Cognitive, Clinical, and Neural Aspects of Drug Addiction

Cognitive, Clinical, and Neural Aspects of Drug Addiction

Ahmed A. Moustafa
Marcs Institute for Brain, Behaviour, and
Development and School of Psychology
Western Sydney University
Penrith, NSW, Australia

ELSEVIER

ACADEMIC PRESS
An imprint of Elsevier

Academic Press is an imprint of Elsevier
125 London Wall, London EC2Y 5AS, United Kingdom
525 B Street, Suite 1650, San Diego, CA 92101, United States
50 Hampshire Street, 5th Floor, Cambridge, MA 02139, United States
The Boulevard, Langford Lane, Kidlington, Oxford OX5 1GB, United Kingdom

Library of Congress Cataloging-in-Publication Data
A catalog record for this book is available from the Library of Congress

British Library Cataloguing-in-Publication Data
A catalogue record for this book is available from the British Library

ISBN 978-0-12-816979-7

For information on all Academic Press publications visit our website at https://www.elsevier.com/books-and-journals

Publisher: Candice Janco
Acquisition Editor: Joslyn Chaiprasert-Paguio
Editorial Project Manager: Sam W. Young
Production Project Manager: Paul Prasad Chandramohan
Cover Designer: Christian J. Bilbow

Typeset by SPi Global, India

Dedication

I dedicate this book to all the substance abuse patients I have worked with in Sydney and elsewhere in Australia. I thank you all for the time and efforts to make this research possible. I very much hope we can provide a treatment for you in the very near future.

Contents

Contributors

Numbers in parentheses indicate the pages on which the authors' contributions begin.

Eid Abohamza (289), Department of Social Sciences, College of Arts and Sciences, Qatar University, Doha, Qatar

Irina Baetu (137), University of Adelaide, Adelaide, SA, Australia

David Best (251), Sheffield Hallam University, Sheffield, United Kingdom; The Australian National University, Canberra, ACT, Australia

Ana-Maria Bliuc (251), School of Social Sciences and Psychology, Western Sydney University, Sydney, NSW, Australia

Peter Casbolt (113), School of Social Sciences and Psychology, Western Sydney University, Sydney, NSW, Australia

Lyndsey E. Collins-Praino (137), University of Adelaide, Adelaide, SA, Australia

Abeer M. Eissa (239), Psychogeriatric Research Center, Institute of Psychiatry, Faculty of Medicine, Ain Shams University, Cairo, Egypt

Mohamad El Haj (187, 221), Psychology Laboratory of Pays de la Loire (EA 4638), University of Angers, Angers; Laboratoire de Psychologie des Pays de la Loire (EA 4638), University of Nantes, Nantes; Geriatric Unit, Centre Hospitalier de Tourcoing, Tourcoing; Institut Universitaire de France, Paris, France

Belinda Favaloro (205), School of Social Sciences and Psychology, Western Sydney University, Sydney, NSW, Australia

Lauren M. Foreman (137), University of Adelaide, Adelaide, SA, Australia

Julia Garami (61, 239), School of Social Sciences and Psychology, Western Sydney University, Sydney, NSW, Australia

Farahnaz Ghafar (113), School of Social Sciences and Psychology, Western Sydney University, Sydney, NSW, Australia

Bruno Kluwe-Schiavon (3), Experimental and Clinical Pharmacopsychology Laboratory, Department of Psychiatry, Psychotherapy, and Psychosomatics, Psychiatric Hospital, University of Zurich, Zurich, Switzerland

Yousif S. Mahdi (267), Qatar University, Doha, Qatar

Justin Mahlberg (85), School of Social Sciences and Psychology, Western Sydney University, Sydney, NSW, Australia

Alejandro N. Morris (187), School of Social Sciences and Psychology, Western Sydney University, Sydney, NSW, Australia

Ahmed A. Moustafa (3,21,41,61,85,113,137,187,205,221,251,267,289), Marcs Institute for Brain, Behaviour, and Development and School of Psychology, Western Sydney University, Penrith, NSW, Australia; Qatar University, Doha, Qatar

Jean-Louis Nandrino (221), Univ. Lille, CNRS, CHU Lille, UMR 9193—SCALab—Cognitive and Affective Sciences Lab, Lille, France

Milen L. Radell (113), Department of Psychology, Niagara University, Lewiston, NY, United States

Janice Rego (137), School of Social Sciences and Psychology, Western Sydney University, Sydney, NSW, Australia

Yuliya Richard (21), School of Social Sciences and Psychology, Western Sydney University, Sydney; Marcs Institute for Brain, Behaviour, and Development and School of Psychology, Western Sydney University, Penrith, NSW, Australia

Daniella M. Saleme (41), School of Social Sciences and Psychology, Western Sydney University, Sydney, NSW, Australia

Breno Sanvicente-Vieira (3), Developmental Cognitive Neuroscience Laboratory, Faculty of Psychology, Pontifical Catholic University of Rio Grande do Sul, Porto Alegre, Brazil

Mohamed Shakir (239), Al-Mamoura Mental Hospital, Alexandria, Egypt

Alaa Solieman (239), Psychogeriatric Research Center, Institute of Psychiatry, Faculty of Medicine, Ain Shams University, Cairo, Egypt

Thiago Wendt Viola (3), Developmental Cognitive Neuroscience Laboratory, Faculty of Psychology, Pontifical Catholic University of Rio Grande do Sul, Porto Alegre, Brazil

Preface

Substance use disorder (or simply drug addiction) is about dependence on a substance or multiple substances without being able to stop using them. While there have tens (if not hundreds) of studies on the psychological and neural mechanism of drug addiction, it is not clear which cognitive errors are related to drug abuse. This, in turn, makes it difficult to find suitable treatment options for patients with drug addiction problems.

The first part of this book addresses these cognitive errors that underlie drug use problems. Specifically, in the first three chapters of the first section of the book, we will discuss traditional views associated with drug addiction problems, including executive dysfunction and the role of prefrontal cortex (Chapter 1). Following that, Chapter 2 will discuss the neural correlates of impulsive behavior in patients with drug use disorders. Risk-taking behavior is a key feature of substance use disorders. Accordingly, Chapter 3 will explain the various forms of risk-taking behavior in patients with drug addiction. Following these chapters, the other chapters in this section will explain less typical forms of cognitive errors associated with drug abuse. Chapter 4 discusses different kinds of discounting tasks and their relationships to drug abuse problems. Chapter 5 reviews prior studies on how the environment (or context associated with drug use) can increase craving for drugs or alcohol. The implications of treatments are also discussed in this chapter. Chapters 6 and 7 will discuss learning theories and processes related to drug abuse in both humans and animals. Specifically, Chapter 6 explain how drug abuse is related to exaggerated acquisition and maintenance of avoidance behavior, and Chapter 7 discusses how compulsive drug use stems from increased incentive salience via sensitization of drugs and drug cues. Chapters 8 and 9 will discuss processes related to thinking about the future in patients with drug addiction. Specifically, Chapter 8 will review prior studies on impaired future-related thinking related to drug abuse, and Chapter 9 explains the concept of intolerance of uncertainty, and explain why patients with drug addiction problems are intolerant of the uncertain future, which leads to increased drug use. Finally, Chapter 10 will explain how social cognitive deficits are related to drug abuse problems.

In the second section of the book, we will discuss clinical and treatment aspects of substance use disorders. Chapter 11 will discuss aspects related to the causes of drug abuse, including the effect of peer pressure. Chapter 12 will compare the benefits and limitations of offline vs. online treatment. Similarly, Chapter 13 will address the benefits and limitations of methadone for the treatment of heroin addiction. Finally, Chapter 14 will review old as well recent studies on the effective role of motivational interviewing for the treatment of drug addiction.

Acknowledgement

I thank all students, research assistants, and collaborators who made this book possible.

I would like to personally thank Justin Mahlberg, Daniella Saleme, Alejandro Morris, Julia Garami, Belinda Favaloro, Mohamad El Haj, Jean-Louis Nandrino, Eid Abu Hamza, Yousif Mahdi, Ana-Maria Bliuc, Milen Radell, Bruno Kluwe-Schiavon, Yuliya Richard, Lauren Foreman, Lyndsey Collins-Praino, and Irina Baetu.

Part I

Cognitive and learning aspects of drug addiction

Chapter 1

Executive functioning and substance use disorders

Bruno Kluwe-Schiavon[a], Breno Sanvicente-Vieira[b], Thiago Wendt Viola[b], Ahmed A. Moustafa[c]

[a]Experimental and Clinical Pharmacopsychology Laboratory, Department of Psychiatry, Psychotherapy, and Psychosomatics, Psychiatric Hospital, University of Zurich, Zurich, Switzerland, [b]Developmental Cognitive Neuroscience Laboratory, Faculty of Psychology, Pontifical Catholic University of Rio Grande do Sul, Porto Alegre, Brazil, [c]Marcs Institute for Brain, Behaviour, and Development and School of Psychology, Western Sydney University, Penrith, NSW, Australia

The concept of executive functioning

Executive functioning has been extensively studied in prior psychological research over the past six decades. Some of the key studies in this area include those of the British experimental psychologist Donald E. Broadbent (1926–93) and the American psychologist Michael I. Posner (1936–) on attentional processing, and the studies of the Russian neuropsychologist Alexander Luria (1902–77) regarding the frontal lobes and the hierarchical organization of brain functioning (Freeman, 1967; Luria, 1965, 1970). What do these authors have in common? Broadly speaking, their work highlighted a variety of behaviors that were accepted to be driven by "high-order cognitive processes" and lately become to be known, for instance, as inhibitory control, attention shifting, working memory, goal-directed behavior, and strategic planning. Although several executive functioning models and theories have emerged since then (Goldstein, Naglieri, Princiotta, & Otero, 2014), such idea remains, and executive functioning has become a multifaceted mental concept that includes more than 30 different components (Barkley, 2012), in which the prefrontal cortex (PFC) and its related structures are still the main neuroanatomical counterpart.

The hierarchical framework of executive functioning and the PFC has its roots on two main fields. Based on a neuroscience view, the role of the PFC in executive functioning emerged with observations of patients who had suffered frontal lobe lesions becoming unable to manipulate, integrate, and respond to internal and external stimulus. In other words, these patients seem to have lost the capacity of prospecting future actions and making reasonable decisions

Cognitive, Clinical, and Neural Aspects of Drug Addiction. https://doi.org/10.1016/B978-0-12-816979-7.00001-7

as before, showing difficulties in moderating social behavior and, ultimately, changing their usual personality expression in detriment of impulsive and disinhibit traits (Harlow, 1868). Based a psychological view, the role of PFC in executives functioning was intimately related to the cognitive revolution in the 1950s as an attempt to understand how information is processed by the mind. In this regard, several experimental models on information processing emerged to explain selective attention and multistore memory processes (e.g. the bottleneck theory of attention and the three-component model). These models, however, were not able to explain how information is deliberatively selected or inhibited during demanding tasks and, therefore, the concept of "cognitive control" was introduced to fulfil this gap (Goldstein et al., 2014), while concurrently phenomenologically explaining behavior attributed to the PFC.

With the advent of functional imaging techniques and neuroscience, the exclusive role of the PFC becomes a matter of ongoing debate over the last 20 years (Kluwe-Schiavon, Viola, Sanvicente-Vieira, Malloy-Diniz, & Grassi-Oliveira, 2016; Miyake et al., 2000), but it has been accepted that "the participation of the frontal lobes in virtually any 'executive process' is probably a necessary, but largely insufficient, requirement" (Alvarez & Emory, 2006). Thus, as described by the Alvarez and Emory (2006), executive functioning involves the coordination of activity among diffuse anatomical and functional brain areas, with the frontal lobes associated with executive functioning sensitivity, rather than specificity. The idea that PFC might contribute to executive functioning is vastly well-accepted, since it has been shown that executive functioning skills reach their peak at age 20–29, in association with the constant myelination of neurons in the prefrontal cortex (Best & Miller, 2010). Such finding also contributes to elucidate another debate, supporting the idea that executive functioning constitutes related, but distinct, multifaceted processes, instead of one single underlying ability that can explain all the components (Jurado & Rosselli, 2007).

The self-regulatory model the Lezak's four-component conceptual model (Lesak, 1982) and the three component conceptual model of Diamond (Diamond, 2013), are some of the most influential multifaceted models of executive functioning. Barkley (2001) focused on the evolutionary role of executive functioning and social contexts, and according to him, it serves "to shift the control of behavior from the immediate context, social others, and the temporal now to self-regulation by internal representations regarding the hypothetical social future". More than that, Barkley pinpoints that executive functioning may have evolved to solve some social adaptive problems such as reciprocal altruism or selfish cooperation, highlighting the crucial role of self-regulation. On the other hand, Lezak proposed simple, but pragmatic, framework where volition, planning, purposeful action, and effective performance are the main core executive functioning processes that enable a person to engage successfully in independent, purposive, and self-serving behavior. This model is broadly used by clinicians and researchers, since it allows us to identify and assess its components. Along the same lines, Diamond (2013) suggested that executive functioning can be described as a family of top-down mental processes recruited when automatic, instinct or intuition would be insufficient to cope with an ongoing demand, emphasizing its role as a self-regulatory process, which is in agreement with Barkley's model. According to Diamond (2013), the main functions are behavioral/cognitive inhibition (including selective attention), working memory, and cognitive flexibility. Several other models have also been reported in the literature. For example, the Miyake and Friedman's model is based on latent variable analysis that suggested both the unity and diversity aspects of executive functioning, emphasizing updating, inhibition, and shifting as it processes (Miyake et al., 2000); or the dual-processing frameworks and its branches that have been extensively used to differentiate emotional "hot" processes from rational "cold" processes and to describe cognitive changes associated with psychiatric disorders and/or neurodevelopment (Kahneman, 2003, 2011; Noël, Brevers, & Bechara, 2013; Zelazo & Carlson, 2012).

Nevertheless, in this chapter, rather than describing different models, we will assume that executive functioning is a dynamic multifaceted construct that, consequently, includes different but intercorrelated processes that aimed to self-regulate (or balance) stress and cognitive engagement, enabling individuals to learn by experience inasmuch as they internalize efficient behaviors and use such information to prospect a set of new behaviors and their most likely consequences (Kluwe-Schiavon, Viola, Sanvicente-Vieira, Malloy-Diniz, & Grassi-Oliveira, 2016). Such broad perspective allows us to go beyond the discussion on which specific executive functioning component some neuropsychological test is depicting, by understanding that any highly complex behavior, such as decision-making and theory of mind, demands an efficient executive functioning performance. Here, therefore, we will discuss how substance related disorders (SUD) might impair executive functioning over a series of different perspectives.

Substance use disorders and executive dysfunction

Nonetheless, one may ask, why should we look for specific deficits of executive functioning associated with addictive disorders? Initially, it is widely well-known that addictive disorders are clinically characterized as the compulsive engagement in rewarding stimuli despite the adverse consequences. Hence, "non-substance related" or behavioral addictions refer to compulsive engagement in rewarding behaviors, for instance, sex, gambling, video-game or internet, and shopping. Here, we will focus on SUD, which refer to a pattern of substance consumption in which the user consumes the substance in quantities or with methods which are harmful to themselves or others. Therefore, phenomenologically speaking, one could characterize SUD as a chronic decision-making and self-regulatory disease, since both the onset of substance use and the maintenance of additive behaviors are characterized by the failure of balancing highly stressful demands (e.g. craving or coping with daily-stressful events) and cognitive/behaviors inhibition, culminating on the engagement in rewarding drug-seeking behaviors despite their adverse consequences (George & Koob, 2010). This understanding is strengthened by studies that suggest that critical cortical and subcortical neural structures related to executive functioning performance are gradually compromised in these pathologies, increasing the frequency of impulsive and compulsive search for the substance (Goldstein & Volkow, 2011; Peterson & Welsh, 2014).

As it is not yet clear what the conceptual borders that distinguish the conceptual definitions of executive functioning and decision-making or how worthful such conceptual differentiation would be, to approach how SUD are related to executive functioning alterations based on the broad executive functioning perspective we assumed, we will also accept that decision-making processes (representation of choices and valuation processing, action selection, outcome evaluation and updating) (Rangel, Camerer, & Montague, 2008), may demand the core executive functioning processes, such as planning, set-shifting, inhibition of non-adaptive choice selection and updating. Therefore, as mentioned above, the coordination of all these processes (e.g. valuation processing, action selection, outcome evaluation and updating) is not mainly done by the PFC, although the PFC plays a crucial role on their efficiency. For instance, two meta-analyses revealed that SUD patients have functional alterations in brain structures highly associated to planning and representation of options, such as the ventral striatum and the anterior cingulate cortex (ACC) (Kuhn & Gallinat, 2011). The ACC lies in a unique position in the brain, connecting the "emotional" limbic system and the "cognitive" PFC, being associated to the ability to control and manage uncomfortable emotions (Stevens, Hurley, & Taber, 2011). In addition, not only because its spatial position, but also because its functions, the ACC plays role in shifting between salience and control networks (Menon & Uddin, 2010). The salience network relates to implicit processing and identification of relevant stimuli and events while the control network relates to explicit executive processing. Although it is not totally clear, together with the insula, the ACC is hypothesized to play as a gatekeeper, implicitly selecting important stimuli for attention and explicitly avoiding irrelevant information (Menon & Uddin, 2010).

Positron emission tomography (PET) studies showed that the right dorsal ACC dopamine neurotransmission—critical for the rewarding effects of drugs and the key neurotransmitter involved in several addiction models (Bickel et al., 2018)—increases significantly during the performance of certain executive processes, such as conflict monitoring and set-shifting (Abenavoli, Greenberg, & Bierman, 2017). Using a unique hair toxicology approach to objectively characterize substance use over the past 6 months and proton magnetic resonance spectroscopy (H-MRS), Hulka et al. (2014) revealed that higher cocaine hair concentrations were associated with lower glutamine ratios in the ACC, indicating that cocaine can be related to changes in glutamate cycling and functional alteration of this brain region. Thus, in a decision-making perspective, it has been hypothesized that the ACC alterations observed in people with SUD are associated with a failure to exert control over the selection of the appropriate behaviors when facing a long and possibly uncertain sequence of actions (Peoples, 2002). More than that, changes in the activity of the ACC, together with the insular cortex may explain, for example, the difficulty of chronic users in representing more adaptive and functional options to replace drug-seeking behaviors, especially during periods of stress, withdrawal symptoms, or severe fissures (Naqvi & Bechara, 2010).

The valuation of choices is a decision-making process that can be closely related to cognitive flexibility and set shifting, as it is crucial for setting goals. In decision behavior theory, we say that during the valuation phase people will attribute a subjective value (known as expected utility) to those options that were previously identified (i.e. represented) and, afterwards, these values will be compared to each other, triggering the behavior of chosen the highest rated option. Interestingly, evidence shows that individuals with SUD are less sensitive to the valuation processing of potentially positive (e.g., earning more money on the task) or potentially negative (e.g., the physical and emotional impact of their use) outcomes in daily life situations (Konova et al., 2012). However, the valuation phase is not directly associated with PFC structures, but mainly to subcortical structures such as the nucleus accumbens (NAc) and the amygdala (Gilman et al., 2014).

If the NAc can be described as the "centre of the pleasure" and reward learning, the amygdala can be described as the "centre of fear" and aversive conditioning in the brain. From a cognitive neuroscience perspective based on a decision behavior theory, we can say that the amygdala encodes and responds to risks and the magnitude of losses. In this regard, some studies have shown that cannabis dependent users have a lower volume of NAc when compared to non-users (Yucel et al., 2008), and that even recreational marijuana users may present morphological and volumetric abnormalities in both NAc and amygdala compared to non-users (Gilman et al., 2014), wherein some of these changes being observable already in adolescent users (Padula, McQueeny, Lisdahl, Price, & Tapert, 2015). In a recent study, Augier et al. (2018) showed that animals that chose alcohol over a high-value alternative (rewarding concentration of the non-caloric sweetener saccharin) despite adverse consequences have lower expression of a specific GABA transporter (GAT-3) in the amygdala. The authors went further, showing that increased GABAergic tone in the amygdala due to reduced GABA uptake by GAT-3 transporters contributes to behaviors that are key for alcohol addiction, advocating for a causal contribution of neuroadaptations on the amygdala to the development of alcohol addiction (Augier et al., 2018).

Such studies highlight the role of different neurotransmitters on SUD, complementing the understanding of it with new theories and hypotheses (Bickel et al., 2018). However, the mesolimbic and the mesocortical projections from ventral tegmental area (VTA) into the basal ganglia and the PFC, respectively, make these regions the highest concentrations of dopamine within the brain and crucial for the valuation of rewarding stimuli and learning. Therefore, before describing how SUD are related to alterations on action selection and motivation, and feedback processing and updating, a brief note about the dopaminergic system and it relations with executive functioning is necessary.

The role of dopamine in executive dysfunction in SUD

A common characteristic of addictive drugs is their capacity to reinforce behavior and to act as powerful rewards when these substances reach the central nervous system. The simple notion that drugs of abuse produce rewarding and reinforcing effects in the brain has been elegantly demonstrated in animal self-administration studies. In this regard, cocaine, amphetamine, nicotine, opiates, cannabis and alcohol have the ability to enhance endogenous DA neurotransmission, primarily in the NAc, that consequently leads to enhanced stimulation of DA receptors located on medium spiny neurons (Asensio et al., 2010; Galaj, Ewing, & Ranaldi, 2018; Kalivas & Volkow, 2005; Satel et al., 1991; Yap & Miczek, 2008). To be noted, the NAc is a core region of the brain's reward system, which is an evolutionarily conserved brain network that underlies basic motivational processes activated by survival- and reproduction-related stimuli (McClure & Bickel, 2014).

Paradigms such as the delay discounting, which is a robust neuropsychological task, has been successfully converted to use in rodent studies (Tedford, Persons, & Napier, 2015), and it has been the basis of some important studies that explored the relation between dopamine, decision-making and executive functioning. The basic premise of the delay discounting task is to choose between obtaining an immediate reward of lesser value in relation to the waiting required to obtain a reward of highest value (Pine, Shiner, Seymour, & Dolan, 2010). In rodents, this task is performed in operant conditioning chamber, allowing the animal to choose between two rewarding options: a single sugar pellet with no presentation delay, or three sugar pellets with a delay interval between 10 and 60 s (Saddoris et al., 2015).

In a remarkable set of experiments, one recent study demonstrated the role of dopaminergic neurotransmission in mesocorticolimbic pathways, defining its causal role for inhibitory control in reward-related situations (Saddoris et al., 2015). Initially, bilateral cannulas were implanted into the NAc of adult rats, and, using fast-scan cyclic voltammetry technology, dopamine release levels were recorded with a time accuracy of millisecond. Before reward collection, particularly during the expectation and anticipation of obtaining a reward, animals had a very high dopaminergic release in the NAc, and such release was even more intense in view of the expectation of obtaining the highest reward. Subsequently, the authors investigated the role of dopaminergic stimulation via the implantation of optogenetic cannulae in the NAc, which allows for a precise and dynamic control of a signaling pathway. They observed that the increase in dopamine release artificially, particularly during the anticipation period, induced rats to choose the highest reward regardless of how long they had to wait for it. Therefore, these experiments provided evidence that cues that predict a reinforcing stimulus during the delay discounting task also modulated extra synaptic dopamine concentrations in the NAc, energizing motivation and modulating decision-making (Volkow, Wise, & Baler, 2017). More than that, it was shown that dopamine has an important role on inhibitory control during decision-making processes.

This mechanism is also implicated in the neurobiology of how drugs of abuse trigger strong behavioral and psychological responses immediately after exposure to drug-related cues. Stimuli (including contextual or environmental) associated with the drug effects become conditioned and, with repeated co-exposure, will trigger dopamine neuronal firing in the NAc. This increased dopaminergic signaling that follows exposure to drug-related cues ensures that an individual will have the necessary motivation to engage in drug-seeking behaviors (Volkow & Morales, 2015). However, neuroplastic changes in midbrain dopamine neurons associated with chronic drug exposure may also result in less inhibitory control, which is extremely necessary to cope with the craving elicited by drug or reward-related cues (Volkow et al., 2017).

These observations led researches to establish a critical role for dopamine during valuation processing, since dopamine release before a goal-directed behavior is a key mechanisms underlying action selection that drives an individual to choose a given option/behavior associated with a potential outcome (Volkow et al., 2017). More than that, altered dopaminergic signaling has been directly related to impulsivity and impairments in executive functioning and decision-making among people with SUD (McClure & Bickel, 2014). For instance, impulsivity in the delay discounting task was observed among individuals who are drug dependents, as well as in those individuals at a high-risk of developing SUD during early stages of drug use behavior (Bickel, Koffarnus, Moody, & Wilson, 2014; Hulka et al., 2014). However, the complexities of the behavioral and neurobiological factors underlying SUD created enormous challenges in efforts to develop effective pharmacological treatments targeting dopaminergic signaling. In this regard, traditional approaches that have focused on blocking selective dopaminergic receptors provided encouraging results in animal models of addiction, but the clinical utility of these agents has been limited due to the side effects they might produce (Galaj et al., 2018).

Action selection, feedback processing and updating

Once the individual has identified a number of options and assigned a subjective value to them, the PFC and its substructures (ventromedial portion, vmPFC, dorsolateral portion, dlPFC, and orbitofrontal cortex, OFC) integrate all information, inhibit inappropriate behaviors and drive goal-oriented ones. Accordingly, a study investigating the neural network involved in the transformation of stimulus values into motor responses found that in healthy individuals the magnitude of costs and benefits are computed in both the vmPFC and the dlPFC and that the resulting signaling of this comparison modulates the activity of the motor cortex and the cerebellum (Wunderlich, Rangel, & O'Doherty, 2009), suggesting that the resulting signaling might be related to motivation and volition. In addition, it is largely accepted that the dlPFC is strongly associated to behavioral inhibition and that this process is independent of the participant's conscious evaluation of the situation they are judging (Büchel et al., 2017). Remarkably, such independency is often reported by individuals with SUD, who usually report not being able to inhibit substance-seeking behaviors despite of being aware of the risks involved. Usually, behavioral disinhibition is observed during delayed gratification tasks in which individuals must inhibit the behavior of choosing an immediate but low value reward in detriment of a late, but higher value reward. Not surprisingly, several studies already show that people with SUD and non-substance related disorders perform worse than people with no SUD (Alvarez & Emory, 2006; Amlung, Vedelago, Acker, Balodis, & MacKillop, 2017; MacKillop et al., 2011). Cocaine-dependent patients who underwent repeated sessions of high-frequency transcranial magnetic stimulation (TMS) in the right dlPFC (which leads to an excitatory effect of this region) had significant reductions in craving and a lower relapse rates when compared to a placebo control group (Terraneo et al., 2016).

As we mentioned above, the concepts of executive functioning also comprise the idea that individuals must be able to learn by experience, in such way that feedback processing and updating are central aspects of many executive functioning models. In this regard, many studies have been using neuropsychological paradigms, such as the Iowa Gambling Task (IGT) (Bechara, Damasio, & Damasio, 2000) to investigate feedback processing and implicit probabilistic learning (Ekhtiari, Victor, & Paulus, 2017). The IGT is a card game that assesses the ability to evaluate immediate gains over long-term losses. In this sense, individuals are instructed that the object of the task is to accumulate as much money as possible by picking one card at a time from any of four decks (A, B, C, and D) until the instruction is given to stop. The task ends when the individual has chosen 100 cards. However, the card decks differ along three dimensions: the immediate gain, the expected long-term gain and the schedule of penalties. All the cards from Decks A and B yield a larger short-term reward ($100 per card) than cards from Decks C and D ($50 per card). Certain cards in all of the decks also carry a penalty, and the accumulated penalties in Decks A and B are larger than in Decks C and D. Over the long run, continued choice from either Deck C or D leads to a net gain ($250/10 cards), whereas choice from either Deck A or B leads to a net loss ($250/10 cards). Therefore, Decks A and B are considered disadvantageous decks, while decks C and D are considered advantageous (Grant, Contoreggi, & London, 2000). The inspiring work done by Bechara et al. (2000) identified the first clues regarding how the human brain performs and mediates the processes of judgment and decision-making, given some important insights on how impairments in these processes might lead to impulsive behaviors. By using the IGT, the authors demonstrated that patients with PFC lesions, particularly in the vmPFC, had poor performance in this task, depicted by frequent choices of cards that generated high reward monetary values, but in the long run, resulted in consistent monetary losses (Verdejo-García & Bechara, 2009).

In the early 2000s, one of the first investigations with the IGT and with individuals with SUD was published (Grant et al., 2000). In this study, 30 polysubstance dependents were compared to a comparison group of 24 individuals who did not use illicit drugs of abuse. Individuals with SUD performed much more poorly on the IGT (net score = 10.2) than controls (net score = 26.0). Therefore, these authors have showed that drug dependents are more likely to make maladaptive decisions in the IGT that result in long-term losses exceeding short-term gains; and they may continue to choose cards from the low yield decks in the IGT because they underestimate the magnitude of their losses occurring over an extended period. These findings contributed to the hypothesis that dysfunctional PFC underlies a kind of "myopia" for the future in people with SUD, and this may be one of the principle mechanisms underlying the transition from casual substance taking to compulsive and uncontrollable behavior (Bechara & Damasio, 2002; Verdejo-García, Bechara, Recknor, & Pérez-García, 2006).

Moreover, this pattern of altered decision-making and PFC functionality was not restricted to neurological conditions but was also evident in psychiatric disorders such as SUD. Remarkably, these results were further replicated using samples including individuals with addiction to smoked/snorted cocaine, cannabis, alcohol or opiate (Bolla, Eldreth, Matochik, & Cadet, 2005; Pirastu et al., 2006; Viola et al., 2012), and the disadvantageous choices on the IGT were associated with higher levels of social dysfunction in people with SUD, suggesting the ecological validity of the IGT for real-life decision-making situations (Cunha, Bechara, de Andrade, & Nicastri, 2011). The IGT has become one of the most used tasks to measure feedback processing, decision-making and, therefore, executive functioning in SUD (Kovacs, Richman, Janka, Maraz, & Ando, 2017), suggesting that impaired performance in this task can be explained, in part, by alterations in the vmPFC that might impair the integration of NAc and amygdala signaling (Basten, Biele, Heekeren, & Fiebach, 2010). Moreover, recent evidence demonstrated that the NAc is also involved with the performance in the classic decision-making paradigm, the IGT (Grassi et al., 2018), once more advocating for the dynamic nature of executive functioning involving the coordination of activity among diffuse neural structures.

Recently, another neuropsychological task has been used to investigate explicit feedback processing on SUD. The Columbia Card Task, CCT, (Figner, Mackinlay, Wilkening, & Weber, 2009) is another card game where participants face a deck with 32 facedown cards. In each trial, participants are informed about how many loss cards are on the deck, the value amount of losing cards and the value amount of the winning cards. Considering this information, participants are asked to choose how many cards they want, seeking the best possible score at the end of the trial. The task is composed by two versions, the no-feedback condition and the feedback condition. In the feedback condition, participants had to choose cards one-by-one until they decided that it was too risky to continue. When selected, the card is signaled as an unknown card until participants voluntarily decide to end the round, and then the cards are turned over according to the order in which they were chosen. In the no-feedback condition, participants have to select the number of cards (from 0 to 32) that they would choose. However, cards are randomly selected by the computer and no-feedback is given until the end of the task (composed by 24 rounds). Because the CCT displays all the information, allowing participants to estimate the risks involved, it can be considered as a decision under risk task with explicit information; differently from the IGT in which participants are not informed on the risks involved and must implicitly learn by association which ones are the best decks.

A study with crack-cocaine dependent users showed that women with SUD perform worse than women without SUD, but similar as adolescents, on the no-feedback condition of the CCT. Interesting, when the feedback was provided, women with SUD showed a reduction of risk-taking behavior, performing similar as women without SUD (Kluwe-Schiavon, Viola, Sanvicente-Vieira, Pezzi, & Grassi-Oliveira, 2016). In another study, we investigated the influences of feedback processing and attention to environmental contingencies on risk-taking in heroin-dependent individuals. However, in contrast with previous findings, this study revealed that heroin-dependent patients pay less attention to environmental contingencies during risk-taking than controls, suggesting either that heroin-dependent patients display different risk -taking behavior to cocaine-dependent individuals, or an influence of opioid replacement therapy on behavior (Saleme et al., 2018). Based on such difference, future studies can specifically investigate how different SUD may affect learning from feedback on risk decision-making scenarios.

SUD, a biopsychosocial executive functioning disorder

Above, we have discussed how different neurobiological mechanisms play a central role in orchestrating executive functioning changes on SUD. In this sense, it is clear that besides the structural and the functional changes in the brain, there are many social and psychological characteristics that are related to SUD and alterations on executive functioning performance. For instance, excessive use of stimulants, such as cocaine, is functionally and clinically associated with impairments in every-day life, in form of health problems, disability, and failure to meet major responsibilities at work, school, or home (Volkow, Baler, & Goldstein, 2011).

As occasional drug use progresses into addictive behavior, individuals typically become increasingly impaired in their ability to function socially (Augier et al., 2018; Heilig, Epstein, Nader, & Shaham, 2016). Those impairments result in social marginalization and exclusion, which is associated with further substance-seeking behaviors (Heilig et al., 2016). Other real-life impairments that reported negative associations between executive functioning and measures of social adjustment, particularly related to family and finances, meaning that as much more executive functioning impairments, more financial and family problems may occur (Cunha et al., 2011). Thus, impairments in abilities such as theory of mind—which is a complex social ability related to understanding what other people is thinking about based on contextual information (e.g., face expression, speech, previous knowledge)—relate to higher symptoms and worse SUD profile (Hulka, Preller, Vonmoos, Broicher, & Quednow, 2013; Preller et al., 2014; Sanvicente-Vieira, Kluwe-Schiavon, Corcoran, & Grassi-Oliveira, 2017).

Moreover, it has been reported that early life experiences play a crucial role during development, impacting cognitive functioning and, therefore, increasing or decreasing the vulnerability for psychiatric conditions. In one prior study, childhood maltreatment was found to be a predictive factor for executive impairments among female cocaine users (Viola, Tractenberg, Pezzi, Kristensen, & Grassi-Oliveira, 2013). In fact, childhood maltreatment plays a key role in the development of SUD, particularly because it interplays with cognitive deficits and negative changes in the prognosis of the drug use (Andersen & Teicher, 2009).

Circumstantial evidence support that early life stress experiences trigger a cascade of changes that anticipates some developmental stages, which in turn, contributes to an increased vulnerability for drug use. Substance users with history of childhood maltreatment have changes in stress-related genes that increase the vulnerability for SUD (Rovaris et al., 2017). Such impact on stress-related genes is in association with disruptions in the immune system (Levandowski et al., 2013), which in turn is negatively related with poor decision making, inhibition and flexibility (Levandowski, Hess, Grassi-Oliveira, & de Almeida, 2016) and in turn drug use severity (Yücel, Lubman, Solowij, & Brewer, 2007). These cognitive dysfunctions are not exclusive to executive functioning, only, but extends to long-term memory (Viola et al., 2015), attention, processing speed (Lundqvist, 2005) and also in social cognitive performance as already shown (Hulka et al., 2013; Preller et al., 2014; Sanvicente-Vieira et al., 2017).

Finally, the question whether all these cognitive impairments described so far might be a cause or consequence of poor executive functioning that leads to SUD, is not yet clearly known. It has been shown that early life experiences are associated with worse cognitive performance and risk behaviors, and that early onset of substance consumption may lead to higher vulnerability to the development of SUD as well (Ernst, Romeo, & Andersen, 2009). Additionally, there is evidence that executive functioning impairments are related to earlier age of regular drug use (Solowij et al., 2012). Furthermore, once the SUD is developed, it has been shown that poorer cognitive performance is related to poorer treatment adherence (Domínguez-Salas, Díaz-Batanero, Lozano-Rojas, & Verdejo-García, 2016). Specifically, executive functioning poorer performance can predict relapse rates in a 3-month period among cocaine users (Verdejo-Garcia et al., 2014). Lastly, one study revealed that cocaine users who substantially decreased the amount of cocaine consumption over 1 year performed as good as non-users controls on attention, working memory, declarative memory, and executive functions (Vonmoos et al., 2014). Together, these findings suggest that, at least, part of these deficits are drug induced.

References

Abenavoli, R. M., Greenberg, M. T., & Bierman, K. L. (2017). Identification and validation of school readiness profiles among high-risk kindergartners. *Early Childhood Research Quarterly*, *38*, 33–43. https://doi.org/10.1016/j.ecresq.2016.09.001.

Alvarez, J., & Emory, E. (2006). Executive function and the frontal lobes: A meta-analytic review. *Neuropsychology Review*, *16*(1), 17–42. https://doi.org/10.1007/s11065-006-9002-x.

Amlung, M., Vedelago, L., Acker, J., Balodis, I., & MacKillop, J. (2017). Steep delay discounting and addictive behavior: A meta-analysis of continuous associations. *Addiction*, *112*(1), 51–62. https://doi.org/10.1111/add.13535.

Andersen, S. L., & Teicher, M. H. (2009). Desperately driven and no brakes: Developmental stress exposure and subsequent risk for substance abuse. *Neuroscience and Biobehavioral Reviews*, *33*(4), 516–524. https://doi.org/10.1016/j.neubiorev.2008.09.009.

Asensio, S., Romero, M. J., Romero, F. J., Wong, C., Alia-Klein, N., Tomasi, D., et al. (2010). Striatal dopamine D2 receptor availability predicts the thalamic and medial prefrontal responses to reward in cocaine abusers three years later. *Synapse*, *64*(5), 397–402. https://doi.org/10.1002/syn.20741.

Augier, E., Barbier, E., Dulman, R. S., Licheri, V., Augier, G., Domi, E., et al. (2018). A molecular mechanism for choosing alcohol over an alternative reward. *Science*, *360*(6395), 1321. https://doi.org/10.1126/science.aao1157.

Barkley, R. A. (2001). The executive functions and self-regulation: An evolutionary neuropsychological perspective. *Neuropsychology Review*, *11*(1). https://doi.org/10.1023/a:1009085417776.

Barkley, R. A. (2012). *Executive functions: What they are, how they work, and why they evolved London*. The Guilford Press.

Basten, U., Biele, G., Heekeren, H. R., & Fiebach, C. J. (2010). How the brain integrates costs and benefits during decision making. *Proceedings of the National Academy of Sciences of the United States of America*, *107*(50), 21767–21772. https://doi.org/10.1073/pnas.0908104107.

Bechara, A., & Damasio, H. (2002). Decision-making and addiction (part I): Impaired activation of somatic states in substance dependent individuals when pondering decisions with negative future consequences. *Neuropsychologia*, *40*(10), 1675–1689.

Bechara, A., Damasio, H., & Damasio, A. R. (2000). Emotion, decision making and the orbitofrontal cortex. *Cerebral Cortex*, *10*(3), 295–307. https://doi.org/10.1093/cercor/10.3.295.

Best, J. R., & Miller, P. H. (2010). A developmental perspective on executive function. *Child Development*, *81*(6), 1641–1660. https://doi.org/10.1111/j.1467-8624.2010.01499.x.

Bickel, W. K., Koffarnus, M. N., Moody, L., & Wilson, A. G. (2014). The behavioral- and neuro-economic process of temporal discounting: A candidate behavioral marker of addiction. *Neuropharmacology*, *76*(Pt B), 518–527. https://doi.org/10.1016/j.neuropharm.2013.06.013.

Bickel, W. K., Mellis, A. M., Snider, S. E., Athamneh, L. N., Stein, J. S., & Pope, D. A. (2018). 21st century neurobehavioral theories of decision making in addiction: Review and evaluation. *Pharmacology Biochemistry and Behavior*, *164*, 4–21. https://doi.org/10.1016/j.pbb.2017.09.009.

Bolla, K. I., Eldreth, D. A., Matochik, J. A., & Cadet, J. L. (2005). Neural substrates of faulty decision-making in abstinent marijuana users. *NeuroImage*, *26*(2), 480–492. https://doi.org/10.1016/j.neuroimage.2005.02.012.

Büchel, C., Peters, J., Banaschewski, T., Bokde, A. L., Bromberg, U., Conrod, P. J., et al. (2017). Blunted ventral striatal responses to anticipated rewards foreshadow problematic drug use in novelty-seeking adolescents. *Nature Communications*, *8*, 14140. https://doi.org/10.1038/ncomms14140.

Cunha, P. J., Bechara, A., de Andrade, A. G., & Nicastri, S. (2011). Decision-making deficits linked to real-life social dysfunction in crack cocaine-dependent individuals. *American Journal on Addictions*, *20*(1), 78–86. https://doi.org/10.1111/j.1521-0391.2010.00097.x.

Diamond, A. (2013). Executive functions. *Annual Review of Psychology*, *64*(64), 135–168. https://doi.org/10.1146/annurev-psych-113011-143750.

Domínguez-Salas, S., Díaz-Batanero, C., Lozano-Rojas, O. M., & Verdejo-García, A. (2016). Impact of general cognition and executive function deficits on addiction treatment outcomes: Systematic review and discussion of neurocognitive pathways. *Neuroscience & Biobehavioral Reviews*, *71*, 772–801.

Ekhtiari, H., Victor, T. A., & Paulus, M. P. (2017). Aberrant decision-making and drug addiction—How strong is the evidence? *Current Opinion in Behavioral Sciences*, *13*, 25–33. https://doi.org/10.1016/j.cobeha.2016.09.002.

Ernst, M., Romeo, R. D., & Andersen, S. L. (2009). Neurobiology of the development of motivated behaviors in adolescence: A window into a neural systems model. *Pharmacology, Biochemistry, and Behavior*, *93*(3), 199–211. https://doi.org/10.1016/j.pbb.2008.12.013.

Figner, B., Mackinlay, R. J., Wilkening, F., & Weber, E. U. (2009). Affective and deliberative processes in risky choice: Age differences in risk taking in the Columbia card task. *Journal of Experimental Psychology. Learning, Memory, and Cognition*, *35*(3), 709–730. https://doi.org/10.1037/a0014983.

Freeman, T. (1967). LURIA, AR—Higher cortical functions in man. *British Journal of Medical Psychology*, *40*, 186.

Galaj, E., Ewing, S., & Ranaldi, R. (2018). Dopamine D1 and D3 receptor polypharmacology as a potential treatment approach for substance use disorder. *Neuroscience and Biobehavioral Reviews*, *89*, 13–28. https://doi.org/10.1016/j.neubiorev.2018.03.020.

George, O., & Koob, G. F. (2010). Individual differences in prefrontal cortex function and the transition from drug use to drug dependence. *Neuroscience and Biobehavioral Reviews*, *35*(2), 232–247. https://doi.org/10.1016/j.neubiorev.2010.05.002.

Gilman, J. M., Kuster, J. K., Lee, S., Lee, M. J., Kim, B. W., Makris, N., et al. (2014). Cannabis use is quantitatively associated with nucleus accumbens and amygdala abnormalities in Young adult recreational users. *Journal of Neuroscience*, *34*(16), 5529–5538. https://doi.org/10.1523/jneurosci.4745-13.2014.

Goldstein, S., Naglieri, J., Princiotta, D., & Otero, T. (2014). Introduction: A history of executive functioning as a theoretical and clinical construct. In S. Goldstein & J. Naglieri (Eds.), *Handbook of executive functioning* (pp. 3–12). New York: Springer.

Goldstein, R. Z., & Volkow, N. D. (2011). Dysfunction of the prefrontal cortex in addiction: Neuroimaging findings and clinical implications. *Nature Reviews. Neuroscience*, *12*(11), 652–669. https://doi.org/10.1038/nrn3119.

Grant, S., Contoreggi, C., & London, E. D. (2000). Drug abusers show impaired performance in a laboratory test of decision making. *Neuropsychologia*, *38*(8), 1180–1187.

Grassi, G., Figee, M., Ooms, P., Righi, L., Nakamac, T., Pallanti, S., et al. (2018). Impulsivity and decision-making in obsessive-compulsive disorder after effective deep brain stimulation or treatment as usual. *CNS Spectrums*, *23*(5), 333–339. https://doi.org/10.1017/S1092852918000846.

Harlow, J. (1868). *Recovery from the passage of an iron bar throught the head. Vol. 2 (3)*. (p. 10). Publications of the Massachusetts Medical Society.

Heilig, M., Epstein, D. H., Nader, M. A., & Shaham, Y. (2016). Time to connect: Bringing social context into addiction neuroscience. *Nature Reviews. Neuroscience*, *17*(9), 592–599. https://doi.org/10.1038/nrn.2016.67.

Hulka, L. M., Eisenegger, C., Preller, K. H., Vonmoos, M., Jenni, D., Bendrick, K., et al. (2014). Altered social and non-social decision-making in recreational and dependent cocaine users. *Psychological Medicine*, *44*(5), 1015–1028. https://doi.org/10.1017/S0033291713001839.

Hulka, L. M., Preller, K. H., Vonmoos, M., Broicher, S. D., & Quednow, B. B. (2013). Cocaine users manifest impaired prosodic and cross-modal emotion processing. *Frontiers in Psychiatry*, *4*, 98. https://doi.org/10.3389/fpsyt.2013.00098.

Hulka, L. M., Scheidegger, M., Vonmoos, M., Preller, K. H., Baumgartner, M. R., Herdener, M., et al. (2014). Glutamatergic and neurometabolic alterations in chronic cocaine users measured with (1) H-magnetic resonance spectroscopy. *Addiction Biology*, https://doi.org/10.1111/adb.12217.

Jurado, M. B., & Rosselli, M. (2007). The elusive nature of executive functions: A review of our current understanding. *Neuropsychology Review*, *17*, 213–233. https://doi.org/10.1007/s11065-007-9040-z.

Kahneman, D. (2003). A perspective on judgment and choice: Mapping bounded rationality. *The American Psychologist*, *58*(9), 697–720. https://doi.org/10.1037/0003-066X.58.9.697.

Kahneman, D. (2011). In T. Gilovich & D. Griffin (Eds.), *Thinking, fast and slow* (1st ed.). New York: Farrar, Straus and Giroux.

Kalivas, P. W., & Volkow, N. D. (2005). The neural basis of addiction: A pathology of motivation and choice. *The American Journal of Psychiatry*, *162*(8), 1403–1413. https://doi.org/10.1176/appi.ajp.162.8.1403.

Kluwe-Schiavon, B., Viola, T. W., Sanvicente-Vieira, B., Malloy-Diniz, L. F., & Grassi-Oliveira, R. (2016). Balancing automatic-controlled behaviors and emotional-salience states: A dynamic executive functioning hypothesis. *Frontiers in Psychology*, *7*, 2067. https://doi.org/10.3389/fpsyg.2016.02067.

Kluwe-Schiavon, B., Viola, T. W., Sanvicente-Vieira, B., Pezzi, J. C., & Grassi-Oliveira, R. (2016). Similarities between adult female crack cocaine users and adolescents in risky decision-making scenarios. *Journal of Clinical and Experimental Neuropsychology*, *38*(7), 795–810. https://doi.org/10.1080/13803395.2016.1167171.

Konova, A. B., Moeller, S. J., Tomasi, D., Parvaz, M. A., Alia-Klein, N., Volkow, N. D., et al. (2012). Structural and behavioral correlates of abnormal encoding of money value in the sensorimotor striatum in cocaine addiction. *The European Journal of Neuroscience*, *36*(7), 2979–2988. https://doi.org/10.1111/j.1460-9568.2012.08211.x.

Kovacs, I., Richman, M. J., Janka, Z., Maraz, A., & Ando, B. (2017). Decision making measured by the Iowa gambling task in alcohol use disorder and gambling disorder: A systematic review and meta-analysis. *Drug and Alcohol Dependence*, *181*, 152–161. https://doi.org/10.1016/j.drugalcdep.2017.09.023.

Kuhn, S., & Gallinat, J. (2011). Common biology of craving across legal and illegal drugs—A quantitative meta-analysis of cue-reactivity brain response. *European Journal of Neuroscience*, *33*(7), 1318–1326. https://doi.org/10.1111/j.1460-9568.2010.07590.x.

Lesak, M. D. (1982). The problem of assessing executive functions. *International Journal of Psychology*, *17*, 281–297.

Levandowski, M. L., Hess, A., Grassi-Oliveira, R., & de Almeida, R. (2016). Plasma interleukin-6 and executive function in crack cocaine-dependent women. *Neuroscience Letters*, *628*, 85–90. https://doi.org/10.1016/j.neulet.2016.06.023.

Levandowski, M. L., Viola, T., Tractenberg, S., Teixeira, A., Brietzke, E., Bauer, M., et al. (2013). Adipokines during early abstinence of crack cocaine in dependent women reporting childhood maltreatment. *Psychiatry Research*, https://doi.org/10.1016/j.psychres.2013.07.007.

Lundqvist, T. (2005). Cognitive consequences of cannabis use: Comparison with abuse of stimulants and heroin with regard to attention, memory and executive functions. *Pharmacology, Biochemistry, and Behavior*, *81*(2), 319–330. https://doi.org/10.1016/j.pbb.2005.02.017.

Luria, A. R. (1965). 2 kinds of motor perseveration in massive injury of frontal lobes. *Brain*, *88*, 1. https://doi.org/10.1093/brain/88.1.1.

Luria, A. R. (1970). Functional organization of brain. *Scientific American*, *222*(3), 66.

MacKillop, J., Amlung, M. T., Few, L. R., Ray, L. A., Sweet, L. H., & Munafo, M. R. (2011). Delayed reward discounting and addictive behavior: A meta-analysis. *Psychopharmacology*, *216*(3), 305–321. https://doi.org/10.1007/s00213-011-2229-0.

McClure, S. M., & Bickel, W. K. (2014). A dual-systems perspective on addiction: Contributions from neuroimaging and cognitive training. *Annals of the New York Academy of Sciences*, *1327*, 62–78. https://doi.org/10.1111/nyas.12561.

Menon, V., & Uddin, L. Q. (2010). Saliency, switching, attention and control: A network model of insula function. *Brain Structure & Function*, *214*(5–6), 655–667. https://doi.org/10.1007/s00429-010-0262-0.

Miyake, A., Friedman, N. P., Emerson, M. J., Witzki, A. H., Howerter, A., & Wager, T. D. (2000). The unity and diversity of executive functions and their contributions to complex "frontal lobe" tasks: A latent variable analysis. *Cognitive Psychology*, *41*(1), 49–100. https://doi.org/10.1006/cogp.1999.0734.

Naqvi, N. H., & Bechara, A. (2010). The insula and drug addiction: An interoceptive view of pleasure, urges, and decision-making. *Brain Structure & Function*, *214*(5–6), 435–450. https://doi.org/10.1007/s00429-010-0268-7.

Noël, X., Brevers, D., & Bechara, A. (2013). A triadic neurocognitive approach to addiction for clinical interventions. *Frontiers in Psychiatry*, *4*, https://doi.org/10.3389/fpsyt.2013.00179.

Padula, C. B., McQueeny, T., Lisdahl, K. M., Price, J. S., & Tapert, S. F. (2015). Craving is associated with amygdala volumes in adolescent marijuana users during abstinence. *American Journal of Drug and Alcohol Abuse*, *41*(2), 127–132. https://doi.org/10.3109/00952990.2014.966198.

Peoples, L. L. (2002). Neuroscience: Will, anterior cingulate cortex, and addiction. *Science*, *296*(5573), 1623–1624. https://doi.org/10.1126/science.1072997.

Peterson, E., & Welsh, M. (2014). The development of hot and cool executive functions: Are we getting warmer? In S. Goldstein & J. A. Naglieri (Eds.), *Handbook of executive functions*. New York, NY: Springer.

Pine, A., Shiner, T., Seymour, B., & Dolan, R. J. (2010). Dopamine, time, and impulsivity in humans. *Journal of Neuroscience*, *30*(26), 8888–8896. https://doi.org/10.1523/JNEUROSCI.6028-09.2010.

Pirastu, R., Fais, R., Messina, M., Bini, V., Spiga, S., Falconieri, D., et al. (2006). Impaired decision-making in opiate-dependent subjects: Effect of pharmacological therapies. *Drug and Alcohol Dependence*, *83*(2), 163–168. https://doi.org/10.1016/j.drugalcdep.2005.11.008.

Preller, K., Herdener, M., Schilbach, L., Staempfli, P., Hulka, L., Vonmoos, M., et al. (2014). Functional changes of the reward system underlie blunted response to social gaze in cocaine users. *Proceedings of the National Academy of Sciences of the United States of America*, *111*(7), 2842–2847. https://doi.org/10.1073/pnas.1317090111.

Preller, K. H., Hulka, L. M., Vonmoos, M., Jenni, D., Baumgartner, M. R., Seifritz, E., et al. (2014). Impaired emotional empathy and related social network deficits in cocaine users. *Addiction Biology*, *19*(3), 452–466. https://doi.org/10.1111/adb.12070.

Rangel, A., Camerer, C., & Montague, P. R. (2008). A framework for studying the neurobiology of value-based decision making. *Nature Reviews Neuroscience*, *9*(7), 545–556. https://doi.org/10.1038/nrn2357.

Rovaris, D. L., Schuch, J. B., Grassi-Oliveira, R., Sanvicente-Vieira, B., da Silva, B. S., Walss-Bass, C., et al. (2017). Effects of crack cocaine addiction and stress-related genes on peripheral BDNF levels. *Journal of Psychiatric Research*, *90*, 78–85. https://doi.org/10.1016/j.jpsychires.2017.02.011.

Saddoris, M. P., Sugam, J. A., Stuber, G. D., Witten, I. B., Deisseroth, K., & Carelli, R. M. (2015). Mesolimbic dopamine dynamically tracks, and is causally linked to, discrete aspects of value-based decision making. *Biological Psychiatry*, *77*(10), 903–911. https://doi.org/10.1016/j.biopsych.2014.10.024.

Saleme, D. M., Kluwe-Schiavon, B., Soliman, A., Misiak, B., Frydecka, D., & Moustafa, A. A. (2018). Factors underlying risk taking in heroin-dependent individuals: Feedback processing and environmental contingencies. *Behavioural Brain Research*, *350*, 23–30. https://doi.org/10.1016/j.bbr.2018.04.052.

Sanvicente-Vieira, B., Kluwe-Schiavon, B., Corcoran, R., & Grassi-Oliveira, R. (2017). Theory of mind impairments in women with cocaine addiction. *Journal of Studies on Alcohol and Drugs*, *78*(2), 258–267.

Satel, S. L., Price, L. H., Palumbo, J. M., McDougle, C. J., Krystal, J. H., Gawin, F., et al. (1991). Clinical phenomenology and neurobiology of cocaine abstinence: A prospective inpatient study. *The American Journal of Psychiatry*, *148*(12), 1712–1716.

Solowij, N., Jones, K. A., Rozman, M. E., Davis, S. M., Ciarrochi, J., Heaven, P. C., et al. (2012). Reflection impulsivity in adolescent cannabis users: A comparison with alcohol-using and non-substance-using adolescents. *Psychopharmacology*, *219*(2), 575–586.

Stevens, F. L., Hurley, R. A., & Taber, K. H. (2011). Anterior cingulate cortex: Unique role in cognition and emotion. *Journal of Neuropsychiatry and Clinical Neurosciences*, *23*(2), 120–125. https://doi.org/10.1176/jnp.23.2.jnp121.

Tedford, S. E., Persons, A. L., & Napier, T. C. (2015). Dopaminergic lesions of the dorsolateral striatum in rats increase delay discounting in an impulsive choice task. *PLoS ONE*, *10*(4), e0122063. https://doi.org/10.1371/journal.pone.0122063.

Terraneo, A., Leggio, L., Saladini, M., Ermani, M., Bonci, A., & Gallimberti, L. (2016). Transcranial magnetic stimulation of dorsolateral prefrontal cortex reduces cocaine use: A pilot study. *European Neuropsychopharmacology*, *26*(1), 37–44. https://doi.org/10.1016/j.euroneuro.2015.11.011.

Verdejo-Garcia, A., Albein-Urios, N., Martinez-Gonzalez, J. M., Civit, E., De la Torre, R., & Lozano, O. (2014). Decision-making impairment predicts 3-month hair-indexed cocaine relapse. *Psychopharmacology*, *231*(21), 4179–4187.

Verdejo-García, A., & Bechara, A. (2009). A somatic marker theory of addiction. *Neuropharmacology*, *56*(Suppl. 1), 48–62. https://doi.org/10.1016/j.neuropharm.2008.07.035.

Verdejo-García, A., Bechara, A., Recknor, E. C., & Pérez-García, M. (2006). Executive dysfunction in substance dependent individuals during drug use and abstinence: An examination of the behavioral, cognitive and emotional correlates of addiction. *Journal of the International Neuropsychological Society*, *12*(3), 405–415.

Viola, T. W., Cardoso, C., Francke, I., Gonçalves, H., Pezzi, J., Araújo, R., et al. (2012). Tomada de decisão em dependentes de crack: um estudo com o Iowa Gambling Task. *Estudos de Psicologia*, *17*(1), 7.

Viola, T. W., Tractenberg, S., Kluwe-Schiavon, B., Levandowski, M., Sanvicente-Vieira, B., Wearick-Silva, L., et al. (2015). Brain-derived neurotrophic factor and delayed verbal recall in crack/cocaine dependents. *European Addiction Research*, *21*(5), 273–278. https://doi.org/10.1159/000430436.

Viola, T. W., Tractenberg, S., Pezzi, J., Kristensen, C., & Grassi-Oliveira, R. (2013). Childhood physical neglect associated with executive functions impairments in crack cocaine-dependent women. *Drug and Alcohol Dependence*, *132*(1–2), 271–276. https://doi.org/10.1016/j.drugalcdep.2013.02.014.

Volkow, N. D., Baler, R. D., & Goldstein, R. Z. (2011). Addiction: Pulling at the neural threads of social behaviors. *Neuron*, *69*(4), 599–602. https://doi.org/10.1016/j.neuron.2011.01.027.

Volkow, N. D., & Morales, M. (2015). The brain on drugs: From reward to addiction. *Cell*, *162*(4), 712–725. https://doi.org/10.1016/j.cell.2015.07.046.

Volkow, N. D., Wise, R. A., & Baler, R. (2017). The dopamine motive system: Implications for drug and food addiction. *Nature Reviews. Neuroscience*, *18*(12), 741–752. https://doi.org/10.1038/nrn.2017.130.

Vonmoos, M., Hulka, L. M., Preller, K. H., Minder, F., Baumgartner, M. R., & Quednow, B. B. (2014). Cognitive impairment in cocaine users is drug-induced but partially reversible: Evidence from a longitudinal study. *Neuropsychopharmacology*, *39*(9), 2200–2210. https://doi.org/10.1038/npp.2014.71.

Wunderlich, K., Rangel, A., & O'Doherty, J. P. (2009). Neural computations underlying action-based decision making in the human brain. *Proceedings of the National Academy of Sciences of the United States of America*, *106*(40), 17199–17204. https://doi.org/10.1073/pnas.0901077106.

Yap, J. J., & Miczek, K. A. (2008). Stress and rodent models of drug addiction: Role of VTA-accumbens-PFC-amygdala circuit. *Drug Discovery Today: Disease Models*, *5*(4), 259–270. https://doi.org/10.1016/j.ddmod.2009.03.010.

Yücel, M., Lubman, D. I., Solowij, N., & Brewer, W. J. (2007). Understanding drug addiction: A neuropsychological perspective. *The Australian and New Zealand Journal of Psychiatry*, *41*(12), 957–968. https://doi.org/10.1080/00048670701689444.

Yucel, M., Solowij, N., Respondek, C., Whittle, S., Fornito, A., Pantelis, C., et al. (2008). Regional brain abnormalities associated with heavy long-term cannabis use. *European Neuropsychopharmacology*, *18*, S545–S546. https://doi.org/10.1016/s0924-977x(08)70828-9.

Zelazo, P. D., & Carlson, S. M. (2012). Hot and cool executive function in childhood and adolescence: Development and plasticity. *Child Development Perspectives*, *6*(4), 354–360. https://doi.org/10.1111/j.1750-8606.2012.00246.x.

Further reading

Barkley, R. A. (1997). Behavioral inhibition, sustained attention, and executive functions: Constructing a unifying theory of ADHD. *Psychological Bulletin*, *121*(1), 65–94. https://doi.org/10.1037/0033-2909.121.1.65.

Bechara, A., Damasio, A. R., Damasio, H., & Anderson, S. W. (1994). Insensitivity to future consequences following damage to human prefrontal cortex. *Cognition*, *50*, 7–15.

Bechara, A., Damasio, H., Damasio, A. R., & Lee, G. P. (1999). Different contributions of the human amygdala and ventromedial prefrontal cortex to decision-making. *Journal of Neuroscience*, *19*(13), 5473–5481.

Bechara, A., Dolan, S., Denburg, N., Hindes, A., Anderson, S. W., & Nathan, P. E. (2001). Decision-making deficits, linked to a dysfunctional ventromedial prefrontal cortex, revealed in alcohol and stimulant abusers. *Neuropsychologia*, *39*(4), 376–389.

Bullmore, E., & Sporns, O. (2009). Complex brain networks: Graph theoretical analysis of structural and functional systems. *Nature Reviews. Neuroscience*, *10*(3), 186–198. https://doi.org/10.1038/nrn2575.

Clarke, T. K., Weiss, A. R., Ferarro, T. N., Kampman, K. M., Dackis, C. A., Pettinati, H. M., et al. (2014). The dopamine receptor D2 (DRD2) SNP rs1076560 is associated with opioid addiction. *Annals of Human Genetics*, *78*(1), 33–39. https://doi.org/10.1111/ahg.12046.

Corbetta, M., & Shulman, G. L. (2002). Control of goal-directed and stimulus-driven attention in the brain. *Nature Reviews Neuroscience*, *3*(3), 201–215.

Levran, O., Randesi, M., da Rosa, J. C., Ott, J., Rotrosen, J., Adelson, M., et al. (2015). Overlapping dopaminergic pathway genetic susceptibility to heroin and cocaine addictions in African Americans. *Annals of Human Genetics*, *79*(3), 188–198. https://doi.org/10.1111/ahg.12104.

Liu, Y.-Y., Slotine, J.-J., & Barabási, A.-L. (2011). Controllability of complex networks. *Nature*, *473*(7346), 167–173.

Mallorqui-Bague, N., Fagundo, A. B., Jimenez-Murcia, S., de la Torre, R., Banos, R. M., Botella, C., et al. (2016). Decision making impairment: A shared vulnerability in obesity, gambling disorder and substance use disorders? *PLoS ONE*, *11*(9). https://doi.org/10.1371/journal.pone.0163901.

Orsini, C. A., Mitchell, M. R., Heshmati, S. C., Shimp, K. G., Spurrell, M. S., Bizon, J. L., et al. (2017). Effects of nucleus accumbens amphetamine administration on performance in a delay discounting task. *Behavioural Brain Research*, *321*, 130–136. https://doi.org/10.1016/j.bbr.2017.01.001.

Orsini, C. A., Moorman, D. E., Young, J. W., Setlow, B., & Floresco, S. B. (2015). Neural mechanisms regulating different forms of risk-related decision-making: Insights from animal models. *Neuroscience and Biobehavioral Reviews*, *58*, 147–167. https://doi.org/10.1016/j.neubiorev.2015.04.009.

Ozburn, A. R., Falcon, E., Twaddle, A., Nugent, A. L., Gillman, A. G., Spencer, S. M., et al. (2015). Direct regulation of diurnal Drd3 expression and cocaine reward by NPAS2. *Biological Psychiatry*, *77*(5), 425–433. https://doi.org/10.1016/j.biopsych.2014.07.030.

Patriquin, M. A., Bauer, I. E., Soares, J. C., Graham, D. P., & Nielsen, D. A. (2015). Addiction pharmacogenetics: A systematic review of the genetic variation of the dopaminergic system. *Psychiatric Genetics*, *25*(5), 181–193. https://doi.org/10.1097/YPG.0000000000000095.

West, R., & Hardy, A. (2006). *Theory of addiction*. Oxford: Blackwell.

White, M. J., Lawford, B. R., Morris, C. P., & Young, R. M. (2009). Interaction between DRD2 C957T polymorphism and an acute psychosocial stressor on reward-related behavioral impulsivity. *Behavior Genetics*, *39*(3), 285–295. https://doi.org/10.1007/s10519-008-9255-7.

Chapter 2

Impulsive behavior in drug addiction: Clinical, cognitive, and neural correlates

Yuliya Richard[a,b], Ahmed A. Moustafa[b]
[a]*School of Social Sciences and Psychology, Western Sydney University, Sydney, NSW, Australia,* [b]*Marcs Institute for Brain, Behaviour, and Development and School of Psychology, Western Sydney University, Penrith, NSW, Australia*

Introduction

Our ability to exercise self-control and to adjust our behavior when evaluating potential negative consequences of our actions is essential for our overall quality of life, development of healthy habits and achievement of our goals. People who are high in impulsivity exhibit poor self-control and emotional instability, which manifest in behaviors that might be risky and inappropriate, and often lead to negative and undesirable consequences (Gay, Schmidt, & Van der Linden, 2011). Thus, these individuals often feel powerless, weak, and ashamed due to their lack of control over their impulsive behaviors (Grant & Kim, 2003). Furthermore, such individuals often experience heightened anxiety, depressive moods, low self-esteem and increased stress associated with their impulsive behaviors (Grant & Kim, 2003). Impulsivity has been shown to play a significant role in the onset, development, and maintenance of drug addiction (Argyriou, Um, Carron, & Cyders, 2018), the development and maintenance of binge eating problems and binge eating disorder, and a greater degree of gambling severity (Kim et al., 2018). It has been associated with hypersexuality (Bőthe et al., 2018), higher frequency of suicide attempts (Colborn et al., 2017), and higher aggression (Hecht & Latzman, 2015).

Impulsivity is a complex aspect of human behavior. Despite numerous studies investigating the clinical and cognitive aspects of impulsivity, it has yet to be properly classified (Parry & Lindsay, 2003). Definitions describing this phenomenon incorporate "*behaviours that are poorly conceived, premature, inappropriate and that frequently result in unwanted and deleterious outcomes*" (Chamberlain & Sahakian, 2007). Impulsivity has been defined as a "*predisposition towards rapid, unplanned reactions to internal and external stimuli*

Cognitive, Clinical, and Neural Aspects of Drug Addiction. https://doi.org/10.1016/B978-0-12-816979-7.00002-9

without regard to the negative consequences of these reactions to the impulsive individual or to others" (Moeller, Barratt, Dougherty, Schmitz, & Swann, 2001). Impulsivity is also defined as "*acting on the spur of the moment, not focusing on the task at hand, and a lack of planning*" (Stratton, 2006).

Individuals who are high in impulsivity exhibit poor self-control and emotional instability, which manifests in behaviors that might be risky and inappropriate, and that often lead to negative and undesirable consequences (Gay et al., 2011). Thus, these individuals often feel powerless, weak, and ashamed due to their lack of control over their impulsive behaviors (Grant & Kim, 2003). Additionally, these individuals often experience heightened anxiety, depressive moods, low self-esteem and increased stress associated with their impulsive behaviors (Grant & Kim, 2003). Individuals with high impulsivity levels place themselves and people around them at risk. For individuals who experience difficulties in effectively managing their impulsive behaviors, it may turn into a lifelong debilitating condition in both clinical and nonclinical populations. Individuals with impulse control disorders are often unable to work and are at risk of social isolation, increased self-injurious behavior, and suicide. Thus, not only do individuals with impulse control disorders suffer the consequences of this condition, but their families and wider communities do as well. Individuals who experience difficulties controlling their impulsive actions do not often seek treatment due to the negative stigma associated with it. Particular difficulties might be present when self-control is moralized in terms of morally good or bad behavior, which in turn could be another potential barrier for people who might be willing to address their impulse control issues (Mooijman et al., 2017).

The biopsychosocial definition of impulsivity incorporates a non-adaptive fast response as well as a lack of planning (Moeller et al., 2001). Impulsivity has been broadly defined as a "*perseverance to the response that is punished and unrewarded; as well as, preference for small immediate rewards rather than larger delayed rewards and making responses that are immature or as inability to withhold a response*" (Moeller et al., 2001). Furthermore, Moeller and his colleagues (2001) suggested that the following factors should be considered when defining impulsivity in individuals: their decreased sensitivity to negative consequences of their behaviors, rapid, unplanned reactions to stimuli before processing information, and lack of regard for long-term consequences (Moeller et al., 2001).

It has been proposed that the following five dispositions underlie impulsivity: positive urgency, negative urgency, sensation seeking, lack of planning, and lack of perseverance. Positive urgency and negative urgency are based on emotions (Dick et al., 2010). Positive urgency is a tendency to act rapidly when experiencing positive mood and negative urgency is the tendency to act quickly when experiencing negative mood (Dick et al., 2010). Individuals high in urgency might experience difficulty resisting temptations and cravings (Johnson & Kim, 2011). For example, it has been reported that in borderline personality disorder, dysfunctional beliefs are associated with negative urgency, even after controlling for age, gender, depression, anxiety, and borderline personality disorder symptomatology (Gagnon, Daelman, McDuff, & Kocka, 2013). Negative urgency has been associated with reactive aggression (Gagnon & Rochat, 2017) and binge eating disorder (Fischer, Peterson, & McCarthy, 2013). It has been further shown that adolescents who might be at risk of larger alcohol consumption might act rashly when experiencing positive and negative urgency (Stautz & Cooper, 2013). Negative urgency in particular has been shown to be a predictor of lifelong alcohol use and risky substance use behavior (Athenour, 2016) and positive urgency was associated with higher levels of nicotine dependency (Spillane, Smith, & Kahler, 2010). Furthermore, urgency as well as lack of premeditation has been associated with problematic drug use in young people (Thomsen et al., 2018) and adults (Gipson, 2012). Higher urgency trait was observed in young people with conduct disorder (Urben, Suter, Pihet, Straccia, & Stephan, 2015).

The other two dispositions underlying impulsivity are based on deficits in conscientiousness (Johnson & Kim, 2011). They are considered to be lack of planning and lack of perseverance (Johnson & Kim, 2011). Lack of planning (or lack of premeditation) is considered to be the tendency to act without thinking about actions and their consequences (Johnson & Kim, 2011). Individuals experiencing deficits in the area of planning often act without thinking about the effects of their behavior on themselves and others (Johnson & Kim, 2011). It is considered that psychological factors associated with heightened impulsivity are a decreased level of control and a lesser degree of planning in impulsive acts (Parry & Lindsay, 2003).

According to Dick et al. (2010), individuals who have difficulties tolerating boredom and have difficulties staying focused when encountering distractions appear to experience a lack of perseverance. Dick et al. (2010) suggested that a lack of perseverance might be related to “resistance to proactive interference” (Dick et al., 2010). It has been demonstrated that individuals displaying drug use disorders appeared to be low in the conscientiousness trait (Zilberman, Yadid, Efrati, Neumark, & Rassovsky, 2018). Along these lines, one study investigating differences in personality traits among opiate-addicted females and those who had developed alcohol dependency showed that opiate-addicted females scored low on a conscientiousness scale (Raketic et al., 2017). Another study has demonstrated that individuals who develop a behavioral addiction, such as internet addiction, tend to show low levels of extraversion, openness to new experiences and conscientiousness (Hwang et al., 2014). It has been shown that negative urgency and a lack of perseverance appear to be associated with non-suicidal self-injury (Riley, Combs, Jordan, & Smith, 2015). Both predispositions, urgency and lack of perseverance have been considered predictive factors for the development of addiction in young people (Thomsen et al., 2018) and indicative of a more severe substance use problem (Tomko, Prisciandaro, Falls, & Magid, 2016). In other words, individuals who experience difficulties in lack of planning and lack of perseverance might not be able to attend to tasks that are perceived as boring or difficult (Johnson & Kim, 2011).

Thus, it is considered that sensation seeking, the tendency to seek out novel or thrilling stimulation, underlies a lack of perseverance (Dick et al., 2010). Often, such stimulating activities might include fast driving or engaging in fights, activities that put the involved individuals and people close to them at risk. It is interesting to note that sensation seeking does not appear to reflect deficits in executive functioning (Romer, 2010). Sensation seeking is a predicting factor of a vulnerability to drug addiction (Wingo, Nesil, Choi, & Li, 2016) and has been associated with online gaming addiction (Hu, Zhen, Yu, Zhang, & Zhang, 2017). Sensation seeking, however, might provide a barrier against high levels of anxiety and dysfunctional avoidance, such as thought suppression (Gay et al., 2011). As reviewed above, impulsivity is a multidimensional construct with a number of underlying cognitive, behavioral and biological dispositions that can have a significant impact on the individual's functioning and wellbeing. The traits underlying impulsivity are being studied and suggestions for further revisions of definitions, including conceptualization within the Diagnostic and Statistical Manual of Mental Disorders (DSM-V) suggest that some of the facets of impulsivity lack specificity (Griffin, Lynam, & Samuel, 2017).

Behavioral definition of impulsivity

Within the behavioral analysis domain, psychologists have mostly used two themes when providing operational definitions of impulsiveness in laboratory-based research. First, it is believed that impulsive individuals show deficient tolerance of the delay of gratification; second, that they have difficulty in delaying or inhibiting voluntary responding. Impulsiveness is considered to be a preference for smaller sooner rewards over larger later rewards (Tobin & Logue, 1994), which is known as delay discounting. It has been suggested that impulsiveness might arise as a difficulty of delay of gratification (Ainslie, 1975). At times, survival can be dependent on the choice of smaller rewards sooner rather than larger ones later. For example, a starving animal might have to take a small amount of food immediately to ensure its survival, rather than wait for larger prey later (Tobin & Logue, 1994). Similarly, people might learn as well to use smaller rewards sooner rather than larger later; for example, it is possible that in countries torn by war, people experiencing economic and political instability might choose immediate small rewards to ensure their existence, rather than larger ones later, as there is a real possibility that they might be worse off from waiting for larger rewards. In human studies, several factors, such as age or the quality of the reward, might impact the individual's ability to delay gratification (Tobin & Logue, 1994).

Three main factors that neuropsychologists consider in the definition of impulsivity are as follows: "inability to suppress powerful, overwhelming, and urgent response; inability to delay gratification by choosing smaller rewards now over larger later rewards; and inability to sustain attention" (Sharma, Markon, & Clark, 2014). The preference of individuals for obtaining immediate smaller rewards and discounting larger delayed rewards has been linked to three factors: their failure to recognize the contingency of the reward; their failure to respond to the possibility of the reward (shown by frequent engagement in compulsions or classical conditioning); and difficulties in conceptualizing the delayed rewards (Ainslie, 1975). The inability to delay gratification is referred to as temporal impulsivity and is measured using the delay discounting questionnaires or computerized tasks (Caswell, Morgan, & Duka, 2013). These tasks involve asking participants to make a choice between immediate or sooner smaller rewards or delayed larger rewards. Individuals high in temporal impulsivity tend to choose smaller rewards instantly rather than larger delayed rewards (Caswell et al., 2013).

The degree of delay discounting is considered a main factor in determining the impulsivity or self-control of an individual (Logue, 1998). A tendency to prefer small immediate rewards is considered a key feature in addictive disorders, as individuals who experience difficulties in this domain prefer to use substances in the present (Dick et al., 2010). A recent study has shown that individuals with heroin addiction demonstrated greater delayed discounting as well as decreased sensitivity in making optimal decisions, which is indicative of deficits in decision making (Scherbaum, Haber, Morley, Underhill, & Moustafa, 2018).

An individual's ability to choose delayed over immediate rewards is of great importance, not only in the recovery framework but in helping the individual achieve an optimal lifestyle in the future (Wittmann & Paulus, 2008). The need for immediate gratification is considered to be one of the facets of entitlement; thus, individuals with a high need for immediate gratification often experience emotional distress if they believe that they are entitled to better treatment than they are receiving (Harrington, 2007; Wittmann & Paulus, 2008). However, this is not limited to addictions only; individuals who display disruptive behaviors or are aggressive, and those attempting suicide tend to choose smaller immediate rewards (Dick et al., 2010).

Gratification could be a learnt behavior; thus, children can learn to act immediately to obtain immediate gratification (Moeller et al., 2001). It is possible that delay discounting evolved as an adaptive trait, as the greater the delay of the event the less likely it is to occur (Logue, 1998). This could be the case in those environments where waiting for a larger reward might be more likely to result in disadvantage; for example, in situations where it might threaten existence and minimize the survival rate (Logue, 1998).

Delay discounting is considered to be a cognitive process that allows the comparison of values of immediate and delayed use (Matta, Gonçalves, & Bizarro, 2012). Individuals with heightened impulsivity not only tend to choose smaller sooner rewards rather than delayed larger ones, but also show preference for larger delayed penalties over smaller immediate ones (Farmer & Golden, 2009). For example, a student might choose to deal with the consequences of poor performance during an examination instead of choosing the discomfort of spending a weekend studying.

Delay discounting is not only at the core of many maladaptive behaviors, such as addictive behaviors and gambling as well as poor academic performance and self-care, but it can be predictive of the likelihood of people who smoke and use drugs to relapse when they attempt to cease such behaviors (Odum, 2011). Delay discounting is observed in major forms of addiction, and it has been suggested that the extent of delay discounting is proportional to amount of drug used (Bickel, Moody, Eddy, & Franck, 2017). Delay discounting as a behavioral marker of addiction has been shown to be a predictor of nicotine smoking cessation (Athamneh, Stein, & Bickel, 2017).

Delay discounting plays an important role in the research of impulsiveness, self-control and decision making (Matta et al., 2012). Research on the role of delay of gratification in behavior contributes to the wealth of knowledge in the areas of addictions, obesity and gambling (Odum, 2011).

While delay discounting tasks help measure an individual's preference for immediate small or larger delayed rewards, one of the potential limitations of delay discounting tasks is the measure of hypothetical rewards. It is important to conduct research using real value rewards (Odum, 2011).

Functional and dysfunctional impulsivity

The consequences of impulsive behavior are not always negative (Dickman, 1990). It has been suggested that high impulsiveness when completing simple tasks under time constraints can lead to more accurate results than low impulsiveness (Dickman, 1990). Thus, impulsivity is not always associated with a negative psychological outcome (Johnson & Kim, 2011). While most of the research has been dedicated to studying the negative consequences of impulsivity, also known as negative impulsivity, it is important to note that the tendency to act quickly to achieve a positive outcome has been referred to as functional impulsivity (Winkel, Wyland, Shaffer, & Clason, 2011). Negative impulsivity occurs despite negative consequences; however, positive impulsivity occurs due to the benefits resulting from this type of information processing style (Johnson & Kim, 2011). Thus, functional impulsivity has been associated with enthusiasm, adventurousness and extraversion (Winkel et al., 2011), idea generation and rapid decision making (Di Milia, 2013).

An example of functional impulsivity can be considered in an individual who spontaneously offers a helping hand to their colleague, offering to induct the new employee and show them around the office, or who volunteers to participate in a charitable event. In contrast, dysfunctional impulsivity leads to negative outcomes and is often associated with aggression, violence and conduct disorder (Winkel et al., 2011). It appears that an individual with dysfunctional impulsivity is more likely to disregard facts when making decisions, subsequently leading to increased displaying of behaviors that might boost morale in the office or benefit the whole organization (Winkel et al., 2011). People who are high in functional impulsivity will act without thinking about the consequences of their behavior when such a cognitive style is optimal; they are thereby considered active, enthusiastic, spontaneous and productive risk takers (Johnson & Kim, 2011). It is considered that enthusiastic and active people are more productive, even though they engage in risk-taking activities like individuals displaying dysfunctional impulsivity (Dickman, 1990).

It appears that an individual with dysfunctional impulsivity is more likely to disregard facts when making decisions, subsequently leading to increased negative outcomes of their impulsive behavior (Dickman, 1990). People high in dysfunctional impulsivity are considered to be inattentive and careless, resulting in planning deficits and inability to delay gratification, which contributes to negative consequences (Johnson & Kim, 2011). Dysfunctional impulsivity is associated with fast but inaccurate decision making, high distractibility and suicidal thoughts (Di Milia, 2013). It is considered that individuals high in dysfunctional impulsivity engage in fast and erroneous decision making due to their inability to use a slower methodical approach (Dickman, 1990). It has been suggested that smokers who are high in dysfunctional impulsivity might require more intensive treatment to address their addiction (Pitts & Leventhal, 2012).

It has been suggested that stressful circumstances might interfere with the individual's ability to engage in a slow and measured approach when making decisions (Dickman, 1990). High dysfunctional impulsivity is associated with higher distractibility while driving and lane crossing (Di Milia, 2013), as well as driving while intoxicated (Eensoo, Paaver, Pulver, Harro, & Harro, 2004) and with high incidents of binge drinking in young males (Adan, 2012).

Dysfunctional impulsivity is also associated with the heightened tendency of individuals diagnosed with schizophrenia to display repetitive violent behavior (Kumari et al., 2009). Dysfunctional impulsivity is considered to be at the core of many psychopathies (Poythress & Hall, 2011). Individuals who scored higher on dysfunctional impulsivity demonstrated poorer responses to treatment for pathological gambling (Maccallum, Blaszczynski, Ladouceur, & Nower, 2007). Furthermore, dysfunctional impulsivity was associated with higher craving as well as smoking without conscious awareness, loss of control, smoking while ill and experiencing difficulties abstaining from non-smoking areas in individuals aiming to cease smoking (Pitts & Leventhal, 2012).

Dickman (1990) has developed an inventory that assesses functional and dysfunctional impulsivity (Dickman, 1990). Both functional and dysfunctional impulsivity correlated positively with extraversion on the Eysenck Personality Questionnaire; however, only functional impulsivity correlated positively with psychoticism and negatively with neuroticism (Brunas-Wagstaff, Bergquist, Richardson, & Connor, 1995). The Dysfunctional scale has been reported to correlate with the Motor Impulsivity subscale of the Barratt Impulsiveness Scale-11 (Caci, Nadalet, Bayle, Robert, & Boyer, 2003).

Impulsivity as a clinical problem

Diagnosis of a psychiatric condition in an individual can have an impact on the manifestation of impulsive behaviors (Parry & Lindsay, 2003). Impulsivity is a prominent feature of many psychiatric and developmental disorders: personality disorders (such as Borderline Personality Disorder [BPD]), eating disorders (such as bulimia), mood disorders (such as bipolar disorder [BD]) and substance addiction (such as drug addiction) (Parry & Lindsay, 2003). The Diagnostic and Statistical Manual of Mental Disorders (DSM-V) addresses a number of Disruptive, Impulse-control and Conduct disorders. It acknowledges problems in self-control of emotions and behavior as a main problem experienced by individuals diagnosed with these disorders (DSM-V). Some of these disorders categorized by DSM-V are Intermittent Explosive Disorder, characterized by "recurrent behavioural outbursts representing a failure to control aggressive impulses" and Kleptomania, characterized by the "recurrent failure to resist impulses to steal items even though they are not needed for personal use or for their monetary value", (DSM-V). The DSM-IV listed six categories of impulse control disorders characterized by a variety of uncontrolled behaviors. The categories are the following: failure to resist aggressive impulses, failure to resist impulses to steal objects not needed for personal use, deliberate and purposeful fire setting for personal gratification or relief, persistent maladaptive gambling behavior, and recurrent pulling out of one's hair, and these are the essential features of intermittent explosive disorder, kleptomania, pyromania, pathological gambling and trichotillomania, respectively. The sixth category of "impulse control disorder not otherwise specified" includes any other repeated failure to resist impulses to carry out a particular behavior; for example, compulsive buying, compulsive sexual behavior and repetitive self-mutilation.

It has been shown that there is a strong association between impulsiveness, anger and violence risk in psychiatric patients, impulsiveness and violence risk in the forensic population and suicide and impulsiveness in individuals diagnosed with Post Traumatic Stress Disorder (PTSD) (Parry & Lindsay, 2003).

Generally impulsivity has been associated with self-injurious and suicidal behaviors (Klonsky & May, 2010). It has been reported that impulsivity plays a significant role in individuals diagnosed with BPD, and it is considered an especially important factor in those who attempt self-harm and suicide (Moeller et al., 2001). Thus, it is considered that self-destructive impulsivity is the most challenging aspect of this psychiatric condition. Similarly, individuals diagnosed with Bipolar Disorder (BD) tend to engage in impulsive or risky behaviors that often lead to undesired outcomes (Kathleen et al., 2009). Thus, impulsivity is considered to be a core feature of BD, irrespective of the presence of co-morbid mood states or alcohol abuse history (Kathleen et al., 2009).

Individuals diagnosed with Obsessive Compulsive Disorder (OCD), similarly to individuals with impulsive control disorders, exercise poor control over their behavior, even though it might be considered that in OCD such behaviors are due to heightened anxiety, whereas in impulsive control disorders individuals it is due to their arousal (Ettelt et al., 2007). At the same time, recent studies have found that individuals with OCD obtained high scores on cognitive impulsiveness, and furthermore the studies found strong associations of cognitive impulsiveness, aggressive obsessions and checking (Ettelt et al., 2007).

Impulsivity has been positively correlated with the intensity and amount of drug use and increased withdrawal (Bankston et al., 2009). With repeated drug or alcohol use, as with many other addictive behaviors, the impulsive system becomes sensitized to the substance and the cues that predict the use of the substance (Wiers & Stacy, 2006). Furthermore, impulsive individuals tend to underestimate the risks and consequences of their health behavior, such as drug use (Ouzir & Errami, 2016). A study evaluating rapid discounting of delayed hypothetical rewards by cocaine-dependent individuals, compared to non-drug using individuals, showed that cocaine-dependent individuals consistently discounted delayed rewards and preferred smaller but immediate rewards (Coffey, Gudleski, Saladin, & Brady, 2003). It is possible as well that this outcome could be due to a "reduction of dopamine receptors in the orbitofrontal lode and cingulate gyrus regions of the brain" in chronic cocaine users (Coffey et al., 2003). Similarly, damage to the prefrontal cortex could be observed as a result of long-term alcohol abuse (Walker, Pena-Oliver, & Stephens, 2011). Thus, impulsivity has been associated with the dependence of drug use and poor cessation rates rather than initial habit formation (Hogarth, 2011). It appears that impulsivity might place an individual at a higher risk of developing addictive behaviors, and at the same time prolonged and repeated engagement in alcohol or drug use might influence individuals by impacting their impulsivity.

It has been suggested that individuals who engage in binge drinking behaviors experience problems associated with difficulties in controlling their impulses (motor impulsivity) and impulsive decision making (cognitive impulsivity) (Field, Schoenmakers, & Wiers, 2008). Thus, such individuals consistently choose rewards that are readily available, despite negative consequences (Field et al., 2008). While the impulsivity trait has been shown to be predictive of alcohol use disorders, at the same time alcohol use can reduce self-control and lead to an increase in impulsive behaviors (Dick et al., 2010). It has been noted that severe impulsiveness was one of the clinical presentations of multiple sclerosis as well (Lopez-Meza, Corona-Vazquez, Ruano-Calderon, & Ramirez-Bermudez, 2005). Furthermore, individuals with frontal lobe injuries have found that it affects their ability to plan and sustain attention, which indicates that impulsivity is more prominent when such injuries are present (Moeller et al., 2001).

Impulsivity, alongside maladaptive coping strategies, hopelessness, cognitive rigidity, problem solving difficulties and hostility, is associated with deliberate self-harm and suicide attempts (Raj, Kumaraiah, & Bhide, 2001). Other factors considered to be in highly impulsive individuals which may lead to further self-harm are poor self-worth, emotional intolerance, entitlement beliefs, and anger (Harrington, 2007). Negative life events and stress might increase vulnerability attempts (Mehlum, 2001). Impulsivity is related to the severity of gambling behaviors and is indicative of a future development of pathological gambling (Forbush et al., 2008). It has been suggested that impairment in executive control, especially stopping, has been associated with impulse control disorders, such as ADHD, substance abuse and pathological gambling (Verbruggen, Adams, & Chambers, 2012). People recovering from drug addictions might benefit from addressing deficits in impulsivity and executive functioning which interfere with their daily functioning and place them at risk of premature treatment cessation and relapse (Ellis et al., 2016). Therefore, the treatment focus should be on response inhibition, thus addressing relapse prevention (Verbruggen et al., 2012).

It is worth noting that impulsivity becomes a prominent feature of some other physical conditions and one of the consequences of endured traumas. Impulsivity often becomes a prominent feature in individuals who have been deprived of sleep. Those who had orbitofrontal cortex lesions have also been shown to be more impulsive on behavioral tests including delay discounting (Wittmann & Paulus, 2008).

Difficulties with controlling impulses might be noted in people who have a diagnosis of intellectual disability or cognitive impairment (Parry & Lindsay, 2003). They often experience difficulty in impulse control alongside anger management, poor socialization and poor coping strategies (Parry & Lindsay, 2003). Furthermore, impulsivity alongside repetitive behaviors has been associated with increased self-injurious behavior in individuals diagnosed with an intellectual disability (Bigham, Daley, Hastings, & Jones, 2013).

Impulsivity in the nonclinical population

Impulsivity has been associated with lower academic grades, higher aggression, interpersonal violence, impulsive shopping, and stealing in clinical and nonclinical populations (Spinella, 2004). Some other behaviors associated with high impulsivity include maladaptive mobile phone use and internet pornography viewing (Sharma, Kohl, Morgan, & Clark, 2013). Overall, for individuals to maintain a healthy lifestyle, they need to engage in self-control to maintain exercise programs and restrain themselves in their food choices (Sutin, Ferrucci, Zonderman, & Terracciano, 2011). Impulsiveness, especially Sensation Seeking, has been consistently associated with weight gain (Sutin et al., 2011). A study investigating decision making in obese individuals indicated that poor and impulsive decision making has been shown to be prevalent in choosing high immediate gain but larger future losses (Davis, Levitan, Muglia, Bewell, & Kennedy, 2004).

Individuals displaying high levels of impulsivity might not plan their meals or allow time for exercise, and fail to resist urges for unhealthy food choices (Lyke & Spinella, 2004). Findings have shown that motor and intentional impulsivity might be contributing to young people's impulsive choices related to their food intake (Lyke & Spinella, 2004). Another study investigating delay discounting in obese individuals demonstrated that obese women, when compared to non-obese women, demonstrated greater discount for delayed rewards, thus preferring smaller immediate rewards rather than delayed larger rewards (Weller, Cook, Avsar, & Cox, 2008). Additionally, it has been shown that there is a strong association between high intentional impulsivity and anxiety (Lyke & Spinella, 2004). According to the theory of emotional eating, some individuals who have learnt that eating reduces their anxiety and fear might feel motivated to continue eating as a response to these feelings (Stroebe, van Koningsbruggen, Papies, & Aarts, 2013). It is interesting to note that individuals with high impulsivity demonstrate higher weight gain across their lifespan (Sutin et al., 2011).

Impulsivity, like some other personality traits, is considered to be a trait or characteristic that remains stable regardless of the individual's situation or circumstances (Wells, Parboteeah, & Valacich, 2011). Thus, if an individual tends to engage in impulsive behavior, such as buying online, this behavior can be similarly observed in an offline environment (Wells et al., 2011). Overall, individuals who have difficulties with exercising any one particular self-control area (such as shopping) often have difficulties in other areas as well (such as gambling) (Sharma et al., 2014). In other words, some people act impulsively rarely and usually only in extreme circumstances (impulsiveness state); whereas other people act impulsively persistently—they present with an impulsivity trait (Parry & Lindsay, 2003).

While the initial urge to buy something is very strong for individuals with an impulsivity trait, not all individuals act when they experience an urge to engage in impulsive behavior (Wells et al., 2011). It is interesting to note that the impulsiveness trait in consumerism has been considered from a cultural perspective, studying individuals from both collectivistic (interdependent self-concept) cultures and individualistic (independent self-concept) cultures (Sharma, Sivakumaran, & Marshall, 2011). Individuals with interdependent self-concept were able to differentiate between self-indulgence and involuntary loss of self-control when making decisions regarding purchases, which has not been the same for individuals with independent self-concept (Sharma et al., 2011). The knowledge of the role of impulsivity in consumerism can be used by advertisement and marketing agencies. This can be of benefit to the economy (Arens & Rust, 2012); however, at the same time it might be detrimental to individuals who experience difficulties in controlling their impulses and end up facing the negative consequences of credit card debts, loss of jobs and relationships.

It has been shown that impulsivity facets such as urgency and lack of perseverance, as well as dysfunctional thought control strategies, were linked to insomnia (Schmidt, Gay, Ghisletta, & Van Der Linden, 2010). Thus, lack of perseverance might exacerbate the occurrence of unwanted thoughts and worries and urgency might be linked to the inability to suppress dominant, automatic responses (Schmidt et al., 2010).

Impulsivity varies across ages and gender. A study investigating deviant behaviors among adolescents has shown such behaviors displayed by boys were greater than those displayed by girls (Esteban & Tabernero, 2011). Related to age, younger adolescents have been shown to be more impulsive, which resulted in higher instances of disruptive behavior (Esteban & Tabernero, 2011). Overall, sensation seeking was detected to be greater in boys aged 14–22 than girls (Romer, 2010). Males in this age group are more likely to take risks, initiate drug use and engage in criminal activity (Romer, 2010). Impulsivity in boys has been considered to place them at a higher risk for substance abuse and behavioral impulsivity (Bankston et al., 2009). It has been found that male substance abusers demonstrate high motor impulsivity and non-planning impulsivity in individuals with substance use issues (Farnell, 2013).

At the same time, it has been suggested that younger individuals are more prone to risk-taking activities and displaying impulsive behaviors, such as "drug use, unintentional injuries and unprotected sexual activities, which might be accounted by still maturing prefrontal cortex" (Romer, 2010). At the same time, youth impulsivity is not considered a universal phenomenon and is considered to be mediated among other factors by the effects of state and peer influence (Romer, 2010). While younger people are consider to be more impulsive than adults, the decision processes that are influenced by sensation seeking are the same as in adults (Romer, 2010).

In particular, negative affect and motivation in school students were predictive of high impulsiveness, whereas positive affect, internal locus of control as well as self-esteem and intrinsic motivation were predicative of non-impulsiveness (Palomo, Beninger, Kostrzewa, & Archer, 2008). It is consistent with the notion of mood-congruency hypothesis: when people are in a negative mood they are more likely to interpret events around them negatively; similarly, when they are in a positive mood they are more likely to interpret events more positively (Rusting, 1998).

It has been shown that acute stress has been contributing to the overestimation of time and might be a contributing factor to exacerbation of impulsive behaviors (Wittmann & Paulus, 2008). Furthermore, stress interacts with impulsiveness, placing individuals at a higher risk for engagement in addictive behaviors (Sheffer et al., 2012). Thus, individuals with high impulsivity who experience heightened stress are more likely to relapse after treatment and initiate smoking again (Sheffer et al., 2012). This might be due to them feeling that they have little control over the main events in their lives, attributing this to luck, destiny and the greater power of other people, thus experiencing heightened stress (Sheffer et al., 2012). Dickman (1990) suggested that stress or stressful circumstances interfere with an individual's ability to process information slowly and accurately in individuals who display dysfunctional impulsivity (Dickman, 1990).

Thus, there is overwhelming evidence of the impact of dysfunctional impulsivity on people in nonclinical populations. Difficulties in controlling behavior, ranging from overspending, overeating, controlling anger, poor school performance and others, not only impact the daily functioning of people high in dysfunctional impulsivity, but the severity of these behaviors is exacerbated by stress and in some cases anxiety and negative mood. Therefore, intervention programs will benefit from incorporating psychoeducational materials about dysfunctional impulsivity, stress, negative mood, and anxiety as well as helping participants learn practical strategies to address their mood, anxiety and stress.

Impulsivity and cognitive distortions

It has been suggested that cognitive distortions are associated with impulsive behavior and impulsivity-related psychopathologies (Gagnon et al., 2013). Beck's investigations with his patients led him to believe that when patients experienced depression, certain cognitive schemas may lead to cognitive distortions in those patients; thus, the depressed patient might feel sad and lonely, and mistakenly think that he is abandoned and excluded by others (Weissman & Beck, 1978). Cognitive distortions are not unique in depressive patients; further cognitive distortions can be found in many other psychiatric conditions. Several cognitive distortions have been identified as contributing to people's impulsive behaviors, such as the illusion of control, and have been found to be at the core of pathological gambling. It has been suggested that people who are aiming to address their gambling addiction might benefit from addressing "chasing losses" cognitive distortions (Chamberlain, Stochl, Redden, Odlaug, & Grant, 2017). Cognitive distortions related to substance use disorder have been shown to be related to lack of skill in coping with problems or unpleasant feelings, low tolerance for frustration, sensation-seeking, difficulties tolerating boredom, diminished future time perspective, catastrophizing, blame, punishment, personalization, and all-or-nothing thinking (Ramirez, 2001). People who have developed substance use disorders might experience a cognitive distortion associated with overestimation of a substance they misuse or have been addicted to; for example, smokers tend to overestimate tobacco and experience low response to non-tobacco rewards (Isomura, Suzuki, & Murai, 2014). People with borderline personality disorders are more likely to negatively evaluate the events in their lives due to cognitive distortions. Furthermore, people with comorbid conditions tend to have more cognitive distortions than those with one diagnosis (Gagnon et al., 2013).

Mobini, Pearce, Grant, Mills, and Yeomans (2006) found that highly impulsive individuals present dysfunctional beliefs that impact their ability to interpret events, as well as their inability to make decisions taking into account the long-term consequences of their behavior (Mobini et al., 2006). A recent study demonstrated that negative evaluations of self, world and the future; tendency to overgeneralize, personalize and catastrophize; and selective negative interpretation of stressful events have been shown to be present in youth victims of dating violence who have developed substance use problems (Miller, Williams, Day, & Esposito-Smythers, 2017).

In relation to impulsivity, some cognitive distortions might impact the delay of gratification, such as short-term thinking or confusing needs and wants, while others might impact impulsive decision making (Mobini et al., 2006).

It has been suggested previously that certain cognitive attributes may operate as cognitive processing associated with some subtypes of impulsivity, such as dysfunctional impulsivity, non-planning and cognitive impulsiveness (Mobini, Grant, Kass, & Yeomans, 2007). It has been shown that delay discount rates were positively correlated with both functional and dysfunctional impulsivity (Mobini et al., 2007). Additionally, delay discounting was positively correlated with three cognitive distortions as measured by the Cognitive Distortion Scale, such as "mind reading", "instant satisfaction" and "short-term thinking" (Mobini et al., 2007). It has been previously suggested that cognitive behavioral interventions might be effective in addressing cognitive distortions specific to impulsivity, thus assisting impulsive people in the effective management of their emotions and behaviors (Mobini et al., 2006).

It has been suggested that individuals tend to be more impulsive if they have an external locus of control; for example, if they tend to believe that luck, fate and the actions of others are responsible for their behaviors (Sheffer et al., 2012). A recent study investigating impulsivity and cognitive distortions among pathological gamblers showed that they demonstrated a higher preference for immediate rewards and the presence of cognitive distortions related to gambling (Michalczuk, Bowden-Jones, Verdejo-Garcia, & Clark, 2011). Similar findings were obtained in another study, confirming a robust relationship between pathological gambling and cognitive distortions. Another recent systematic review study has highlighted that increased impulsivity and dysfunctional decision making appear to be strongly associated with general drug use (Betzler, Viohl, Romanczuk-Seiferth, & Foxe, 2017).

As mentioned previously, Mobini et al. (2006) suggests that cognitive behavioral interventions might be of benefit to highly impulsive individuals (Mobini et al., 2006). It has been demonstrated that cognitive behavioral therapy is effective in treating pathological gambling, especially when incorporating cognitive restructuring and so addressing cognitive distortions (Fortune & Goodie, 2012). Thus, in developing an intervention program for dysfunctional impulsive behavior, we need to focus on understanding the cognitive distortions involved in impulsive behavior and the ways in which we can modify these cognitive distortions (Mobini et al., 2007).

High impulsivity relates to a lack of consideration for negative future consequences, and as a result has been shown to have negative effects on many different aspects of life. Dysfunctional impulsivity can be present in any population and leads to a variety of unhealthy life choices, but those who have high impulsivity (including individuals with drug use disorders) are often reluctant to seek treatment for this issue due to social stigma. In particular, impulsivity has been consistently shown to be a factor in the development of a number of psychiatric conditions, particularly addictive and compulsive behaviors.

Those with drug use disorders are often deficient in the ability to employ delay discounting and delay of gratification, both of which are directly related to high impulsivity. In some cases, immediate gratification may be a learned behavior if a person has lived in conditions of scarcity. The extended use of substances can have negative effects on the areas of the brain related to reward centers and executive function, which further decreases self-control and increases impulsivity. Due to this, high impulsivity is also correlated with longer periods of substance abuse and high risk of relapse.

Cognitive distortions contribute to high levels of impulsivity; those with substance use disorders commonly lack strategies to cope with negative emotions and are prone to short-term thinking. These cognitive distortions often correlate with the deficits in delay discounting and delay of gratification seen in people with drug use disorders. However, it has been shown that addressing these cognitive distortions can have positive effects on those suffering from a pathological addiction or compulsion. From this basis, treatments can be developed for those with drug use disorders that modify the cognitive distortions related to high impulsivity. Such treatments are likely to address many of the underlying causes of impulsivity and prevent relapse in those with substance abuse disorders.

References

Adan, A. (2012). Functional and dysfunctional impulsivity in young binge drinkers. *Adicciones, 24*(1), 17–22.

Ainslie, G. (1975). Specious reward: A behavioral theory of impulsiveness and impulse control. *Psychological Bulletin, 82*(4), 463.

Arens, Z. G., & Rust, R. T. (2012). The duality of decisions and the case for impulsiveness metrics. *Journal of the Academy of Marketing Science, 40*(3), 468–479.

Argyriou, E., Um, M., Carron, C., & Cyders, M. A. (2018). Age and impulsive behavior in drug addiction: A review of past research and future directions. *Pharmacology, Biochemistry, and Behavior, 164*, 106–117.

Athamneh, L. N., Stein, J. S., & Bickel, W. K. (2017). Will delay discounting predict intention to quit smoking? *Experimental and Clinical Psychopharmacology, 25*(4), 273–280. https://doi.org/10.1037/pha0000129.

Athenour, D. R. (2016). Substance use across adolescence: The role of negative urgency, school bonding, and perceptions of peer use. *Dissertation Abstracts International: Section B: The Sciences and Engineering, 77*(2-B(E)).

Bankston, S. M., Carroll, D. D., Cron, S. G., Granmayeh, L. K., Marcus, M. T., Moeller, F. G., et al. (2009). Substance abuser impulsivity decreases with a nine-month stay in a therapeutic community. *American Journal of Drug & Alcohol Abuse, 35*(6), 417–420.

Betzler, F., Viohl, L., Romanczuk-Seiferth, N., & Foxe, J. (2017). Decision-making in chronic ecstasy users: A systematic review. *European Journal of Neuroscience, 45*(1), 34–44. https://doi.org/10.1111/ejn.13480.

Bickel, W. K., Moody, L. N., Eddy, C. R., & Franck, C. T. (2017). Neurocognitive dysfunction in addiction: Testing hypotheses of diffuse versus selective phenotypic dysfunction with a classification-based approach. *Experimental and Clinical Psychopharmacology, 25*(4), 322–332. https://doi.org/10.1037/pha0000115.

Bigham, K., Daley, D. M., Hastings, R. P., & Jones, R. S. P. (2013). Association between parent reports of attention deficit hyperactivity disorder behaviours and child impulsivity in children with severe intellectual disability. *Journal of Intellectual Disability Research*, *57*(2), 191–197.

Bőthe, B., Tóth-Király, I., Potenza, M. N., Griffiths, M. D., Orosz, G., & Demetrovics, Z. (2018). Revisiting the role of impulsivity and compulsivity in problematic sexual behaviors. *Journal of Sex Research*, 1–14.

Brunas-Wagstaff, J., Bergquist, A., Richardson, P., & Connor, A. (1995). The relationships between functional and dysfunctional impulsivity and the Eysenck personality questionnaire. *Personality and Individual Differences*, *18*(5), 681–683.

Caci, H., Nadalet, L., Bayle, F. J., Robert, P., & Boyer, P. (2003). Functional and dysfunctional impulsivity: Contribution to the construct validity. *Acta Psychiatrica Scandinavica*, *107*(1), 34–40.

Caswell, A. J., Morgan, M. J., & Duka, T. (2013). Inhibitory control contributes to "motor"-but not "cognitive"-impulsivity. *Experimental Psychology*.

Chamberlain, S. R., & Sahakian, B. J. (2007). The neuropsychiatry of impulsivity. *Current Opinion in Psychiatry*, *20*(3), 255–261.

Chamberlain, S. R., Stochl, J., Redden, S. A., Odlaug, B. L., & Grant, J. E. (2017). Latent class analysis of gambling subtypes and impulsive/compulsive associations: Time to rethink diagnostic boundaries for gambling disorder? *Addictive Behaviors*, *72*, 79–85. https://doi.org/10.1016/j.addbeh.2017.03.020.

Coffey, S. F., Gudleski, G. D., Saladin, M. E., & Brady, K. T. (2003). Impulsivity and rapid discounting of delayed hypothetical rewards in cocaine-dependent individuals. *Experimental and Clinical Psychopharmacology*, *11*(1), 18–25.

Colborn, V. A., LaCroix, J. M., Neely, L. L., Tucker, J., Perera, K., Daruwala, S. E., et al. (2017). Motor impulsivity differentiates between psychiatric inpatients with multiple versus single lifetime suicide attempts. *Psychiatry Research*, *253*, 18–21.

Davis, C., Levitan, R. D., Muglia, P., Bewell, C., & Kennedy, J. L. (2004). Decision-making deficits and overeating: A risk model for obesity. *Obesity Research*, *12*(6), 929–935.

Di Milia, L. (2013). A revised model of Dickman's dysfunctional impulsivity scale. *Journal of Individual Differences*.

Dick, D. M., Smith, G., Olausson, P., Mitchell, S. H., Leeman, R. F., O'malley, S. S., et al. (2010). Understanding the construct of impulsivity and its relationship to alcohol use disorders. *Addiction Biology*, *15*(2), 217–226.

Dickman, S. (1990). Functional and dysfunctional impulsivity: Personality and cognitive correlates. *Journal of Personality and Social Psychology*, *58*(1), 95.

Eensoo, D., Paaver, M., Pulver, A., Harro, M., & Harro, J. (2004). Low platelet MAO activity associated with high dysfunctional impulsivity and antisocial behavior: Evidence from drunk drivers. *Psychopharmacology*, *172*(3), 356–358.

Ellis, C., Hoffman, W., Jaehnert, S., Plagge, J., Loftis, J. M., Schwartz, D., et al. (2016). Everyday problems with executive dysfunction and impulsivity in adults recovering from methamphetamine addiction. *Addictive Disorders & Their Treatment*, *15*(1), 1–5. https://doi.org/10.1097/ADT.0000000000000059.

Esteban, A., & Tabernero, C. (2011). Relationship between impulsiveness and deviant behavior among adolescents in the classroom: Age and sex differences. *Psychological Reports*, *109*(3), 703–717.

Ettelt, S., Ruhrmann, S., Barnow, S., Buthz, F., Hochrein, A., Meyer, K., et al. (2007). Impulsiveness in obsessive-compulsive disorder: Results from a family study. *Acta Psychiatrica Scandinavica*, *115*(1), 41–47.

Farmer, R. F., & Golden, J. A. (2009). The forms and functions of impulsive actions: Implications for behavioral assessment and therapy. *International Journal of Behavioral Consultation and Therapy*, *5*(1), 12.

Farnell, E. (2013). Gender differences in impulsivity and decision making. *Dissertation Abstracts International: Section B: The Sciences and Engineering*, *73*(7-B(E)).

Field, M., Schoenmakers, T., & Wiers, R. W. (2008). Cognitive processes in alcohol binges: A review and research agenda. *Current Drug Abuse Reviews*, *1*(3), 263–279.

Fischer, S., Peterson, C. M., & McCarthy, D. (2013). A prospective test of the influence of negative urgency and expectancies on binge eating and purging. *Psychology of Addictive Behaviors*, *27*(1), 294.

Forbush, K. T., Shaw, M., Graeber, M. A., Hovick, L., Meyer, V. J., Moser, D. J., et al. (2008). Neuropsychological characteristics and personality traits in pathological gambling. *CNS Spectrums*, *13*(4), 306–315.

Fortune, E. E., & Goodie, A. S. (2012). Cognitive distortions as a component and treatment focus of pathological gambling: A review. *Psychology of Addictive Behaviors*, *26*(2), 298–310.

Gagnon, J., Daelman, S., McDuff, P., & Kocka, A. (2013). UPPS dimensions of impulsivity. *Journal of Individual Differences*.

Gagnon, J., & Rochat, L. (2017). Relationships between hostile attribution bias, negative urgency, and reactive aggression. *Journal of Individual Differences*, *38*(4), 211.

Gay, P., Schmidt, R. E., & Van der Linden, M. (2011). Impulsivity and intrusive thoughts: Related manifestations of self-control difficulties? *Cognitive Therapy and Research*, *35*(4), 293–303.

Gipson, C. D. (2012). A translational model of mood-based rash action. *Dissertation Abstracts International: Section B: The Sciences and Engineering*, *73*(6-B), 3996.

Grant, J. E., & Kim, S. W. (2003). *Stop me because i can't stop myself: Taking control of impulsive behavior*. New York: McGraw-Hill.

Griffin, S. A., Lynam, D. R., & Samuel, D. B. (2017). Dimensional conceptualizations of impulsivity. *Personality Disorders*, *9*(4), 333–345.

Harrington, N. (2007). Frustration intolerance as a multidimensional concept. *Journal of Rational-Emotive & Cognitive-Behavior Therapy*, *25*(3), 191–211.

Hecht, L. K., & Latzman, R. D. (2015). Revealing the nuanced associations between facets of trait impulsivity and reactive and proactive aggression. *Personality and Individual Differences*, *83*, 192–197.

Hogarth, L. (2011). The role of impulsivity in the aetiology of drug dependence: Reward sensitivity versus automaticity. *Psychopharmacology*, *215*(3), 567–580.

Hu, J., Zhen, S., Yu, C., Zhang, Q., & Zhang, W. (2017). Sensation seeking and online gaming addiction in adolescents: A moderated mediation model of positive affective associations and impulsivity. *Frontiers in Psychology*, *8*, 699.

Hwang, J. Y., Choi, J.-S., Gwak, A. R., Jung, D., Choi, S.-W., Lee, J., et al. (2014). Shared psychological characteristics that are linked to aggression between patients with Internet addiction and those with alcohol dependence. *Annals of General Psychiatry*, *13*(2014), 6. https://doi.org/10.1186/1744-859X-13-6.

Isomura, T., Suzuki, J., & Murai, T. (2014). Paradise lost: The relationships between neurological and psychological changes in nicotine-dependent patients. *Addiction Research & Theory*, *22*(2), 158–165. https://doi.org/10.3109/16066359.2013.793312.

Johnson, J. L., & Kim, L. M. (2011). The role of impulsivity in forgiveness. *Individual Differences Research*, *9*(1), 12–21.

Kathleen, H. M., Bearden, C. E., Barguil, M., Fonseca, M., Serap Monkul, E., Nery, F. G., et al. (2009). Conceptualizing impulsivity and risk taking in bipolar disorder: Importance of history of alcohol abuse. *Bipolar Disorders*, *11*(1), 33–40.

Kim, H. S., Cassetta, B. D., Hodgins, D. C., Tomfohr-Madsen, L. M., McGrath, D. S., & Tavares, H. (2018). Assessing the relationship between disordered gamblers with psychosis and increased gambling severity: The mediating role of impulsivity. *The Canadian Journal of Psychiatry*, *63*(6), 370–377.

Klonsky, E. D., & May, A. M. A. (2010). Rethinking impulsivity in suicide. *Suicide and Life-threatening Behavior*, *40*(6), 612–619.

Kumari, V., Barkataki, I., Goswami, S., Flora, S., Das, M., & Taylor, P. (2009). Dysfunctional, but not functional, impulsivity is associated with a history of seriously violent behaviour and reduced orbitofrontal and hippocampal volumes in schizophrenia. *Psychiatry Research*, *173*(1), 39–44. https://doi.org/10.1016/j.pscychresns.2008.09.003.

Logue, A. W. (1998). Laboratory research on self-control: Applications to administration. *Review of General Psychology*, *2*(2), 221.

Lopez-Meza, E., Corona-Vazquez, T., Ruano-Calderon, L. A., & Ramirez-Bermudez, J. (2005). Severe impulsiveness as the primary manifestation of multiple sclerosis in a young female. *Psychiatry and Clinical Neurosciences*, *59*(6), 739–742.

Lyke, J. A., & Spinella, M. (2004). Associations among aspects of impulsivity and eating factors in a nonclinical sample. *International Journal of Eating Disorders*, *36*(2), 229–233.

Maccallum, F., Blaszczynski, A., Ladouceur, R., & Nower, L. (2007). Functional and dysfunctional impulsivity in pathological gambling. *Personality and Individual Differences*, *43*(7), 1829–1838.

Matta, A.d., Gonçalves, F. L., & Bizarro, L. (2012). Delay discounting: Concepts and measures. *Psychology & Neuroscience*, *5*(2), 135–146.

Mehlum, L. (2001). Suicidal behaviour and personality disorder. *Current Opinion in Psychiatry*, *14*(2), 131–135.

Michalczuk, R., Bowden-Jones, H., Verdejo-Garcia, A., & Clark, L. (2011). Impulsivity and cognitive distortions in pathological gamblers attending the UK National Problem Gambling Clinic: A preliminary report. *Psychological Medicine*, *41*(12), 2625–2635.

Miller, A. B., Williams, C., Day, C., & Esposito-Smythers, C. (2017). Effects of cognitive distortions on the link between dating violence exposure and substance problems in clinically hospitalized youth. *Journal of Clinical Psychology*, *73*(6), 733–744. https://doi.org/10.1002/jclp.22373.

Mobini, S., Grant, A., Kass, A. E., & Yeomans, M. R. (2007). Relationships between functional and dysfunctional impulsivity, delay discounting and cognitive distortions. *Personality, & Differences, Individual*, *43*(6), 1517–1528.

Mobini, S., Pearce, M., Grant, A., Mills, J., & Yeomans, M. R. (2006). The relationship between cognitive distortions, impulsivity, and sensation seeking in a non-clinical population sample. *Personality and Individual Differences*, *40*(6), 1153–1163.

Moeller, F., Barratt, E. S., Dougherty, D. M., Schmitz, J. M., & Swann, A. C. (2001). Psychiatric aspects of impulsivity. *The American Journal of Psychiatry*, *158*(11), 1783–1793. https://doi.org/10.1176/appi.ajp.158.11.1783.

Mooijman, M., Meindl, P., Oyserman, D., Monterosso, J., Dehghani, M., Doris, J. M., et al. (2017). Resisting temptation for the good of the group: Binding moral values and the moralization of self-control. *Journal of Personality and Social Psychology*, *115*(3), 585–599.

Odum, A. L. (2011). Delay discounting: I'm ak, you're ak. *Journal of the Experimental Analysis of Behavior*, *96*(3), 427–439.

Ouzir, M., & Errami, M. (2016). Etiological theories of addiction: A comprehensive update on neurobiological, genetic and behavioural vulnerability. *Pharmacology, Biochemistry and Behavior*, *148*, 59–68. https://doi.org/10.1016/j.pbb.2016.06.005.

Palomo, T., Beninger, R. J., Kostrzewa, R. M., & Archer, T. (2008). Focusing on symptoms rather than diagnoses in brain dysfunction: Conscious and nonconscious expression in impulsiveness and decision-making. *Neurotoxicity Research*, *14*(1), 1–20.

Parry, C. J., & Lindsay, W. R. (2003). Impulsiveness as a factor in sexual offending by people with mild intellectual disability. *Journal of Intellectual Disability Research*, *47*(6), 483–487.

Pitts, S. R., & Leventhal, A. M. (2012). Associations of functional and dysfunctional impulsivity to smoking characteristics. *Journal of Addiction Medicine*, *6*(3), 226–232. https://doi.org/10.1097/ADM.0b013e31825e2a67.

Poythress, N. G., & Hall, J. R. (2011). Psychopathy and impulsivity reconsidered. *Aggression and Violent Behavior*, *16*(2), 120–134.

Raj, M. A. J., Kumaraiah, V., & Bhide, A. V. (2001). Cognitive-behavioural intervention in deliberate self-harm. *Acta Psychiatrica Scandinavica*, *104*(5), 340–345.

Raketic, D., Barisic, J. V., Svetozarevic, S. M., Gazibara, T., Tepavcevic, D. K., & Milovanovic, S. D. (2017). Five-factor model personality profiles: The differences between alcohol and opiate addiction among females. *Psychiatria Danubina*, *29*(1), 74–80. http://dx.doi.org/10.24869/psyd.2017.74.

Ramirez, J. (2001). Cognitive distortions in adolescents with substance-related disorders. *Dissertation Abstracts International: Section B: The Sciences and Engineering*, *62*(2-B), 1060.

Riley, E. N., Combs, J. L., Jordan, C. E., & Smith, G. T. (2015). Negative urgency and lack of perseverance: Identification of differential pathways of onset and maintenance risk in the longitudinal prediction of nonsuicidal self-injury. *Behavior Therapy*, *46*(4), 439–448.

Romer, D. (2010). Adolescent risk taking, impulsivity, and brain development: Implications for prevention. *Developmental Psychobiology: The Journal of the International Society for Developmental Psychobiology*, *52*(3), 263–276.

Rusting, C. L. (1998). Personality, mood, and cognitive processing of emotional information: Three conceptual frameworks. *Psychological Bulletin*, *124*(2), 165–196.

Scherbaum, S., Haber, P., Morley, K., Underhill, D., & Moustafa, A. A. (2018). Biased and less sensitive: A gamified approach to delay discounting in heroin addiction. *Journal of Clinical and Experimental Neuropsychology*, *40*(2), 139–150. https://doi.org/10.1080/13803395.2017.1324022.

Schmidt, R. E., Gay, P., Ghisletta, P., & Van Der Linden, M. (2010). Linking impulsivity to dysfunctional thought control and insomnia: A structural equation model. *Journal of Sleep Research*, *19*(1 (Pt 1)), 3–11.

Sharma, L., Kohl, K., Morgan, T. A., & Clark, L. A. (2013). "Impulsivity": Relations between self-report and behavior. *Journal of Personality and Social Psychology*, *104*(3), 559–575.

Sharma, L., Markon, K. E., & Clark, L. A. (2014). Toward a theory of distinct types of "impulsive" behaviors: A meta-analysis of self-report and behavioral measures. *Psychological Bulletin*, *140*(2), 374.

Sharma, P., Sivakumaran, B., & Marshall, R. (2011). Deliberate self-indulgence versus involuntary loss of self-control: Toward a robust cross-cultural consumer impulsiveness scale. *Journal of International Consumer Marketing*, *23*(3–4), 229–245.

Sheffer, C., MacKillop, J., McGeary, J., Landes, R., Carter, L., Yi, R., et al. (2012). Delay discounting, locus of control, and cognitive impulsiveness independently predict tobacco dependence treatment outcomes in a highly dependent, lower socioeconomic group of smokers. *The American Journal on Addictions*, *21*(3), 221–232.

Spillane, N. S., Smith, G. T., & Kahler, C. W. (2010). Impulsivity-like traits and smoking behavior in college students. *Addictive Behaviors*, *35*(7), 700–705. https://doi.org/10.1016/j.addbeh.2010.03.008.

Spinella, M. (2004). Neurobehavioral correlates of impulsivity: Evidence of prefrontal involvement. *International Journal of Neuroscience*, *114*(1), 95–104.

Stautz, K., & Cooper, A. (2013). Impulsivity-related personality traits and adolescent alcohol use: A meta-analytic review. *Clinical Psychology Review*, *33*(4), 574–592. https://doi.org/10.1016/j.cpr.2013.03.003.

Stratton, K. J. (2006). Mindfulness-based approaches to impulsive behaviors. *The New School Psychology Bulletin*, *4*(2), 49–71.

Stroebe, W., van Koningsbruggen, G. M., Papies, E. K., & Aarts, H. (2013). Why most dieters fail but some succeed: A goal conflict model of eating behavior. *Psychological Review*, *120*(1), 110–138.

Sutin, A. R., Ferrucci, L., Zonderman, A. B., & Terracciano, A. (2011). Personality and obesity across the adult life span. *Journal of Personality and Social Psychology*, *101*(3), 579–592.

Thomsen, K. R., Callesen, M. B., Hesse, M., Kvamme, T. L., Pedersen, M. M., Pedersen, M. U., et al. (2018). Impulsivity traits and addiction-related behaviors in youth. *Journal of Behavioral Addictions*, *7*(2), 317–330. https://doi.org/10.1556/2006.7.2018.22.

Tobin, H., & Logue, A. W. (1994). Self-control across species (*Columba livia*, *Homo sapiens*, and *Rattus norvegicus*). *Journal of Comparative Psychology*, *108*(2), 126.

Tomko, R. L., Prisciandaro, J. J., Falls, S. K., & Magid, V. (2016). The structure of the UPPS-R-Child impulsivity scale and its relations with substance use outcomes among treatment-seeking adolescents. *Drug and Alcohol Dependence*, *161*, 276–283. https://doi.org/10.1016/j.drugalcdep.2016.02.010.

Urben, S., Suter, M., Pihet, S., Straccia, C., & Stephan, P. (2015). Constructive thinking skills and impulsivity dimensions in conduct and substance use disorders: Differences and relationships in an adolescents' sample. *Psychiatric Quarterly*, *86*(2), 207–218. https://doi.org/10.1007/s11126-014-9320-8.

Verbruggen, F., Adams, R., & Chambers, C. D. (2012). Proactive motor control reduces monetary risk taking in gambling. *Psychological Science*, *23*(7), 805–815.

Walker, S. E., Pena-Oliver, Y., & Stephens, D. N. (2011). Learning not to be impulsive: Disruption by experience of alcohol withdrawal. *Psychopharmacology*, *217*(3), 433–442.

Weissman, A., & Beck, A. (1978). Development and validation of the dysfunctional attitude scale: A preliminary investigation. In *Proceedings of the annual meeting of the American Educational Research Association,* Toronto, 27–31 March (pp. 1–13).

Weller, R. E., Cook, E. W., 3rd, Avsar, K. B., & Cox, J. E. (2008). Obese women show greater delay discounting than healthy-weight women. *Appetite*, *51*(3), 563–569.

Wells, J. D., Parboteeah, V., & Valacich, J. S. (2011). Online impulse buying: Understanding the interplay between consumer impulsiveness and website quality. *Journal of the Association for Information Systems*, *12*(1), 32.

Wiers, R. W., & Stacy, A. W. (2006). Implicit cognition and addiction. *Current Directions in Psychological Science*, *15*(6), 292–296.

Wingo, T., Nesil, T., Choi, J.-S., & Li, M. D. (2016). Novelty seeking and drug addiction in humans and animals: From behavior to molecules. *Journal of Neuroimmune Pharmacology*, *11*(3), 456–470.

Winkel, D. E., Wyland, R. L., Shaffer, M. A., & Clason, P. (2011). A new perspective on psychological resources: Unanticipated consequences of impulsivity and emotional intelligence. *Journal of Occupational and Organizational Psychology*, *84*(1), 78–94.

Wittmann, M., & Paulus, M. P. (2008). Decision making, impulsivity and time perception. *Trends in Cognitive Sciences*, *12*(1), 7–12.

Zilberman, N., Yadid, G., Efrati, Y., Neumark, Y., & Rassovsky, Y. (2018). Personality profiles of substance and behavioral addictions. *Addictive Behaviors*, *82*, 174–181. https://doi.org/10.1016/j.addbeh.2018.03.007.

Chapter 3

The multifaceted nature of risk-taking in drug addiction

Daniella M. Saleme[a], Ahmed A. Moustafa[b]

[a]*School of Social Sciences and Psychology, Western Sydney University, Sydney, NSW, Australia,* [b]*MARCS Institute for Brain, Behaviour, and Development and School of Psychology, Western Sydney University, Penrith, NSW, Australia*

Introduction

Drug addiction is a chronic and recurring disorder characterized by drug-seeking and drug-taking behavior that individuals continuously engage in despite dire consequences (Camí & Farré, 2003). Drug addiction has been recognized as a neuropsychiatric disorder with biopsychosocial consequences for the individual, the community and the healthcare system (Robbins & Everitt, 1999). The fifth edition of the Diagnostic and Statistical Manual of Mental Disorders (DSM-5; American Psychiatric Association, 2013) includes the diagnostic category of Substance-Related and Addictive Disorders, with specifiers for opioid (e.g., heroin); stimulant (e.g., cocaine); and alcohol intoxication, dependence, and withdrawal.

In neuropsychiatry, animal models of drug dependence have revealed that there are shared and also distinct neurobiological pathways involved in the reinforcing and addictive effects of different drugs of abuse: the addictive effects of cocaine are dependent upon the release of dopamine in the mesocorticolimbic dopamine system; the dopaminergic and dopamine-independent systems are involved in acute reinforcement of opiates such as heroin; dopaminergic and opioid peptidergic systems seem to be involved in nicotine addiction; and the reinforcing nature of ethanol seems to occur as a result of a number of neurobiological pathways involving, but not limited to, the release of dopamine, opioid, and serotonin neurotransmitters (Koob & Nestler, 1997). In addition to the neuropsychiatric facets of drug addiction and how it forms, there are neurocognitive and psychological factors to be considered in the decision-making of drug-dependent individuals. Individuals with drug dependence have consistently shown impaired decision-making when compared to healthy controls (Myers et al., 2016). Studies of risk-taking have shown that chronic drug use and dependence tends to leave individuals vulnerable to poorer decision-making (e.g., continually choosing to

Cognitive, Clinical, and Neural Aspects of Drug Addiction. https://doi.org/10.1016/B978-0-12-816979-7.00003-0

use illicit drugs such as heroin and cocaine despite the adverse effects on social, legal, and health areas of life; choosing to steal money to buy drugs despite being on parole from jail; or leaving infants at home on their own to meet a drug dealer and make a purchase; Torres, Megias, Catena, Candido, & Maldonado, 2017). Research has shown that drug-dependent individuals' ability to perceive and evaluate the outcomes of a decision is impaired (e.g., heroin-dependent individuals' abilities to consider the different consequences of a situation or action and then making a decision based on the best scenario are impaired), and so they tend to make riskier decisions than those without drug dependence (Wittwer, Hulka, Heinimann, Vonmoos, & Quednow, 2016).

Risk-taking

Risk-taking refers to refers to behavior known to possibly lead to danger, harm or loss and where decisions are made due to the perceived benefits being greater than the costs. Risk-taking behavior can occur across many different domains, including social risk-taking such as disagreeing with a friend or questioning an acquaintance's religious beliefs; financial risk-taking such as engaging in impulsive spending or investing one's life savings in an unfamiliar investment; recreational risk-taking such as skydiving or bungee jumping; health risks such as neglecting to visit a general practitioner regarding ongoing symptoms or sharing used needles during drug administration; ethical risk-taking such as cheating on a spouse or stealing from another person; and general risk-taking tendencies across domains overall. Risk-taking is a multifaceted construct as it is influenced by several different factors such as the decision-making abilities of the individual, the type of situation, and the attention that the individual pays to clues about the different consequences of their decisions. Risk-taking is a type of decision-making that can be adaptive (e.g., taking a risk by investing in a stock option that will yield a large return) or maladaptive (e.g., taking heroin despite the probable legal, health, and social ramifications). When risk-taking is maladaptive, it is at its core a poor decision that is made despite the dangers of the risk outweighing its benefits. Risk-taking is influenced by individuals' abilities to process feedback about previous experience (e.g., considering that previous drug use resulted in hospitalization and so not using drugs again) and consider information about situational outcomes when making decisions (e.g., considering the probability of getting caught buying heroin from a dealer by police; Torres et al., 2017). As taking risks is a type of decision-making, it is also influenced by decision-making biases such as intolerance of uncertainty which is a biased perception of uncertain or ambiguous events as having negative outcomes (e.g., not being able to tolerate the unknown outcome of a court appearance, so using heroin to reduce anxiety; Garami et al., 2017). The propensity of individuals to engage in such risk-taking behavior can be assessed using self-report scales that ask individuals to rate their likelihood of engaging in risky behavior; computerized decision-making programs in which participants make decisions in hypothetical situations designed to simulate real-world scenarios. Given the multifaceted nature of risk-taking, and the fact that individuals with drug dependence tend to be vulnerable to greater risk-taking than healthy populations, the factors influencing risk-taking in both populations will now be reviewed.

Dual-systems theory of decision-making

Risk-taking is a type of decision-making where the benefits of taking a risk are perceived to outweigh the potential consequences (Brockhaus, 1980). Recent theories have suggested a dual-systems model of risk-taking involving (a) affective or emotional processes or (b) cognitive and deliberative processes. Affective processes underlie actions that tend to be more spontaneous, and are influenced by the emotions an individual is experiencing, which can bias the interpretation of the probability of an outcome and the associated gains and losses associated with an action, whereas cognitive or deliberative processes underlie actions that are intentional and are planned based on a consideration of the situational information presented (Figner, Mackinlay, Wilkening, & Weber, 2009). Decisions based on the desire to fill a void, for example, may be made under the affective system due to the influence of emotions (e.g., despair or longing). One study investigated the applicability of the risk-sensitive foraging (RSF) theory—the theory that the level of risk undertaken to obtain food will depend on the level of hunger for food (Caraco, Martindale, & Whittam, 1980)—to decision-making in heroin addiction. Bickel, Giordano, and Badger (2004) assessed heroin-dependent hospital patients ($N=30$) on RSF measures (i.e., scales that measure participant's willingness to take risks given their hunger for food) with scenarios adapted to heroin use. Participants were asked to choose between constant (low risk) and variable (high risk) sources of heroin. Heroin-dependents chose the constant (low risk) heroin source more often when the scenario was of heroin satiation and the variable (high risk) heroin source more often when the scenarios were of hunger for heroin. The transferability of hypothetical scenarios to real-world behavior is questionable, and the reliability and validity of adaptations of the RSF scenarios are unknown, however the results suggest that RSF theory may be applicable to heroin use where hunger for heroin may result in higher risk-taking, driven by the affective decision-making system. During satiation, fewer risks are taken, as more deliberative processes can be engaged when making decisions.

Outside of the laboratory, the significance of the findings of Bickel et al. (2004) can be seen in the example of the goal of methadone maintenance therapy for heroin addiction. The provision of methadone is intended to induce satiation for heroin, decrease withdrawal symptoms, facilitate sobriety from heroin use, and therefore reduce the risk of patients seeking heroin illegally (Gorzelańczyk, Fareed, Walecki, Feit, & Kunc, 2014). However, when decision-making processes are disrupted, an individual's ability to consider the short- and long-term consequences of their actions and decide on an advantageous choice becomes impaired (Verdejo-Garcia, Chong, Stout, Yücel, & London, 2017). This is true of drug-dependent individuals who tend to repeatedly engage in drug use despite the negative consequences they face, including criminal charges, ill-health, and in some cases, death (Verdejo-Garcia et al., 2017; Voon et al., 2015). The affective system tends to override the cognitive control system in states of emotional arousal, which may explain increased risk-taking in stressful or emotional situations (e.g., taking heroin during methadone maintenance therapy in response to a stressful situation; Figner et al., 2009). Adherence to intervention programs is also influenced by such decisions, whereby the continued use of illicit drugs lowers adherence to treatment plans such as methadone maintenance (Gorzelańczyk et al., 2014). Therefore, research on how affective and deliberative systems influence decision-making in heroin-dependent individuals is crucial to informing adherence to, and positive outcomes of, treatment programs (e.g., increased health and daily functioning and reduced crime rates).

Risk-taking in drug addiction and healthy populations

Risk-taking and feedback and environmental contingency processing

Decision-making has been proposed to be made up of three cognitive stages that individuals cycle between: (a) preference formation, where individuals develop a preference for one option over another; (b) choice implementation, where the preferred choice is selected; and (c) feedback processing, where behavioral outcomes from previous actions are considered and guide future decision-making and behavior (Verdejo-Garcia et al., 2017). In laboratory studies of decision-making, drug-dependent individuals consistently exhibit impaired performance when compared to non-drug-using individuals because of deficits in feedback processing and neglect of environmental contingencies (Ekhtiari, Victor, & Paulus, 2017). Feedback processing concerns an individual's ability to use information that is available to them about their past behavior to guide future decisions (e.g., learning that in the past possession of heroin lead to being arrested and so considering this before being in possession of heroin again), and attending to environmental contingencies concerns the attention an individual pays to information about a situation before making a decision (e.g., observing that an alley way in a dangerous part of town is not well-lit and deciding not to walk through it alone at night; Verdejo-Garcia et al., 2017). The decision-making deficits of poor feedback processing and neglect of environmental contingencies may undermine patients' efforts to remain abstinent from drug-use (Dominguez-Salas, Diaz-Batanero, Lozano-Rojas, & Verdejo-Garcia, 2016). Drug use further exacerbates these decision-making deficits, leading to a vicious cycle of negative consequences, making abstinence from drug use increasingly difficult (Verdejo-Garcia et al., 2017). As such, poor clinical outcomes such as drug relapse during methadone maintenance programs for heroin addiction are very common, inhibiting heroin-dependent individuals' long-term recovery and placing increased strain on the healthcare and criminal justice system (Verdejo-Garcia, 2016). Therefore, it is important to investigate the way that decision-making is influenced by the processing of feedback and the attention paid to environmental contingencies. Empirical evidence into the role of these factors in risk-taking in drug-dependent populations will now be reviewed.

Healthy populations

The influence of affective and deliberative processes, feedback processing, and environmental contingencies on risk-taking has been investigated using the dynamic Columbia Card Task (Figner et al., 2009). The CCT is a computerized behavioral task providing measures of risk-taking across conditions with and without feedback. During the CCT, participants are shown 32 cards and given information about the probability of a hidden loss card (1 or 3), the gain amount for winning cards (10 or 30) and the loss amount for losing cards (250 or 750). Turning over a higher number of cards is associated with greater outcome variability and so is a higher-risk strategy than overturning fewer cards (Figner et al., 2009). The average number of cards overturned represents a measure of risk-taking. In a study of feedback and environmental contingencies and risk-taking in adolescents using the CCT, Figner et al. (2009) found that adolescent risk-taking differed across the feedback and no-feedback conditions more than adult risk-taking did, suggesting greater balance between adult affective and deliberative decision-making processes. Participant's information use regarding the probability of a loss, gain amount, and loss amount on each CCT trial revealed that in adults, all three information factors influenced the extent to which they took risks; however, in the two younger age groups only the probability of a loss influenced greater risk-taking. These results suggest that risky decision-making may be influenced by the performance feedback that is available and whether participants consider environmental contingencies in their decision-making. Additionally, the authors suggest that the results support the dual-systems model for explaining risk-taking as a product of the competition between affective and deliberative processes.

Risk-taking was also measured in the context of affective and deliberative processes, feedback, and environmental contingencies using the CCT by Buelow (2015). Undergraduate university students' ($N=489$) performance was assessed with two different samples assigned to the feedback and the no-feedback CCT conditions. Participants in the no-feedback condition took more risks (i.e., selected more cards) than participants in the feedback condition. By investigating information use, it was found that participants tended to display greater risk-taking on the no-feedback condition when the probability of a loss card was one, when the loss amount was 250, and when the gain amount was 30 (i.e., when the situation was the most advantageous). On the feedback condition, however, participants only took greater risks when the probability of a loss was one and gain and loss amounts did not influence risk-taking. These results suggest that when feedback about performance is unavailable, healthy individuals tend to pay more attention to all the information factors present when taking more risks, however when feedback about performance is available, individuals tend to pay more attention to the probability of a loss, during risk-taking.

Drug-dependent populations

Feedback processing or integrating information about previous experiences to guide future decisions, is a stage of decision-making in which individuals with drug-dependence tend to experience significant impairment compared with healthy individuals (Verdejo-Garcia et al., 2017). The influence of feedback processing on risky decision-making has been investigated in drug-dependent populations. Wittwer et al. (2016) investigated decision-making under risky information with no feedback in cocaine-dependent individuals. Participants were grouped based on cocaine use 1 year after baseline measures, indicated by hair concentrations of cocaine ($n=10$ with a strong increase in hair concentration of cocaine; $n=12$ with a strong decrease; $n=9$ with no change in hair concentration) and compared to healthy controls ($n=26$). Participants completed the Randomized Lottery Task (RALT) measuring risky behavior when presented with either a lottery with an expected value or a payoff that is guaranteed. Cocaine users were more disposed to making riskier decisions in the lottery task compared to healthy controls. Cocaine dependents with increased hair concentrations of cocaine (i.e., with an indication of greater cocaine use) made riskier choices with high loss probabilities, suggesting that higher impairments in processing information about risk, and that deficits in attending to environmental contingencies may mediate the risk-taking behavior of drug-dependent individuals. The study is limited by the questionable reliability of measuring hair concentrations of cocaine, and that risk-taking was only tested one time of the study, not at baseline, so claims about the effect of 1-year increases in cocaine use on decision-making need to be further investigated using longitudinal studies.

The attention an individual pays to environmental contingencies such as the probability of an event or outcome occurring influences their approach towards decision-making. Lower probabilities of a rewarding outcome tend to be overvalued, increasing risk-taking to achieve gains (e.g., buying lottery tickets) and to avoid losses (e.g., foregoing health insurance to save money), and higher probabilities tend to be undervalued, resulting in risk avoidance for gains (e.g., gambling a small amount of money to ensure a small but probable gain) and risk-taking for high probability losses (e.g., investing a large sum of money for a high return that is unlikely; Voon et al., 2015). Stimulant- and alcohol-dependent individuals as well as those with gambling addictions have been shown to exhibit poorer decision-making than non-users when environmental contingencies are made explicit to them in risky situations (Lawrence, Luty, Bogdan, Sahakian, & Clark, 2009). However, the effects of situational probability, gain, and loss information on decision-making have been less researched. Voon et al. (2015) therefore investigated the influence of probability, gain and loss, and value of reward on risk-taking and pathological behavior in abstinent alcohol- ($n=30$) and methamphetamine-dependent ($n=23$) subjects. Participants had to choose between a risky and sure choice and were presented with the probability of winning and the value of gains for each probability. Probability, gain and loss, and reward value were found to influence attitudes towards risk-taking in drug-dependent participants. Methamphetamine-dependent subjects tended to take greater risks for larger more unlikely rewards and alcohol-dependent subjects had greater risk-taking for smaller but more likely rewards. The authors suggest that probability may mediate the choices that drug-dependent individuals make when chasing the high from drug rewards. An important limitation of this study is that it could not be distinguished whether participants' risk attitudes reflected state factors (i.e., whether risk attitudes in the study were temporary and related external influences, e.g., stress or drug use) or trait factors (i.e., more permanent risk attitudes that would tend to be more stable over time, e.g., long-standing views of drug use as high-risk behavior). Nonetheless, Voon et al., 2015 demonstrated that attention paid towards probability, gain and loss, and their value influence the risk-taking of drug-dependent individuals.

The influence of feedback processing and attention to environmental contingencies on making risky decisions in individuals with substance addiction has been investigated in cocaine users. Kluwe-Schiavon, Viola, Sanvicente-Vieira, Pezzi, and Grassi-Oliveira (2016) tested participants using the feedback and no feedback conditions of the Columbia Card Task (CCT) measure of risky decision-making. The CCT (Figner et al., 2009) is a computerized behavioral task assessing risk-taking across conditions with and without feedback, in which participants are shown 32 cards and provided information regarding the trial's probability of experiencing a loss, the gain amount, and loss amount. Turning over a higher number of cards is associated with greater outcome variability and so is a higher-risk strategy than overturning fewer cards (i.e., the average cards overturned provides a measure of risk-taking) (Figner et al., 2009). The feedback CCT condition activates affective decision-making processes and the no feedback CCT condition activates deliberative decision-making processes (Figner et al., 2009; Figner & Murphy, 2011; Figner & Weber, 2011). The CCT is also decomposable as attention paid to environmental contingencies (i.e., probability of a loss, gain amount, and loss amount) during risk-taking (Schonberg, Fox, & Poldrack, 2011) can be analyzed, allowing for motivations underlying risk-taking to be investigated (Figner & Weber, 2011). Cocaine users ($n=27$) demonstrated reduced risk-taking behavior when offered delayed feedback about their choices, and both cocaine users and controls ($n=18$) showed increased risk-taking without feedback. Even though the sample size was relatively small, and participants were not matched for demographic factors, the results suggest that feedback about performance may function to reduce risk-taking behavior in cocaine addicts.

To determine whether similar effects are seen in heroin-dependent individuals, Saleme et al. (2018) investigated the influences of feedback processing and attention towards environmental contingencies (probability of a loss, gain amount, and loss amount), and their relationship to risk-taking, in heroin-dependent patients receiving opioid-replacement therapy ($n=25$) and healthy individuals ($n=27$) Patients undergoing treatment, and controls, completed the feedback and no-feedback conditions of the Columbia Card Task (CCT), which is described above. Analyses of covariance, controlling for education and task design (the order in which the CCT conditions were completed) as covariates revealed a significant interaction between (a) probability, gain and loss amount, and group, such that patients undergoing treatment took greater risks than controls across all information factors of both CCT conditions (Fig. 1). An interaction was also found between (b) group and probability, such that patients took greater risks than controls when the probability of overturning a loss card was higher (Fig. 2). The findings suggested that heroin-dependent patients pay less attention to environmental contingencies such as the probability of negative consequences and the feedback from performance during risk-taking than controls. This is an important factor to consider when evaluating the multifaceted nature of risk-taking in drug addiction and the factors that influence how and why patients take greater risks than controls do.

Risk-taking behavior has also been studied in users of other drugs such as cocaine. Previous research has shown chronic cocaine users to exhibit impairments in attention, memory, and in executive functioning areas of decision-making and impulsivity (Wittwer et al., 2016). Reduced ability to incorporate feedback into decision-making has also been associated with cocaine dependence (Kluwe-Schiavon et al., 2016). Wittwer et al. (2016) investigated risky decision-making without feedback in cocaine users considering hair concentrations of cocaine (i.e., cocaine use) 1 year after baseline measures ($n=10$ with a

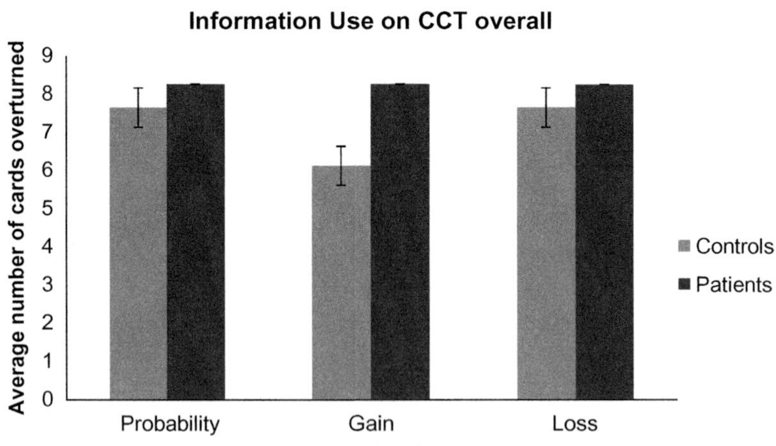

FIG. 1 Information use by controls and patients on the CCT overall.

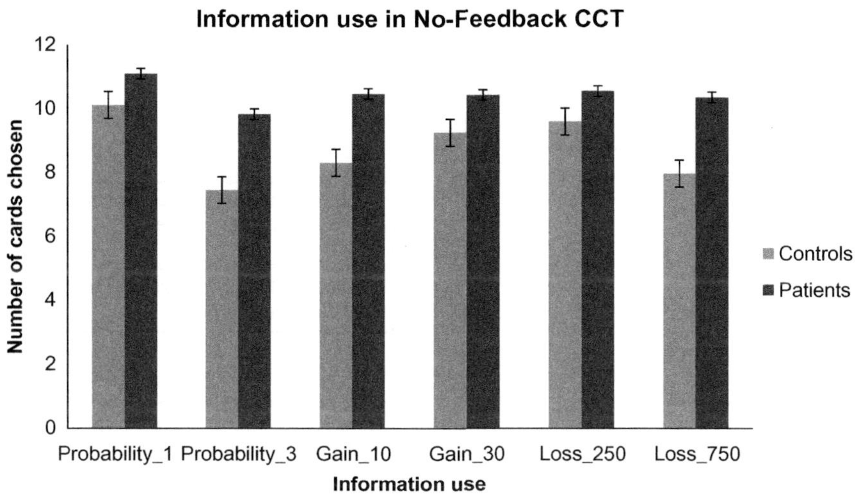

FIG. 2 Information use by controls and patients on the no-feedback CCT.

strong increase in hair concentration of cocaine; $n=12$ with a strong decrease; $n=9$ with no change in hair concentration) and compared to non-users ($n=26$). Participants completed the Randomized Lottery Task (RALT) measuring risky behavior when presented with either a lottery with an expected value or guaranteed payoff. Cocaine users made riskier decisions in the lottery task compared to healthy controls. Cocaine-dependents with increased cocaine use made riskier choices with high loss probabilities, suggesting that neglect of environmental contingencies may mediate risk-taking behavior in drug dependence. The effects of 1-year increases in cocaine use on decision-making need to be further investigated using longitudinal studies.

Previous research therefore suggests that in stimulant and opioid-dependent individuals, poor decision-making tends to be a result of greater attention being directed towards gains, neglect of losses, and inconsistency in the use of feedback from previous experiences to guide future behavior (Figner & Weber, 2011; Saleme et al., 2018; Voon et al., 2015; Wittwer et al., 2016). The results of studies of the influences on risk-taking in healthy and drug-dependent population samples suggest that the different systems underlying decision-making, feedback processing, and attention paid towards environmental contingencies, may work together to influence an individual's risk-taking (Fig. 3). Thus, the empirical evidence on the influence of these factors together on risk-taking in healthy and drug-dependent populations will now be reviewed.

Domain-specific risk-taking

Risky decision making is influenced by several different factors, one of which being the domain in which the behavior is occurring (Smith, Ebert, & Broman-Fulks, 2016). Risky decisions can be made across a number of different contexts

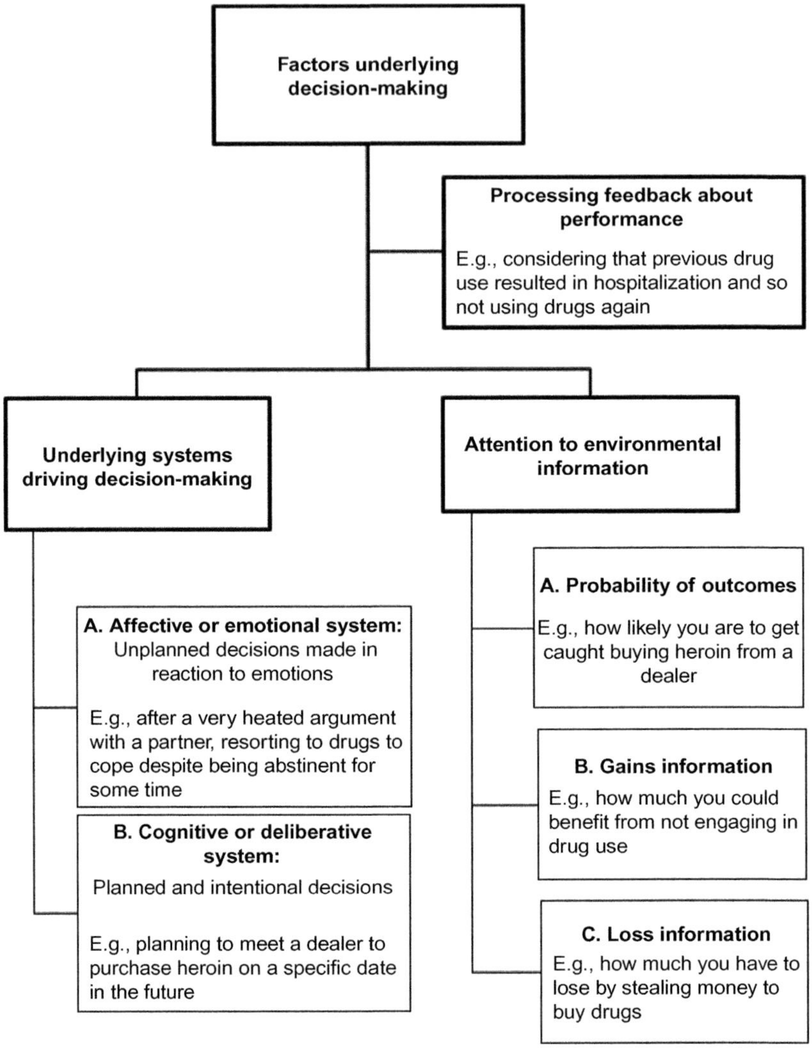

FIG. 3 Factors influencing decision-making.

or domains which can include (a) social risk-taking such as disagreeing with a friend or questioning an acquaintance's religious beliefs; (b) financial risk-taking such as engaging in impulsive spending or investing one's life savings in an unfamiliar investment; (c) recreational risk-taking such as skydiving or bungee jumping; (d) health risks such as neglecting to visit a general practitioner regarding ongoing symptoms or sharing used needles during drug administration; (e) ethical risk-taking such as cheating on a spouse or stealing from another person; and (f) general risk-taking tendencies across domains overall. The propensity of individuals to engage in such risk-taking behavior can be assessed using self-report scales that ask individuals to rate their likelihood of engaging in risky behavior; computerized decision-making programs in which participants make decisions in hypothetical situations designed to simulate real-world scenarios (Ekhtiari et al., 2017). When measured using such tools, individuals afflicted by substance addictions tend to display greater engagement in risk-taking behavior in some of these domains than healthy populations (Wittwer et al., 2016). Furthermore, there can be other forms of risk-taking that can be time-based (e.g., preferring a smaller reward today rather than a bigger reward in 1 month, as measured by delay discounting tasks) or certainty-based (e.g., preferring a proximal but uncertain reward rather than a certain but distant reward, as measured by probability discounting tasks).

Healthy populations

Risk-taking propensity ha's been found to remain relatively stable across the lifespan (Seaman et al., 2016). The domains in which healthy populations tend to take risks have also been investigated. Josef et al. (2016) investigated the propensity for risk-taking behavior in participants aged 18–85 years longitudinally by examining the results of a German national census. Risk-taking tendency was found to be moderately stable across the life span. Participants displayed variations in risk-taking across the different risk domains (generalized risk-taking—the total risk-taking across all domains; social risk-taking—risk-taking behavior in social interactions such as disagreeing with a friend; and financial risk-taking—risk-taking behavior regarding financial decisions such as making investments or making impulsive purchases) suggesting that healthy individuals differ in their risk-taking behavior dependent on the risk domain. Risk domains were significantly correlated with each other suggesting that risk-taking in some domains may impact risk-taking in other domains (e.g., propensity to take risks in the safety domain may influence propensity to take risks in the health domain). Differential stability—the preservation of individual differences over time—demonstrated an inverted U-curved trend where risk-taking tendency was highest from youth to middle age and then declined with maturity, suggesting that risk-taking decreases with increasing age. Mean-level changes revealed decreasing risk-taking with age and showed that domains such as social risk-taking are stable across adult years. Individual changes in risk-taking revealed no significant association with income; a moderate positive relationship between the Big Five personality factors of extraversion and openness to experience and risk-taking, and a negative relationship between within-person consciousness, neuroticism, and agreeableness and risk-taking tendency. These results suggest that risk-taking propensity remains stable across adulthood but tends to decrease upon approaching older age, that individual differences in risk-taking exist across different risk-taking domains, and that personality factors may influence the tendency to take risks.

Similarly, in community members ($N=92$), Seaman et al. (2016) found that older individuals tended to choose rewards with shorter temporal delays, increased social interactions and increased health benefits. What these findings suggest is that the domains in which individuals take risks may be influenced by factors salient to their circumstance. In older age, health and social interaction tend to be of greater importance to the individual than recreational and financial factors, for example, and so older adults tend to take risks with maximized social and health benefits. This finding has implications for prevention and intervention programs targeted for heroin dependent individuals—if the factors that are salient to the individual's circumstances can be assessed and then their propensity to take risks in these domains may be prevented. The association between risk-taking behavior and personality factors found by Josef et al. (2016) has also been examined in relation to creativity. Tyagi, Hanoch, Hall, Runco, and Denham (2017) measured risk-taking domains (ethical, financial and gambling, health and safety, recreational, and social) using the Domain-Specific Risk-Taking Scale in university students ($n=64$) and community members ($n=417$), the Roulette Betting Task (RBT) and measures of creativity. It was found that creativity is related to increased tendencies for social risk-taking only, suggesting that certain forms of risk-taking may be adaptive, and not necessary signs of psychological impairment.

The factors that influence an individual's likelihood of engaging in risk-taking behavior have also been investigated. Zou and Scholer (2016) investigated the relationship between promotion (opportunity for growth) and prevention (safety maintenance) motivations in predicting risk-taking in community participants ($n=302$, 235). The domains of health, moral, leisurely, gambling, financial, and social risk-taking were examined across three studies, using the Regulatory Focus Questionnaire measuring prevention and promotion focus, and the Domain-Specific Risk-Taking scale of risk-taking behavior. Results demonstrated that promotion and prevention motivation can be considered as predictive of the propensity for risk-taking in health, ethical, leisurely, gambling, financial, and social risk domains. Prevention motivation was negatively associated with risk-taking in all domains and promotion motivation was positively associated with risk-taking in only the domains where true growth was possible. Finally, the perceived gains mediated risk-taking and promotion focus only, suggesting that perceived gains are important for promotional and not prevention motivation. Despite the use of a non-clinical sample the results reveal that promotion and prevention motivations are predictive of risk-taking across different domains. If different motivations underlie an individuals' tendency to engage in risk-taking behavior and their existence is predictive of risk-taking propensity, assessment of these motivations may be implemented in intervention programs for heroin dependent individuals to identify the domains in which they are likely to take risks.

The influences of genetic differences and environmental factors on risk-taking behavior have also been investigated in healthy populations across risk domains. A mock scale, named the DOSPERT-7, was created by Wang, Zheng, Xuan, Chen, and Li (2016) by combining the DOSPERT, DOSPERT-revised, the Chinese adaptation of the DOSPERT (DOSPERT-C), and a scale measuring the tendency for risk-taking in evolutionarily salient domains. The risk domains measured by the mock scale included competition, safety, procreation, physical, ethical, financial, and gambling risks. The DOSPERT-7 was used to assess the estimated influence of genes and unshared environmental factors in monozygotic and dizygotic twins on risk-taking tendency, and then reviewed the literature on risk-taking in twin studies in the risk domains. Results demonstrated that individual variations in tendencies to take risks and the consistency of these differences were moderated by genetic and unique environmental factors. Genetics explained the most variance in risk-taking tendency in the safety and reproductive domains and least in the financial risk domain. Genetics also explained a significant portion of the variance in ethical risk-taking and were significantly associated with risk-taking tendency in moral, financial, and physical risk. Individual differences in tendency for risk-taking were found to stem from unshared environmental factors and experiences unique to the individual. These results suggest tendencies for risk-taking can be domain-specific, genetic influences on risk-taking in different domains, and that genetic and environmental factors play a significant role in risk-taking propensity.

Drug-dependent populations

Domain-specific risk-taking was investigated for the first time in heroin-dependent patients undergoing opioid-replacement treatment ($n=25$) and healthy controls ($n=27$) by Saleme et al. (2018). Heroin-dependent patients were administered the Domain Specific Risk-taking Scale to measure their levels of self-reported risk-taking in the areas of health and safety, ethical, recreational, financial (gambling and investment areas), and social risk domains. Data was analyzed using several ANCOVAs whilst controlling for the covariate of education. Significant differences were found between patients and controls such that patients took greater risks than controls on the ethical, health and safety, social, and recreational risk-taking domains (Fig. 4). Patient and control differences in risk-taking in the financial domain approached significance, however when the financial domain was decomposed into its subscales, patients exhibited significantly greater risk-taking in the gambling domain than controls (Fig. 5). Additionally, Saleme et al. (2018) conducted correlational analyses to investigate any relationship between domain specific risk-taking and IU, finding that IUS was significantly correlated with risk-taking in the ethical risk-taking domain, such that higher levels of IU were related to higher ethical risk-taking. This suggests that IU mediates domain-specific risk-taking in drug-dependent individuals to the extent that the level of risk taking may increase as IU increases.

As such, further research into the extent of the effect of decision-making biases on risk-taking in drug addiction is necessary to inform treatment programs such as opioid maintenance therapy as behavioral therapy to reduce risk-taking may be a useful adjunct to opioid replacement for reduced rates of relapse or recidivism. Future research should also include more behavioral measures

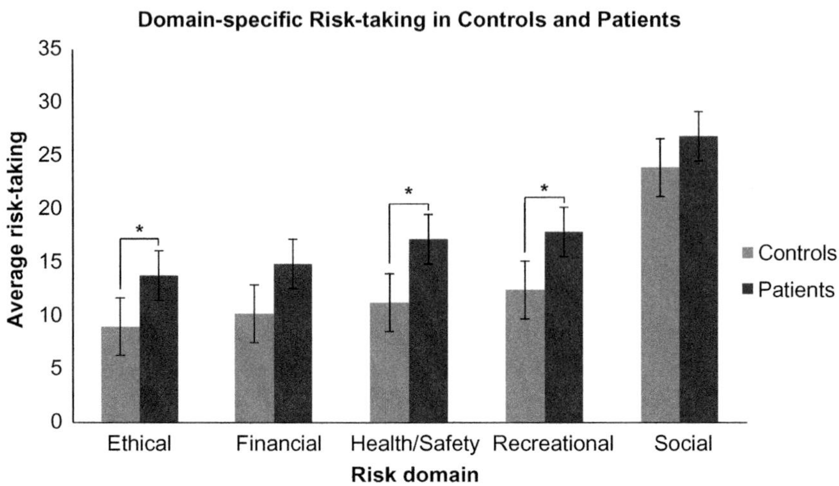

FIG. 4 Domain-specific risk-taking in controls and patients undergoing opioid replacement therapy.

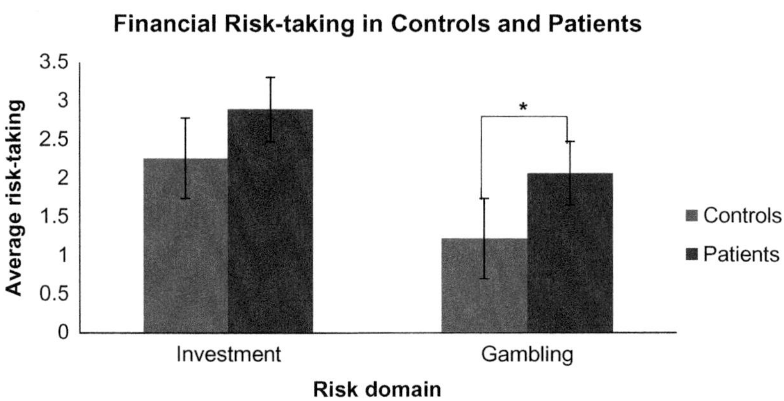

FIG. 5 Financial risk-taking in controls and patients undergoing opioid maintenance therapy.

of domain-specific risk-taking with less of a reliance on self-report measures employed thus far. Further to this, the transferability of laboratory-based risk-taking measures, including self-reported risk-taking, to risk-taking in real life scenarios is unknown and requires research in the future.

Decision-making biases and risk-taking

Risk-taking is an example of decision-making that can be maladaptive, and so an individual's tendency for risk-taking may be influenced by the presence of any decision-making biases. Decision-making biases can alter the way that individuals perceive situational information and consider it when making decisions (Dalley & Robbins, 2017). Intolerance of uncertainty (IU) is a decision-making bias towards interpreting and responding to uncertain situations negatively due to distorted views about their outcomes (Garami et al., 2017). Intolerance of uncertainty and the behaviors associated with addiction including poor decision-making are suggested to be underpinned by neuropsychological mechanisms shared with reward and loss processes (Dalley & Robbins, 2017; Ortega, Solano, Torres, & Papini, 2017). Gorka et al. (2016) investigated brain region activation using the Intolerance of Uncertainty Scale (IUS; Carleton, Norton, & Asmundson, 2007) and functional magnetic resonance imaging (fMRI). Community members ($N=37$) with higher IU demonstrated increased anterior insula activation than low IU community members overall, and more so for the prospective than the inhibitory IU subscale items. Despite the fact that there were differences in the predictability of the rewards in the fMRI conditions, the results suggest that neurobiological processes underlie IU in healthy adults, such that different areas of cortical activation may influence decision-making and contribute to the multifaceted nature of risk-taking. Decision-making biases such as intolerance of uncertainty have been studied in healthy and some drug-dependent populations.

Intolerance of uncertainty in healthy populations

The manifestation of IU in healthy populations has been studied by several researchers. Radell, Myers, Beck, Moustafa, and Allen (2016) tested IU in students ($N=88$), reporting high IU individuals showed a higher tendency to chase rewards. Similarly, Carleton et al. (2016) investigated the influence of IU on decision-making in community members ($n=115$) and university students ($n=98$) using the computerized Risky-Gains task and modified Iowa Gambling Task (IGT). Despite threat levels insufficient to test clinical IU, and the use of non-clinical samples, the results suggest that the effects of IU on behavior are greater in high-risk situations and are moderated by the perceived threat. Similar constructs to IU, ambiguity and need for cognitive closure (NCC)—a strong aversion to uncertainty and motivation to seek answers to questions (Schumpe et al., 2017)—may also have a role in moderating risk-taking behavior. Schumpe et al. (2017) investigated NCC in community members ($n=139$), finding that risk perception mediated the NCC and risk-taking relationship; that high NCC results in less risk-taking in university students ($n=151$); and that university students' ($n=113$) willingness to take risks decreased as NCC increased, with high NCC individuals preferring smaller but more certain and proximal rewards. NCC may therefore be related to behaviors directed at minimizing uncertainty and risk-taking is engaged in only to reduce uncertainty. Similarly, Smith et al. (2016) investigated behavior under ambiguity in university students ($N=124$). Higher anxiety predicted reduced risk-taking in high ambiguity, and anxiety and risk-taking were not related in low ambiguity conditions. Despite the use of a non-clinical convenience sample, results suggest that individuals high in anxiety are more risk-averse in highly ambiguous situations and so when uncertainty is high, risk-taking decreases. Investigations of IU in healthy populations therefore suggest that an aversion to, or an inability to cope with, ambiguity can have a moderating impact on risk-taking behavior such that as intolerance of uncertainty increases, an individual's willingness to take risks decreases, suggestive of a fear of further uncertainty from risk-taking.

Intolerance of uncertainty in drug-dependent populations

IU has only been investigated in drug-dependent samples in a few studies. The study of IU in drug dependence has mainly focused on its occurrence in alcohol-dependent individuals where it fits into the coping model of drug abuse as avoidance behavior to cope with uncertainty and worry about future events (Garami et al., 2017). When sober, individuals with alcohol dependence display exaggerated reactivity to threats that are uncertain (U-threat). Since this aversive affective state can be alleviated via acute alcohol intoxication, it has been posited that individuals who exhibit heightened reactivity to U-threat at baseline are motivated to use alcohol as a means of avoidance-based coping, setting the stage for excessive drinking. To date, however, no study has attempted to characterize the dispositional nature of exaggerated reactivity to U-threat and test whether it is a vulnerability factor or exclusively a disease marker of problematic alcohol use. Gorka et al. (2016) therefore used a family study design to address these gaps by examining whether (1) reactivity to U-threat is associated with risk for problematic alcohol use, defined by family history of alcohol use disorder (AUD) and (2) reactivity to U-threat is correlated amongst adult biological siblings. 157 families ($n = 458$) participated in the study and two biological siblings completed a threat-of-shock task designed to probe reactivity to U-threat and predictable threat (P-threat). Startle potentiation was collected as an index of aversive responding. Within biological siblings, startle potentiation to U-threat ($ICC=0.35$) and P-threat ($ICC=0.63$) was significantly correlated. In addition, independent of an individuals' own AUD status, startle potentiation to U-threat, but not P-threat, was positively associated with risk for AUD (i.e. AUD family history). This suggests that heightened reactivity to U-threat may be a familial vulnerability factor for problematic drinking and a novel prevention target for AUD.

In marijuana-dependent individuals, Hefner, Starr, and Curtin (2016) found that heavy users who had been deprived of marijuana ($N=104$), showed increased aversion to ambiguity and decreased valuation of uncertain rewards for fear of uncertain outcomes. Results suggest that uncertainty may mediate the way that drug-dependent individuals perceive rewards and their risk-taking behavior. Garami et al. (2017) tested IU for the first time in opioid-dependent patients ($n = 114$) and controls ($n=69$) as well as anxiety and impulsivity. Opioid dependent individuals demonstrated higher IU, anxiety, and impulsivity than controls and the measures were highly inter-correlated, suggesting a moderator variable underlying the results in the patient group. These results were replicated in a study by Saleme et al. (2018) who found that heroin-dependent individuals undergoing opioid-replacement therapy ($n=25$) demonstrated significantly higher levels

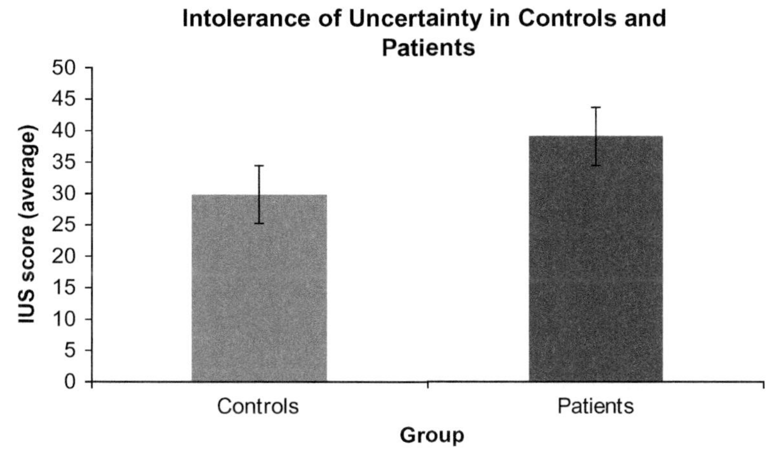

FIG. 6 Self-reported levels of intolerance of uncertainty in controls and patients.

of IU than healthy individuals ($n=27$) on the Intolerance of Uncertainty Scale (Carleton et al., 2007), a self-report measure of an individual's ability to tolerate uncertainty (Fig. 6). Research on IU in drug-dependent populations therefore suggests that such individuals demonstrate higher levels of IU than controls. There is limited evidence to suggest that marijuana-dependent individuals' higher IU may reduce their willingness to engage in risk-taking for fear of increased uncertainty. In heroin-dependent individuals, however, there is limited evidence to explain the increased incidence of IU and risk-taking behavior. Given these reported differences in drug-dependent and healthy individuals in IU, further research is crucial to discern its presence in drug-dependent individuals. Research in this field is important, as it is not known whether IU occurs because of drug dependence or if drug abuse is a method of coping with pre-existing IU. The role of IU in the risk-taking behavior of drug-dependent individuals is also unknown. Future research into the effect of IU on risk-taking is therefore crucial to inform interventions targeted to poor decision-making resulting from feedback processing deficits and decision-making biases in drug dependent populations.

Future studies on risk-taking in drug dependence

Further investigation of the multifaceted nature of risk-taking in drug-dependent population and the factors that influence risk-taking in such individuals would allow for more targeted intervention and prevention programs. Future studies should employ behavioral measures of risk-taking and intolerance of uncertainty to reduce the reliance on self-report measures. Future studies should also focus on examining real-world risk-taking rather than laboratory-based measures of risk-taking. This would increase the generalizability of risk-taking measures and results. Future studies should also further investigate domain-specific risk-taking in heroin-dependent and other drug-dependent individuals. The effectiveness of intervention programs in eliciting positive treatment outcomes may be strengthened by the integration of decision-making training so that patients can learn to evaluate decisions before taking risks. Additionally, adherence to treatments such as opioid-replacement therapy may be strengthened with improvement in patients' decision-making abilities, as they may be better able to evaluate the costs and benefits of taking illicit drugs during treatment. This would also serve to reduce rates of relapse.

References

American Psychiatric Association. (2013). *Diagnostic and statistical manual of mental disorders (DSM-5®)*. American Psychiatric Pub.

Bickel, W. K., Giordano, L. A., & Badger, G. J. (2004). Risk-sensitive foraging theory elucidates risky choices made by heroin addicts. *Addiction*, *99*(7), 855–861. https://doi.org/10.1111/j.1360-0443.2004.00733.x.

Brockhaus, R. H., Sr. (1980). Risk taking propensity of entrepreneurs. *Academy of Management Journal*, *23*(3), 509–520.

Buelow, M. T. (2015). Predicting performance on the Columbia card task: Effects of personality characteristics, mood, and executive functions. *Assessment*, *22*(2), 178–187. https://doi.org/10.1177/1073191114539383.

Camí, J., & Farré, M. (2003). Drug addiction. *The New England Journal of Medicine*, *349*, 975–986. https://doi.org/10.1056/NEJMra023160.

Caraco, T., Martindale, S., & Whittam, T. S. (1980). An empirical demonstration of risk-sensitive foraging preferences. *Animal Behaviour*, *28*, 820–830. Available from: http://www.stewartcalculus.com/data/BIOCALCULUS/BB/BB_Chapter4/E4.2.6/Caraco%20et%20al%201980.pdf (Accessed August 27, 2017).

Carleton, R. N., Duranceau, S., Shulman, E. P., Zerff, M., Gonzales, J., & Mishra, S. (2016). Self-reported intolerance of uncertainty and behavioural decisions. *Journal of Behavior Therapy and Experimental Psychiatry*, *51*, 58–65.

Carleton, R. N., Norton, M. A., & Asmundson, G. J. G. (2007). Fearing the unknown: A short version of the intolerance of uncertainty scale. *Journal of Anxiety Disorders*, *21*, 105–117. https://doi.org/10.1016/j.janxdis.2006.03.014.

Dalley, J. W., & Robbins, T. W. (2017). Fractioning impulsivity: Neuropsychiatric implications. *Nature*, *18*, 158–171. https://doi.org/10.1038/nrn.2017.8.

Dominguez-Salas, S., Diaz-Batanero, C., Lozano-Rojas, O. M., & Verdejo-Garcia, A. (2016). Impact of general cognition and executive function deficits on addiction treatment outcomes: Systematic review and discussion of neurocognitive pathways. *Neuroscience and Biobehavioral Reviews*, *71*, 772–801. https://doi.org/10.1016/j.neubiorev.2016.09.030.

Ekhtiari, H., Victor, T. A., & Paulus, M. P. (2017). Aberrant decision-making and drug addiction—How strong is the evidence? *Current Opinion in Behavioral Sciences*, *13*, 25–33. https://doi.org/10.1016/j.cobeha.2016.09.002.

Figner, B., Mackinlay, R. J., Wilkening, F., & Weber, E. U. (2009). Affective and deliberative processes in risky choice: Age differences in risk taking in the Columbia card task. *Journal of Experimental Psychology: Learning, Memory, and Cognition*, *35*(3), 709–730. https://doi.org/10.1037/a0014983.

Figner, B., & Murphy, R. O. (2011). *Using skin conductance in judgment and decision making research.* In *A handbook of process tracing methods for decision research* (pp. 163–184).

Figner, B., & Weber, E. U. (2011). Who takes risk when and why? Determinants of risk taking. *Current Directions in Psychological Science*, *20*, 211–216. https://doi.org/10.1177/0963721411415790.

Garami, J., Haber, P., Myers, C. E., Allen, M. T., Misiak, B., Frydecka, D., et al. (2017). Intolerance of uncertainty in opioid dependency: Relationship with trait anxiety and impulsivity. *PLoS ONE*, *12*(7), e0181955https://doi.org/10.1371/journal.pone.0181955.

Gorka, S. M., Hee, D., Lieberman, L., Mittal, V. A., Phan, K. L., & Shankman, S. A. (2016). Reactivity to uncertain threat as a familial vulnerability factor for alcohol use disorder. *Psychological Medicine*, *46*(16), 3349–3358. https://doi.org/10.1017/S0033291716002415.

Gorzelańczyk, E. J., Fareed, A., Walecki, P., Feit, J., & Kunc, M. (2014). Risk behavior in opioid-dependent individuals after the administration of a therapeutic dose of methadone. *The American Journal on Addictions*, *23*(6), 608–612. https://doi.org/10.1111/j.1521-0391.2014.12154.x.

Hefner, K. R., Starr, M. J., & Curtin, J. J. (2016). Altered subjective reward valuation among drug-deprived heavy marijuana users: Aversion to uncertainty. *Journal of Abnormal Psychology*, *125*(1), 138–150. https://doi.org/10.1037/abn0000106.

Josef, A. K., Richter, D., Samanez-Larkin, G. R., Wagner, G. G., Hertwig, R., & Mata, R. (2016). Stability and change in risk-taking propensity across the adult life span. *Journal of Personality and Social Psychology*, *111*(3), 430–450. https://doi.org/10.1037/pspp0000090.

Kluwe-Schiavon, B., Viola, T. W., Sanvicente-Vieira, B., Pezzi, J. C., & Grassi-Oliveira, R. (2016). Similarities between adult female crack cocaine users and adolescents in risky decision-making scenarios. *Journal of Clinical and Experimental Neuropsychology*, *38*(7), 795–810. https://doi.org/10.1080/13803395.2016.1167171.

Koob, G. F., & Nestler, E. J. (1997). The neurobiology of drug addiction. *Journal of Neuropsychiatry and Clinical Neurosciences*, *9*(3), 482–497. https://doi.org/10.1176/jnp.9.3.482.

Lawrence, A. J., Luty, J., Bogdan, N. A., Sahakian, B. J., & Clark, L. (2009). Problem gamblers share deficits in impulsive decision-making with alcohol-dependent individuals. *Addiction*, *104*(6), 1006–1015. https://doi.org/10.1111/j.1360-0443.2009.02533.x.

Myers, C. E., Shenyin, J., Baldson, T., Luzardo, A., Beck, K. D., Hogarth, L., et al. (2016). Probabilistic reward- and punishment-based learning in opioid addiction: Experimental and computational data. *Behavioural Brain Research*, *296*, 240–248. https://doi.org/10.1016/j.bbr.2015.09.018.

Ortega, L. A., Solano, J. L., Torres, C., & Papini, M. R. (2017). Reward loss and addiction: Opportunities for cross-pollination. *Pharmacology Biochemistry and Behavior*, *154*, 39–52. https://doi.org/10.1016/j.pbb.2017.02.001.

Radell, M. L., Myers, C. E., Beck, K. D., Moustafa, A. A., & Allen, M. T. (2016). The personality trait of intolerance to uncertainty affects behavior in a novel computer- based conditioned place preference task. *Frontiers in Psychology*, *7*(1175), https://doi.org/10.3389/fpsyg.2016.01175.

Robbins, T. W., & Everitt, B. J. (1999). Drug addiction: Bad habits add up. *Nature*, *398*, 567–570. https://doi.org/10.1038/19208.

Saleme, D. M., Kluwe-Schiavon, B., Soliman, A., Misiak, B., Frydecka, D., & Moustafa, A. A. (2018). Factors underlying risk taking in heroin-dependent individuals: Feedback processing and environmental contingencies. *Behavioural Brain Research*, *350*, 23–30. https://doi.org/10.1016/j.bbr.2018.04.052.

Schonberg, T., Fox, C. R., & Poldrack, R. A. (2011). Mind the gap: bridging economic and naturalistic risk-taking with cognitive neuroscience. *Trends in Cognitive Sciences*, *15*(1), 11–19.

Schumpe, B. M., Brizi, A., Giacomantonio, M., Panno, A., Kopetz, C., Kosta, M., et al. (2017). Need for cognitive closure decreases risk taking and motivates discounting of delayed rewards. *Personality and Individual Differences*, *107*, 66–71. https://doi.org/10.1016/j.paid.2016.11.039.

Seaman, K. L., Gorlick, M. A., Vekaria, K. M., Hsu, M., Zald, D. H., & Samanez-Larkin, G. R. (2016). Adult age differences in decision making across domains: Increased discounting of social and health-related rewards. *Psychology and Aging*, *31*(7), 737–746. https://doi.org/10.1037/pag0000131.

Smith, A. R., Ebert, E. E., & Broman-Fulks, J. J. (2016). The relationship between anxiety and risk taking is moderated by ambiguity. *Personality and Individual Differences*, *95*, 40–44. https://doi.org/10.1016/j.paid.2016.02.018.

Torres, M. A., Megias, A., Catena, A., Candido, A., & Maldonado, A. (2017). Opposite effects of feedback contingency on the process of risky decision-making. *Transportation Research Part F*, *45*, 147–156. https://doi.org/10.1016/j.trf.2016.12.007.

Tyagi, V., Hanoch, Y., Hall, S. D., Runco, M., & Denham, S. L. (2017). The risky side of creativity: Domain specific risk taking in creative individuals. *Frontiers in Psychology*, *8*, 145. https://doi.org/10.3389/fpsyg.2017.00145.

Verdejo-Garcia, A. (2016). Cognitive training for substance use disorders: Neuroscientific mechanisms. *Neuroscience and Biobehavioural Reviews*, *68*, 270–281. https://doi.org/10.1016/j.neubiorev.2016.05.018.

Verdejo-Garcia, A., Chong, T.T.-J., Stout, J. C., Yücel, M., & London, E. D. (2017). Stages of dysfunctional decision-making in addiction. *Pharmacology, Biochemistry, and Behavior*, https://doi.org/10.1016/j.pbb.2017.02.003.

Voon, V., Morris, L. S., Irvine, M. A., Ruck, C., Worbe, Y., Derbyshire, K., et al. (2015). Risk-taking in disorders of natural and drug rewards: Neural correlates and effects of probability, valence, and magnitude. *Neuropsychopharmacology*, *40*, 804–812. https://doi.org/10.1038/npp.2014.242.

Wang, X. T., Zheng, R., Xuan, Y.-H., Chen, J., & Li, S. (2016). Not all risks are created equal: A twin study and meta-analyses of risk taking across seven domains. *Journal of Experimental Psychology: General*, *145*(11), 1548–1560. https://doi.org/10.1037/xge0000225.

Wittwer, A., Hulka, L. M., Heinimann, H. R., Vonmoos, M., & Quednow, B. B. (2016). Risky decisions in a lottery task are associated with an increase of cocaine use. *Frontiers in Psychology*, *7*, 640. https://doi.org/10.3389/fpsyg.2016.00640.

Zou, X., & Scholer, A. A. (2016). Motivational affordance and risk-taking across decision domains. *Personality and Social Psychology Bulletin*, *42*(3), 275–289. https://doi.org/10.1177/0146167215626706.

Chapter 4

Delay, probability, and effort discounting in drug addiction

Julia Garami[a], Ahmed A. Moustafa[b]
[a]*School of Social Sciences and Psychology, Western Sydney University, Sydney, NSW, Australia,* [b]*Marcs Institute for Brain, Behaviour, and Development and School of Psychology, Western Sydney University, Penrith, NSW, Australia*

Introduction

Drug addiction has been conceptualized as a dysregulation of multiple brain circuits and dysfunction in brain areas responsible for self-regulation and cognitive control (Bechara, 2005; Everitt & Robbins, 2005; Goldstein & Volkow, 2002). The sensitization of the mesolimbic dopamine system has traditionally thought to be the primary mechanism underlying transition from drug use to addiction (Hyman, Malenka, & Nestler, 2006), in which addictive substances activate the dopaminergic reward system responsible for the hedonic effect of drugs (Di Chiara & Imperato, 1988; Koob & Bloom, 1988; Volkow, Fowler, & Wang, 2003). The drug's reinforcing effect is thought to become so great that it overshadows natural reinforcers, thus motivating drug use (Volkow et al., 2003) at the expense of other rewarding activities. Drugs become more desirable, and "wanting" the drug develops into craving and compulsive drug taking, which occurs relatively independently of desired or undesired consequences (Robinson & Berridge, 2001).

Damage to the prefrontal cortex (PFC), a brain region largely responsible for decision making and self-control, has been associated with impaired decision-making in substance dependent individuals (Bechara et al., 2001). Insensitivity to future consequences, as demonstrated by problematic drug use, has been shown to also be related to damage to the PFC (Bechara, Damasio, Damasio, & Anderson, 1994). Substance abusers exhibit impaired decision making similar to those with damage to the PFC (Bechara et al., 2001; Rogers et al., 1999). A combination of sensitization to drug-related rewards, associative learning, and lack of behavioral control manifest in the maladaptive choices observed in drug addiction (Jentsch & Taylor, 1999; Lyvers et al., 2014). Over time, neural changes can contribute to the transition to a loss of control over one's drug use. However, the pathway from drug taking to drug addiction is unclear

Cognitive, Clinical, and Neural Aspects of Drug Addiction. https://doi.org/10.1016/B978-0-12-816979-7.00004-2

and highly complex, as some regular drug users never develop an addiction. The National Comorbidity Survey, composed of a large representative US sample, revealed that only 15% of those who use drugs develop a substance use disorder (SUD), and 7.5% of analgesic users (e.g. opioids) reported dependency (Anthony, Warner, & Kessler, 1994). This survey also revealed that among those who use drugs sporadically, only some of them go on to start using regularly, and even a smaller subset develop an addiction. An estimated 3% of chronic pain patients receiving opioid medication management develop an addiction to opioids (Portenoy et al., 2007), and around 20.4% exhibit aberrant drug-related behaviors that often precede addiction (Fishbain, Cole, Lewis, Rosomoff, & Rosomoff, 2008). Animal models also support the supposition of individual differences in addition vulnerability. When rodents are freely able to self-administer drugs, only some rats increase their drug intake (Mantsch, Yuferov, Mathieu-Kia, Ho, & Kreek, 2004; Piazza, Deminiere, Le Moal, & Simon, 1989) or consistently take high doses of the drug (Piazza, Deroche-Gamonent, Rouge-Pont, & Le Moal, 2000). Finally, addiction can develop to many other hedonic activities such as sex, food or gambling, which clearly indicates that there are non-drug exposure factors that underlie substance abuse.

One key personality characteristic associated with substance abuse and addictive behaviors is impulsivity. In the broadest terms, impulsivity is a multifaceted construct delineated by the tendency to act without considerable forethought or inhibition (Bari & Robbins, 2013; Gullo, Loxton, & Dawe, 2014). Impulsivity can have both positive and negative implications. For example, spontaneity and risk taking are fundamental to scientific progress and to the creative arts, and the ability to make quick, error-free decisions is considered to be an advantageous quality in many fast-paced work environments. However, impulsivity is widely conceptualized as a form of maladaptive decision making when the outcomes of acting upon impulses are counterproductive for achieving one's best interests. Thoughtlessly grabbing a candy bar at the grocery checkout counter can undermine the otherwise carefully considered food choices in one's basket, particularly if one is on a diet. Likewise, taking up a friend's offer to attend a party the night before a big exam most likely will be detrimental to one's final grade. These decisions are made on the basis of desire and inhibition, in which the consequences are not thoroughly considered or are willfully ignored in favor of pleasure.

Such maladaptive decisions are a fundamental element to chronic substance use, in which the user continues to take drugs despite the potential for great harm to herself and/or to others. Impulsivity in the context of substance abuse manifests as devaluing negative future consequences in favor of immediate gratification and the short-term rewarding effects of the drug. Furthermore, substance abusers demonstrate a lack of behavioral inhibition to repress drug-related urges in favor of healthy, pro-social pursuits that yield greater rewards in the long run. The literature largely supports impulsivity as a stable personality trait that correlates well with psychopathologies such as SUD, problem gambling and violent criminal offending (see Stanford et al., 2009 for a review). It has been theorized that heightened approach behavior and difficulty inhibiting behavior combine as a risk factor for drug addiction (Gullo et al., 2014; Loxton & Dawe, 2001). In other words, a pre-morbid predisposition towards risk-taking may incline one to initially try drugs, and deficits in impulse control may impede regulation of the intensity of drug taking. Once addicted, a lack of inhibitory control may also interfere with attempts to quit or impede adherence to a treatment program.

Impulsivity is multi-dimensional, and a variety of self-report measures have been developed to evaluate its diverse manifestations, but have many inherent limitations that restrict the interpretation of their results, such as intentional misreporting and the respondent's ability for self-assessment. Consequently, behavioral measures have been created to more precisely evaluate impulsivity. Of the many methods, discounting paradigms are of the most widely used assessment of maladaptive decision-making in addiction research. Discounting is based in behavioral economics and is indicative of how one subjectively devalues a reward according to a variety of factors such as delay, probability or effort. Unlike economic theories of decision-making that are based in logical calculation of optimal choices, behavioral economics attempts to describe our behavior in relation to psychosocial factors that impact our decisions. The influence of such factors on optimal decision making is believed to be a reflection of impulsivity, and will be the focus of this chapter.

Delay discounting

One of the most widely used empirical ways of assessing impulsive behavior is the preference for smaller, immediate rewards over larger, delayed ones—otherwise known as delay discounting. Delay discounting is observed in both animals and humans, and is generally considered to be a naturally occurring phenomenon in which one devalues (i.e. discounts) future rewards as the delay until its receipt increases (Green & Myerson, 2004). Delay discounting is a cost/benefit decision that accounts for the inherent uncertainty of receiving a future outcome according to the possible interfering factors during the delay. Receiving $100 today may be more attractive than receiving $105 in a year because the person offering the money may, for instance, lose your contact details, go bankrupt or even pass away. However, discounting future rewards becomes maladaptive when it fails to maximize one's overall gains in the long run. It is commonly accepted that higher levels of delay discounting reflects a desire for immediate gratification at the expense of greater long-term benefits (Green & Myerson, 2010). Such an attitude can be observed in lottery payouts, as typically the full sum of money won is paid out over a number of years, with an option to receive a lesser lump sum immediately. Rationally, receiving the total dollar amount of winnings is preferable to the smaller immediate amount, however humans deviate from rationality.

Delay discounting studies typically utilize an amount adjusting procedure consisting of a series of dichotomous choices between a smaller, immediate outcome and a larger, delayed outcome (e.g. would you prefer $100 today or $150 in a year?) Depending on whether the participant chooses the delayed or immediate outcome, the immediate outcome is either increased or decreased (respectively) on the next choice. The values continue to be adjusted until the delayed outcome is equally likely to be selected as the immediate outcome, otherwise known as the indifference point. At this point, the participant subjectively values the immediate outcome equally to the objective value of the delayed outcome. The procedure is repeated then for different delay lengths (e.g. 1 day, 1 week, 1 month, 6 months, 1 year, 2 years) to obtain a series of indifference points used to calculate one's level of discounting. While the titrating procedure is the most commonly used in research, a linear method has also been used in which the delayed amount is fixed while the immediate amount steadily increases/decreases—at which point the respondent "switches" from selecting the immediate option to the delayed option is taken as the indifference point (e.g. Frank & Triplett, 2009). The fill-in-the-blank method is also used in which a single question is posed for each delay length (e.g. "How much money would you rather receive today rather than wait one year for $100?"). However, Smith and Hantula (2008) found that the binary choice method produced steeper discounting rates than the fill-in-the-blank method, indicating that the discounting method is important to consider when comparing results between studies.

Delay discounting is largely accepted to be hyperbolic in nature, and discounting data has been described by the following hyperbolic function (Rachlin, Raineri, & Cross, 1991):

$$V = A / (1 + kD)$$

The indifference point is the subjective value (V) of the objective outcome (A) to be delivered after a delay (D). The discounting parameter (k) specifies the degree of discounting, with greater values indicating a tendency to discount future outcomes at a higher rate, which results in a hyperbolic function that declines steeply. Steep delay discounting functions are used as an indication of impulsivity, in which a preference for immediate rewards is preferred rather than waiting for a larger reward.

Another method of quantifying discounting is calculating the area under the curve (AUC). This method was proposed by Myerson, Green, and Warusawitharana (2001) as a response to the limitations of the hyperbolic function. The AUC method does not depend on a theoretical function and has been shown to be a valid way to compare impulsivity between a variety of samples. The hyperbolic discounting parameter is skewed, which necessitates non-parametric statistical analysis that often has a diminished power. The AUC is calculated by normalizing the subjective values of the delayed outcome and plotting these values as a function of the delay. The area of the trapezoids beneath the data points are summed, which can range from 0.00 (greatest discounting) to 1.00 (no discounting). It has been suggested that the AUC method is preferable when comparing sensitivity to delay between groups as discounting is derived from actual data points rather than estimated from an equation that may not necessarily be a good fit to the obtained data (Myerson et al., 2001). In fact, it is rare for the hyperbolic equation to provide an adequate fit to every participant's data and often can range from 10% to 30% (Madden, Begotka, Raiff, & Kastern, 2003; Patak & Reynolds, 2007; Rasmussen, Lawyer, & Reilly, 2010), often necessitating the exclusion of otherwise meaningful data from the statistical analysis. The AUC method is not without limitations, as it may be disproportionally influenced by subjective values at longer delays while being less impacted by those of shorter delays (Borges, Kuang, Milhorn, & Yi, 2016; Yoon et al., 2017).

Delay discounting has shown consistent correlations with other impulsivity measures such as the Eysenk Impulsiveness Questionnaire (Andrade & Petry, 2012; Madden, Petry, Badger, & Bickel, 1997; Malesza & Ostaszewski, 2016) and the Barratt Impulsivity Scale non-planning (de Wit, Flory, Acheson, McCloskey, & Manuck, 2007; Mobini, Kass, Yeomans, & Grant, 2007) and motor subtraits (Malesza & Ostaszewski, 2016). Cognitive distortions regarding one's right to immediate satisfaction and focus on short-term consequences have also been positively correlated with discounting delayed outcomes (Mobini et al., 2007). Behavioral indices of impulsivity have also been associated with delay discounting such as the Stop-Signal task (Malesza & Ostaszewski, 2016), and Reimers, Maylor, Stewart, and Chater (2009) found that delay discounting was related to risky, impulsive behaviors such as early sexual activity and infidelity.

Substance abuse

The delay discounting paradigm has been so frequently used in the addiction literature because of its potential ability to predict hallmark facets of substance abuse. For example, the immediate reward of the drug's desired effects is chosen at the expense of the larger, delayed rewards of abstinence (e.g. increased career prospects, better family relationships or health improvements). As a measure of impulsivity, delay discounting can also shed light on why addicts are aware of the negative consequences of their drug use, have a desire to quit or seek treatment, but often turn to drugs before they start. The link between delay discounting and addictive behaviors is well documented. A meta-analysis conducted by Mackillop et al. (2011) reported that problematic users of alcohol, nicotine, cocaine, amphetamines, marijuana, heroin and pathological gamblers consistently exhibit greater levels of delay discounting compared to non-drug using controls. High levels of delay discounting can also predict smoking relapse (Krishnan-Sarin et al., 2007; Mackillop & Kahler, 2009; Sheffer et al., 2012, 2014; Yoon et al., 2007) and fewer days of cocaine abstinence after treatment (Washio et al., 2011). The knowledge base for delay discounting and its links to addiction is extensive and has been reviewed in great depth by a number of authors (see Green & Myerson, 2010; Madden & Johnson, 2010). Consequently, the current chapter will focus primarily on more recent developments in discounting research as they relate to substance abuse and addiction.

Contemporary studies have replicated and reinforced the connection between the devaluation of delayed rewards and addiction. Robles, Huang, Simpson, and McMillan (2011) observed that methadone maintained opioid addicts appeared to discount delayed rewards more so than non-drug users, and that current illicit drug using methadone patients demonstrated similar discounting to abstinent patients. Stimulant and/or alcohol dependent adults have exhibited greater discounting of delayed rewards significantly more than non-drug abusing controls, regardless of comorbid depression or antisocial personality disorder (Moody, Franck, & Bickel, 2016), which indicates that excessive discounting may exist independently from impulsivity related psychopathologies. Hofmeyr et al. (2017) considered whether smokers delay discount differently than never-smokers, and whether a hyperbolic, exponential or mixed model was the best in describing participants' data. Overall, smokers discounted delayed rewards more steeply, but the groups were comparable when the hyperbolic model was used, suggesting that a mixed discounting model may better describe discounting behavior in smokers. Such findings may influence the way in which we interpret previous discounting research and/or inform future discounting analyses.

Recent research has also supported that delay discounting becomes more exaggerated as addiction severity increases, as a meta-analysis conducted by Amlung, Vedelago, Acker, Balodis, and Mackillop (2017) reported. Discounting is more pronounced in the literature for those with more intense levels of addiction, such as a clinical diagnosis of substance abuse disorder. Delay discounting has also shown to predict the number of cannabis dependence symptoms, but does not predict the frequency of marijuana use (McKetin, Parasu, Cherbuin, Eramudugolla, & Anstey, 2016). In other words, a desire for immediate gratification is associated with the severity of dependence independent of the actual amount of drug used. Furthermore, Lim, Cservenka, and Ray (2017) found that alcohol dependence severity correlates with neural reward system dysfunction during delay discounting tasks. However, discounting has not consistently been was correlated with addiction severity symptoms in opioid users Robles et al. (2011), which suggests that attitudes towards delayed rewards may be a consequence of long-term opioid use.

There are lines of evidence that support delay discounting as a pre-existing addiction vulnerability trait. Longitudinal studies have revealed that delay discounting rates in early adolescence can predict future alcohol and drug use (Fernie et al., 2013; Khurana, Romer, Betancourt, & Hurt, 2017). Alcohol, tobacco and marijuana use, as well as drug experimentation have been correlated with greater delay discounting, which has shown to mediate the relationship between a family history of substance misuse and current substance use (VanderBroek, Acker, Palmer, de Wit, & MacKillop, 2016). Growing up in an environment with parental drug abuse may have an impact on the development of self-control in children, which subsequently heightens their vulnerability to addiction in adulthood. It seems that parental substance abuse may be related to impulsivity and delay discounting, although there have been mixed results. A first-degree family history of alcoholism predicts alcohol use disorder risk in non-heavy drinkers (Smith, Steel, Parrish, Kelm, & Boettiger, 2015). Herting, Schwartz, Mitchell, and Nagel (2010) found a nonsignificant positive trend of DD among adolescents with a family history of alcoholism, while Petry, Kirby, and Kranzler (2002) found higher discount rates only in women with a family history of alcohol dependence but not men. Athamneh, Stein, Quisenberry, Pope, and Bickel (2017) attempted to clarify the relationship between parental drug abuse and offspring delay discounting, and found that individuals with two parents with a history of drug addiction exhibit greater delay discounting compared to individuals who have only one or none. In fact, there may be genetic basis for delay discounting that is related to hereditary dopaminergic abnormalities (see Mackillop, 2013 for a review).

Delay discounting research has expanded in recent years to adapt to the perpetual creation of novel drugs and methods of administration. One development is the electronic cigarette, which is an instrument that vaporizes nicotine and is promoted as a safer alternative to traditional cigarettes. Białaszek, Marcowski, and Cox (2017) found that both types of cigarette smokers discounted delayed rewards greater than non-smokers, and that the smoking groups were similar in discounting. Given that many smokers begin using electronic cigarette as a smoking cessation method, while others report using them as a safer alternative to traditional cigarettes, it is interesting that both groups exhibit similar impulsivity. The rise of prescription opioid abuse in the last decade necessitates further enquiry as we seek to identify vulnerability factors before prescribing such powerful medications. Karakula et al. (2016) found that among treatment-seeking patients with a diagnosed opioid use disorder, those who were heroin users discounted delayed rewards more than those who used prescription opioids exclusively, suggesting that the there may be a separate set of vulnerability factors for illicit opioid use. Identifying these factors is of great importance given the health hazards of heroin, such as potency unpredictability and risk of disease transmission through intravenous use. Further investigation will need to include a demarcation between individuals whose trajectory to opioid addiction began with prescribed opioid medication versus heroin or other illicit opioids.

Researchers have also recently sought to better understand the mechanisms underlying discounting in substance abusers. Scherbaum, Haber, Morley, Underhill, and Moustafa (2018) created a virtual delay discounting "game" to assess differences in optimal decision making between addicts and to those without a history of addiction. The authors confirmed that methadone maintained opioid patients delay discount more steeply than non-drug using controls, and that this pattern of decision making resulted in suboptimal choices that could be attributed to difficulty discriminating suboptimal options. Therefore, delay discounting may reflect not only deficits in self-control but also cognitive deficits in decision making and information processing.

Probability discounting

Delay discounting has been the primary measure of impulsivity in the addiction literature, and far less attention has been paid to probability discounting. Probability discounting refers to our tendency to devalue an outcome as the probability of its occurrence becomes less likely. When presented with a choice between a certain reward and a larger probabilistic reward, the subjective value of the probabilistic reward diminishes as the odds against receiving it increase (McKerchar & Renda, 2012). In other words, we are more likely to stick with a safer, certain reward rather than take a risky gamble for a larger reward if the odds of that gamble are poor. For example, the chance to win \$100 with 95% chance may seem more appealing compared to an assured \$20, but the same \$100 becomes less attractive when its probability drops to only 5% The less likely one is to discount the probabilistic reward (i.e. chooses to gamble for the \$100 regardless of low odds) is thought to be a marker for risk taking (Green & Myerson, 2010; Patton, Stanford, & Barratt, 1995). Risky choices often yield great rewards, such as investing in the stock market or pursuing a romantic interest. However, risks that fail to maximize benefits in the long run are suboptimal and a potential sign of impulsivity.

The typical probability discounting procedure is very similar to that of delay discounting, in that the respondent is presented with dichotomous choices between two outcomes. The choices are between an assured reward and a larger, uncertain reward at varying odds. The amount of the assured reward is adjusted until an indifference point is reached, in which the respondent is equally as likely to select the assured reward as the probabilistic reward. Inverse to delay discounting, greater probability discounting indicates less risk taking. That is, the more the probabilistic outcome is discounted, the more likely the respondent is to stick with the safer, but smaller, reward. Probability discounting is well described by the same hyperbolic function, however k is replaced as h as the indicator of discounting, and Θ is calculated as the probability against the reward $(p - 1/p)$.

$$V = A / (1 + h\Theta)$$

The same caveats exist for the hyperbolic probability discounting function as the delay function, and therefore AUC is also a valid and theoretically neutral way to assess probability discounting.

The links between probability discounting and other measures of impulsivity are tenuous, leading researchers to conclude that the paradigm is assessing a particular facet of impulsivity not otherwise captured by self-report and behavioral measures. The literature has failed to consistently find correlations between probability discounting and other impulsivity measures such as the BIS-11 (Baumann & Odum, 2012; Mitchell, 1999), the Eysenk Impulsiveness Questionnaire (Andrade & Petry, 2012; Crean, de Wit, & Richards, 2000; Reynolds, Richards, Horn, & Karraker, 2004), the Zuckerman Sensation

Seeking Scale (Mitchell, 1999; Reynolds et al., 2004) and dopamine-related impulse control (Mobini, Chiang, Ho, Bradshaw, & Szabadi, 2000). Furthermore, positive correlations have frequently been observed between delay and probability discounting (Baumann & Odum, 2012; Crean et al., 2000; Johnson, Johnson, Herrmann, & Sweeney, 2015; Ohmura, Takahashi, & Kitamura, 2005; Richards, Zhang, Mitchell, & Wit, 1999), a counterintuitive result that suggests those who prefer the immediacy of a reward (i.e. are impulsive) simultaneously place a high value on the certainty of a reward (i.e. are risk-averse). It is possible that the two discounting paradigms are capturing different impulsivity facets or that delay discounting is actually a measure of risk taking rather than delayed gratification, given that delay inherently involves risk.

A meta-analysis of neuroimaging studies conducted by Gowin, Mackey, and Paulus (2013) revealed that individuals with SUD exhibit neural activity while engaging in risk taking that is distinctly different than the activity witnessed in non-drug abusing control groups. The meta-analysis also found that the neural activation observed in SUD occurs in brain areas that are responsible for dopaminergic regulation and executive functioning. There is also extensive evidence for neural correlates with risk taking and desensitization to risk in areas of the brain affected by drug abuse discussed in Chapter 1, such as the OFC (Hsu, Bhatt, Adolphs, Tranel, & Camerer, 2005) and the PFC (Manes et al., 2002; Rahman, Sahakian, Hodges, Rogers, & Robbins, 1999).

As there has been a dearth of probability discounting research in relation to addiction, the current chapter aims to review the existing knowledge across various types of addictive substances in order to provide a comprehensive review of what is known.

Nicotine

The study of PD in the addiction context is gaining momentum in recent years. Early studies were limited to cigarette smokers and found contradictory results. Reynolds et al. (2004) found that heavy smokers (20+ cigarettes per day) discounted a probabilistic $10 more than never-smokers, although delay discounting was a stronger predictor of smoking status. Similarly, Yi, Chase, and Bickel (2007) found that heavy cigarette smokers displayed greater PD than non-smokers, but this result was limited to higher probability magnitudes (95%, 75% and 50%) and greater values of probabilistic money ($100 and $1000), which the authors interpreted a floor effect. Taken together, these two studies suggest that cigarette smokers have more risk averse attitudes towards uncertain rewards than non-smokers, which is surprising given that smokers tend to be risk-taking and sensation seeking (Munafò, Zetteler, & Clark, 2007).

However, there are a variety of studies that have failed to replicate these results. Mitchell (1999) compared smokers (M = 18.15 cigarettes per day) to never-smokers and found no meaningful difference in the discounting of a probabilistic monetary reward. Ohmura et al. (2005) compared moderate smokers to never-smokers (M = 14.38 cigarettes per day), and did not observe differences for a probabilistic amount of money. It has been argued that the smokers in theses studies were more moderate in their addiction, and that shallower discounting rates are likely to be found among those with more severe habits, based on similar findings in the delay discounting field, such as Kollins (2003); Ohmura et al. (2005), who found that delay discounting was more apparent in heavy smokers and those with greater drug use. More recent research has recruited even lighter smokers, and found no probability discounting differences between never-smokers and cigarette smokers, in which study half of the participants smoked 5 cigarettes or less daily (Białaszek et al., 2017).

Alcohol

There is similarly an undefined picture of probability discounting in relation to alcohol consumption and abuse, as alcoholics (Myerson, Green, Berk-Clark, & Grucza, 2015) or everyday drinkers (Ida & Goto, 2009) were found to discount similarly to non-drinkers or light drinkers. To date, there is no association between drinking frequency and probability discounting (Takahashi, Ohmura, Oono, & Radford, 2009). However, Bernhardt et al. (2017) observed that treatment-seeking alcoholic dependent participants demonstrated a shallower probability discounting curve than non-problematic drinkers for monetary rewards, which suggested that alcoholics may be risk-seeking. However, discounting did not predict problematic alcohol use in a non-clinical sample, a finding that bolsters the proposition that inflated discounting is only found in higher levels of substance abuse. Risk-aversion for probabilistic losses was predictive of relapse to hazardous drinking in the first year after treatment. This result is intriguing because patients who were inclined to risk incurring great losses had better long-term success in their rehabilitation. A contradictory pattern between alcohol and nicotine abuse has thus far emerged, in that abusers of nicotine demonstrate risk-aversion and those dependent on alcohol exhibit risk taking when deciding between assured and probabilistic rewards.

Illicit drugs

Nicotine and alcohol are highly addictive drugs that carry immense health risks and can occasion death if taken in too large of a quantity. However, each addictive substance is unique in its psychotropic effects as well as its risk factor for addiction, dependency and health outcomes. Probability discounting research branched out to illicit substances and have had mixed results. Primarily, illicit drug users appear to discount probabilistic outcomes similarly to control groups. For example, Andrade and Petry (2012) observed that problem gamblers with or without a history of substance abuse problems (including alcohol or any other drug used for non-medicinal purposes) discounted probabilistic monetary similarly. Comparable findings have been obtained for cocaine and marijuana dependent individuals receiving in-patient treatment, although in the study conducted by Mejía-Cruz, Green, Myerson, Morales-Chainé, and Nieto (2016), the exclusion criteria for control participants was limited to a history of drug use consistent with a substance use disorder, rather than a history of any drug use. Control participants who had used cocaine and/or marijuana recreationally may have been too similar to the drug-dependent patients for differences in discounting to be apparent. However, Johnson et al. (2015) also did not find that those with cocaine use disorder discounted probabilistic money or sexual intercourse any differently to never-users of cocaine.

In contrast, there are two studies to date that have found that drug abusers tend towards smaller, certain monetary rewards more so than non-users, such as methamphetamine users (Yi, Carter, & Landes, 2012). The authors attributed this result not to risk aversion, but to a limited capacity to consider the future impact of probability on one's choices. They postulated that one with such a restricted future perspective would be indifferent to the probabilistic nature of one option, and place equal values on an uncertain and an assured reward. In the most recent study, Garami and Moustafa (2019) found that opioid dependent patients receiving maintenance therapy also displayed a preference for smaller, certain monetary rewards over greater, probabilistic ones compared to non-drug dependent controls.

While the literature is far from conclusive, there is a pattern that emerges between non-alcohol substance abusers and non-abusing control groups. When statistically meaningful differences occur, addicted cohorts tend to choose safer, certain rewards and are less willing to take risks for comparatively larger rewards than controls. Problematic drug taking behavior is reflected by this pattern, as users seem to prefer the smaller, certain desired outcomes of the drug rather than take a risk on the less certain (and arguably more obscure) benefits of abstinence. Repeated drug administration has shown to cause significant changes to stress and reward neural pathways (Koob & Kreek, 2007; Sinha, 2008) and can result in neurobiological adaptations that impair decision-making by altering the subjective value of uncertain rewards and impeding the ability to predict rewards (Berridge & Aldridge, 2008). Reduced accuracy of predicting reward value and an exaggerated asymmetry between the subjective and objective value of a reward may result in suboptimal decision-making (Schultz, 2011). As a result, a drug-addicted individual may become biased towards certain rewards, such as a drug's desired effects, and away from relatively uncertain non-drug rewards. It is worth noting that Bernhardt et al. (2017) found the opposite effect of alcohol use disorder, potentially signifying inherent differences in alcohol abuse compared to other addictive substances. As with all the probability discounting literature, many replication studies are needed before we can begin to make informed conclusions about the nature of probability discounting in addictive individuals.

It is possible that probability discounting is not actually a measure of risk-taking, and Yi et al. (2012) posit an explanation for substance abusing individuals' preference for certain rewards that meshes more elegantly with popular conceptualizations of risk in addictive behaviors. Given the apparent contradiction to psychobiological research evidencing greater impulsivity and risk taking in substance abusers, discounting probabilistic outcomes may be indicative of insensitivity to the impact of risk on the consequences of one's decisions. Yi et al. (2012) argue that the substance abuser does not differentiate between high and low probability outcomes, and chooses the certain outcome because it is certain, without factoring in the probability of a larger reward. This interpretation is limited in that if probabilities are not considered, the respondent would be equally likely to favor the probabilistic choice.

Effort discounting

One important factor that has not been addressed thus far is the amount of effort required in making the choice between the immediacy/certainty of the drug's desired effect and the delayed/uncertain benefits of abstinence. Delay and probability discounting have the potential to help us understand impulsivity as a general trait present in addictive behaviors, but the application of the discounting paradigm to drug taking behavior is limited due to the host of additional variables contributing to drug use. Addiction to drugs (particularly to those that cause physical dependency) involves strong motivating factors such as the prevention of withdrawal symptoms, escape from painful memories, and improvements in negative affect. Operant conditioning and alterations in the dopaminergic system also work in tandem to amplify the reward value of drugs and promote compulsive drug use. Undoubtedly, there is a considerable amount of effort required to overcome such strong psychobiological drives, as well as to maintain abstinence and pursue non-drug activities. The way in which one discounts effortful tasks is particularly applicable to rehabilitation, as relapse occurs despite the absence of physiological effects of the drug. In the best-case scenarios, the psychological desire for the drug has also been reduced and new associations have been made with non-drug reinforcers. Despite successful rehabilitation, relapse is incredibly common and many consider addiction to be a lifelong illness that requires constant effort to resist temptation, avoid drug-related scenarios, and engage in non-drug activities. There are clear implications of attitudes towards effort in recovering from addiction, yet there is a dearth of investigation on the topic.

Stated simply, effort discounting is behavioral economic decision-making process in which an outcome is devalued as the effort required to obtain it increases (Bijleveld, Custers, & Aarts, 2012). We make effort based decisions frequently in our every-day lives when we weigh the payoff of our actions in relation to the necessary exertion of effort. For example, the reward of eating at your favorite restaurant most likely diminishes in subjective value if your car is out of service and you have to walk instead. Effort can also be operationalized in terms of cognitive and emotional effort. Splitting the check among many friends at a restaurant is simpler than calculating exactly what everyone ate and drank, even though you may pay more than your share. The subjective value of a close friendship may also devalue according to the amount of emotional effort you have to invest supporting a friend through a difficult time. Effort discounting has been described well by the hyperbolic discounting function (Mitchell, 1999, 2004; Ostaszewski, Bąbel, & Swebodziński, 2013; Prevost, Pessiglione, Metereau, Clery-Melin, & Dreher, 2010; Sugiwaka & Okouchi, 2004), although Hartmann, Hager, Tobler, and Kaiser (2013) found that a concave parabolic model best fit physical effort discounting data. They argued that the parabolic model makes more logical sense than the hyperbolic, in that the impact of increasing increments are more powerful as one gets closer to their effortful maximum. This is illustrated in running a marathon, where each successive kilometer gets more difficult as more distance has already been run.

Three human studies have provided evidence that effort discounting fits the hyperbolic discounting function, operationalizing effort in a number of ways. Sugiwaka and Okouchi (2004) found that their participants were more willing to forgo payment for cleaning bathrooms as the number of bathrooms to be cleaned increased and preferred to accept a smaller amount of money for not engaging in any effort at all. Effort was defined more nebulously in a study conducted by Nishiyama (2014), in which participants mentally compiled a list of 100 physical/psychological effortful tasks in order of difficulty. As the tasks increased in difficulty, the more likely the participant chose be to be paid a smaller sum of money to avoid engaging in the task. A follow-up study was conducted by Nishiyama (2014) that used the same procedure, but delineated effort in terms of physical, cognitive and emotional effort. The hyperbolic equation described that data for each type of effort adequately, and the discounting rates across effort types were closely correlated. It appears that not only can we assess effort discounting in a similar way to delay and probability discounting, but that individuals have a cost analysis process that values effort as a general construct.

Impulsivity

Effort discounting falls under the umbrella of behavioral inhibition and is thought to underpin impulsivity (Bari & Robbins, 2013) in that restraining behaviors based on what we want to do rather than what we feel like doing clearly requires effort. Impulsivity is apparent in the predisposition towards effortless rewards that are ultimately suboptimal compared to channeling energy towards more productive behaviors, but effort discounting as a method for assessing impulsivity has not been well studied in human research. In one of the earliest studies, Mitchell (1999) found that greater levels of effort discounting were correlated with higher thrill and novelty seeking personality traits. Impulsive persons may prefer new and exciting options rather than working on a tedious, yet important, task. Malesza and Ostaszewski (2013) also found moderate correlations between discounting sums of money paid for climbing flights of stairs and higher scores on Harm Avoidance and Reward Dependency subscales of the Temperament and Character Inventory (Cloninger, Svrakic, & Przybeck, 1993). It appeared that those who tend to shy away from negative situations and respond strongly to reward are more likely to be enticed by a smaller, effortless reward. Contrary to the authors' expectations, the Novelty Seeking subscale did not have a meaningful correlation with discounting, although discounting was predictive of lower persistence (i.e. perseverance to complete effortful, non rewarding tasks). However, effort discounting has also shown no relationship to other measures of self-control, such as the Redressive-Reformative Self-Control Scale or the delay discounting paradigm (Sugiwaka & Okouchi, 2004). There has not been sufficient investigation to date to elucidate the relationship between impulsivity and effort discounting, and much more work is clearly necessary in the near future.

Substance abuse

The implications for substance abuse can be found in non-human animal research with dopamine and brain regions associated with decision-making and impulsivity. Rodent studies have revealed that dopamine antagonists increase choices for low reward, low effort options while dopamine agonists increase high effort/high reward choices (see Bailey, Simpson, & Balsam, 2016 for a review). Furthermore, dopamine is involved in effort-guided behaviors, particularly those that are reward-driven (Miller, Thome, & Cowen, 2013). The role of dopamine dysfunction in drug addiction is well established (Hyman et al., 2006; Nestler & Aghajanian, 1997; Robinson & Berridge, 2001), and there is evidence of a hypodopaminergic state resulting from substance abuse that may reduce goal driven behavior for non-drug reinforcers (Martinez et al., 2011; Melis, Spiga, & Diana, 2005). There is impetus for the study of effort discounting in substance abuse, as dopamine deficits may account for the difficulties that recovering addicts have in remaining abstinent during and after rehabilitation. There is only one human study to date comparing cigarette smokers to non-smokers, which failed find any meaningful differences in the amount of physical effort expended on squeezing a hand dynometer for money (Mitchell, 1999). The current literature, while scant, provides a necessary foundation for future substance abuse research.

Future directions

There is a large database of theoretical knowledge on the way in which substance abusers discount delayed outcomes which should be applied to the development of more effective treatment. For the most insidious and dangerous class of addictive substances, the relapse rate of opioid addiction is around 56% by 1 week post-treatment and can be as high as 90% after 6-weeks

(Bentzley, Barth, Back, & Book, 2015; Smyth, Barry, Keenan, & Ducray, 2010). A greater devaluation of delayed rewards predict greater substance abuse treatment attrition (Stevens, Verdejo-García, Roeyers, Goudriaan, & Vanderplasschen, 2015), which highlights the potential importance of improving delay of gratification during addiction therapy. Preliminary research has evidenced that working memory training can reduce delay discounting in cocaine and methamphetamine abusers (Bickel, Yi, Landes, Hill, & Baxter, 2011; Zhu et al., 2018), as well as a money management-based intervention designed to improve long-term goal planning (Black & Rosen, 2011). The intervention not only reduced discounting, but also predicted greater cocaine abstinence use over time. Episodic future thinking (EFT) defined as mentally projecting an autobiographical future event in vivid detail, is a technique which has shown to reduce discounting of delayed rewards along with substance use (Atance & O'Neill, 2001). Engaging in an EFT task has also been successful in improving both discounting and alcohol consumption in alcohol dependent patients (Bulley & Gullo, 2017; Snider, Laconte, & Bickel, 2016), and there is evidence that EFT can be effective in reducing cigarette use in nicotine dependent smokers (Stein et al., 2016). The future events imagined in EFT tasks are unrelated to substance use in addiction studies, suggesting that generally orientating individuals to a positive future can decrease a bias towards immediate rewards and improve long-term decision making.

The devaluation of future rewards has shown associations with poor executive control in adolescents (Stanger et al., 2013), and predicts substance abuse onset and poorer treatment outcomes in young people (Audrain-Mcgovern et al., 2004; Stanger et al., 2012). Targeting attitudes towards delayed rewards at an early age is an important avenue for the application of discounting interventions. There are indications that pharmaceutical drugs can improve delay discounting as well as illicit drug use, particularly monoamine transporters in the treatment stimulant addiction (see Perkins & Freeman, 2018 for a review). However, this line of enquiry has predominantly utilized rodents as test subjects and replication in human studies are needed.

There has yet to be any investigation into the efficacy of probability discounting interventions and the applicability to addiction treatment. There is tangential evidence that modulating attitudes towards uncertainty using cognitive bias modification methods and cognitive behavioral therapy is possible and can yielded lasting improvements in the acceptability of uncertainty (Boswell, Thompson-Hollands, Farchione, & Barlow, 2013; Hui & Zhihui, 2017; Mahoney & McEvoy, 2012; Oglesby, Allan, & Schmidt, 2017; Talkovsky & Norton, 2016). Changing perceptions of uncertain rewards may improve maladaptive discounting of probabilistic outcomes and positively impact treatment success. The same can be said about the effort discounting paradigm and relapse prevention, in that increasing the value of effort in achieving positive outcomes has a great possibility of managing cravings and temptations after rehabilitation.

Conclusion

Drug addiction is very complex and involves an interplay between psychological, biological and social influences. In this chapter, we have reviewed the literature relevant to cognitive behavioral underpinnings of drug abuse and the subjective discounting of rewards in relation to temporal distance, probability and effort exertion. Discounting measures are widely used to assess different facets of impulsivity, and delay discounting has received the bulk of research efforts. We can conclude that the tendency to prefer sooner rewards rather than wait for more beneficial ones has been significantly associated with addiction, and may increase according to the severity of drug abuse. What is yet unclear is the role of risk taking and exerted effort on drug taking behavior as assessed by the discounting paradigm. Whether drug addicted individuals exhibit unique patterns of probability discounting is far from conclusive, and a counterintuitive trend has appeared in that drug abusers to be risk-averse when deciding about probabilistic rewards. Because of the sparse number of studies, we are not able meaningfully infer attitudes towards uncertain rewards and researchers have largely omitted interpretations of their non-significant results in terms of impulsivity. We also are quite in the dark about how discounting effort applies to drug addiction; despite the obvious influence effort has on reward-driven behavior. Future research focusing on lesser understood discounting processes is necessary in order to obtain a fuller picture of the behavioral economic choices involved in drug addiction.

References

Amlung, M., Vedelago, L., Acker, J., Balodis, I., & Mackillop, J. (2017). Steep delay discounting and addictive behavior: A meta-analysis of continuous associations. *Addiction*, *112*, 51–62.

Andrade, L., & Petry, N. (2012). Delay and probability discounting in pathological gamblers with and without a history of substance use problems. *Psychopharmacology*, *219*(2), 491–499.

Anthony, J. C., Warner, L. A., & Kessler, R. C. (1994). Comparative epidemiology of dependence on tobacco, alcohol, controlled substances, and inhalants: Basic findings from the National Comorbidity Survey. *Experimental and Clinical Psychopharmacology*, *2*(3), 244–268.

Atance, C. M., & O'Neill, D. K. (2001). Episodic future thinking. *Trends in Cognitive Sciences*, *5*(12), 533–539.

Athamneh, L. N., Stein, J. S., Quisenberry, A. J., Pope, D., & Bickel, W. K. (2017). The association between parental history and delay discounting among individuals in recovery from addiction. *Drug and Alcohol Dependence*, *179*, 153–158.

Audrain-Mcgovern, J., Rodriguez, D., Tercyak, K., Epstein, L. H., Goldman, P., & Wileyto, E. (2004). Applying a behavioral economic framework to understanding adolescent smoking. *Psychology of Addictive Behaviors*, *18*(1), 64–73.

Bailey, M. R., Simpson, E. H., & Balsam, P. D. (2016). Neural substrates underlying effort, time, and risk-based decision making in motivated behavior. *Neurobiology of Learning and Memory*, *133*, 233–256.

Bari, A., & Robbins, T. (2013). Inhibition and impulsivity: Behavioral and neural basis of response control. *Progress in Neurobiology*, *108*, 44–79.

Baumann, A. A., & Odum, A. L. (2012). Impulsivity, risk taking, and timing. *Behavioural Processes*, *90*(3), 408–414.

Bechara, A. (2005). Decision making, impulse control and loss of willpower to resist drugs: A neurocognitive perspective. *Nature Neuroscience*, *8*(11), 1458–1463.

Bechara, A., Damasio, A. R., Damasio, H., & Anderson, S. W. (1994). Insensitivity to future consequences following damage to human prefrontal cortex. *Cognition*, *50*(1), 7–15.

Bechara, A., Dolan, S., Denburg, N., Hindes, A., Anderson, S. W., & Nathan, P. E. (2001). Decision-making deficits, linked to a dysfunctional ventromedial prefrontal cortex, revealed in alcohol and stimulant abusers. *Neuropsychologia*, *39*(4), 376–389.

Bentzley, B. S., Barth, K. S., Back, S. E., & Book, S. W. (2015). Discontinuation of buprenorphine maintenance therapy: Perspectives and outcomes. *Journal of Substance Abuse Treatment*, *52*, 48–57.

Bernhardt, N., Nebe, S., Pooseh, S., Sebold, M., Sommer, C., Birkenstock, J., et al. (2017). Impulsive decision making in young adult social drinkers and detoxified alcohol-dependent patients: A cross-sectional and longitudinal study. *Alcoholism: Clinical and Experimental Research*, *41*(10), 1794–1807.

Berridge, K., & Aldridge, J. W. (2008). Decision utility, the brain, and pursuit of hedonic goals. *Social Cognition*, *26*(5), 621–646.

Białaszek, W., Marcowski, P., & Cox, D. (2017). Differences in delay, but not probability discounting, in current smokers, e-cigarette users, and never smokers. *The Psychological Record*, *67*(2), 223–230.

Bickel, W. K., Yi, R., Landes, R. D., Hill, P. F., & Baxter, C. (2011). Remember the future: Working memory training decreases delay discounting among stimulant addicts. *Biological Psychiatry*, *69*(3), 260–265.

Bijleveld, E., Custers, R., & Aarts, H. (2012). Adaptive reward pursuit: How effort requirements affect unconscious reward responses and conscious reward decisions. *Journal of Experimental Psychology: General*, *141*(4), 728–742.

Black, A. C., & Rosen, M. I. (2011). A money management-based substance use treatment increases valuation of future rewards. *Addictive Behaviors*, *36*(1), 125–128.

Borges, A. M., Kuang, J., Milhorn, H., & Yi, R. (2016). An alternative approach to calculating area-under-the-curve (AUC) in delay discounting research. *Journal of the Experimental Analysis of Behavior*, *106*(2), 145–155.

Boswell, J. F., Thompson-Hollands, J., Farchione, T. J., & Barlow, D. H. (2013). Intolerance of uncertainty: A common factor in the treatment of emotional disorders. *Journal of Clinical Psychology*, *69*(6), 630–645.

Bulley, A., & Gullo, M. J. (2017). The influence of episodic foresight on delay discounting and demand for alcohol. *Addictive Behaviors*, *66*, 1–6.

Cloninger, C. R., Svrakic, D. M., & Przybeck, T. R. (1993). A psychobiological model of temperament and character. *Archives of General Psychiatry*, *50*(12), 975–990.

Crean, J. P., de Wit, H., & Richards, J. (2000). Reward discounting as a measure of impulsive behavior in a psychiatric outpatient population. *Experimental and Clinical Psychopharmacology*, *8*(2), 155–162.

de Wit, H., Flory, J. D., Acheson, A., McCloskey, M., & Manuck, S. B. (2007). IQ and nonplanning impulsivity are independently associated with delay discounting in middle-aged adults. *Personality and Individual Differences*, *42*(1), 111–121.

Di Chiara, G., & Imperato, A. (1988). Drugs abused by humans preferentially increase synaptic dopamine concentrations in the mesolimbic system of freely moving rats. *Proceedings of the National Academy of Sciences of the United States of America*, *85*(14), 5274–5278.

Everitt, B. J., & Robbins, T. W. (2005). Neural systems of reinforcement for drug addiction: From actions to habits to compulsion. *Nature Neuroscience*, *8*(11), 1481–1489.

Fernie, G., Peeters, M., Gullo, M. J., Christiansen, P., Cole, J. C., Sumnall, H., et al. (2013). Multiple behavioural impulsivity tasks predict prospective alcohol involvement in adolescents. *Addiction*, *108*(11), 1916–1923.

Fishbain, D. A., Cole, B., Lewis, J., Rosomoff, H. L., & Rosomoff, R. S. (2008). What percentage of chronic nonmalignant pain patients exposed to chronic opioid analgesic therapy develop abuse/addiction and/or aberrant drug-related behaviors? A structured evidence-based review. *Pain Medicine*, *9*(4), 444–459.

Garami, J., & Moustafa, A. A. (2019). Probability discounting of monetary gains and losses in opioid-dependent adults. *Behavioural Brain Research*, *364*, 334–339.

Goldstein, R., & Volkow, N. (2002). Drug addiction and its underlying neurobiological basis: Neuroimaging evidence for the involvement of the frontal cortex. *American Journal of Psychiatry*, *159*(10), 1642–1652.

Gowin, J. L., Mackey, S., & Paulus, M. P. (2013). Altered risk-related processing in substance users: Imbalance of pain and gain. *Drug and Alcohol Dependence*, *132*(1–2), 13–21.

Green, L., & Myerson, J. (2004). A discounting framework for choice with delayed and probabilistic rewards. *Psychological Bulletin*, *130*(5), 769–792.

Green, L., & Myerson, J. (2010). Experimental and correlational analyses of delay and probability discounting. In G. J. Madden & W. K. Bickel (Eds.), *Impulsivity: The behavioral and neurological science of discounting*: American Psychological Association.

Gullo, M. J., Loxton, N. J., & Dawe, S. (2014). Impulsivity: Four ways five factors are not basic to addiction. *Addictive Behaviors*, *39*(11), 1547–1556.

Hartmann, M. N., Hager, O. M., Tobler, P. N., & Kaiser, S. (2013). Parabolic discounting of monetary rewards by physical effort. *Behavioural Processes*, *100*(C), 192–196.

Herting, M. M., Schwartz, D., Mitchell, S. H., & Nagel, B. J. (2010). Delay discounting behavior and white matter microstructure abnormalities in youth with a family history of alcoholism. *Alcoholism: Clinical and Experimental Research*, *34*(9), 1590–1602.

Hofmeyr, A., Monterosso, J., Dean, A. C., Morales, A. M., Bilder, R. M., Sabb, F. W., et al. (2017). Mixture models of delay discounting and smoking behavior. *The American Journal of Drug and Alcohol Abuse*, *43*(3), 271–280.

Hsu, M., Bhatt, M., Adolphs, R., Tranel, D., & Camerer, C. F. (2005). Neural systems responding to degrees of uncertainty in human decision-making. *Science*, *310*(5754), 1680–1683.

Hui, C., & Zhihui, Y. (2017). Group cognitive behavioral therapy targeting intolerance of uncertainty: A randomized trial for older Chinese adults with generalized anxiety disorder. *Aging and Mental Health*, *21*(12), 1294–1302.

Hyman, S., Malenka, R., & Nestler, E. (2006). Neural mechanisms of addiction: The role of reward-related learning and memory. *Annual Review of Neuroscience*, *29*, 565–598.

Ida, T., & Goto, R. (2009). Interdependency among addictive behaviours and time/risk preferences: Discrete choice model analysis of smoking, drinking, and gambling. *Journal of Economic Psychology*, *30*(4), 608–621.

Jentsch, J. D., & Taylor, J. R. (1999). Impulsivity resulting from frontostriatal dysfunction in drug abuse: Implications for the control of behavior by reward- related stimuli. *Psychopharmacology*, *146*(4), 373–390.

Johnson, M. W., Johnson, P., Herrmann, E., & Sweeney, M. (2015). Delay and probability discounting of sexual and monetary outcomes in individuals with cocaine use disorders and matched controls. *PLoS ONE*, *10*(5).

Karakula, S. L., Weiss, R. D., Griffin, M. L., Borges, A. M., Bailey, A. J., & McHugh, R. K. (2016). Delay discounting in opioid use disorder: Differences between heroin and prescription opioid users. *Drug and Alcohol Dependence*, *169*, 68–72.

Khurana, A., Romer, D., Betancourt, L. M., & Hurt, H. (2017). Working memory ability and early drug use progression as predictors of adolescent substance use disorders. *Addiction*, *112*(7), 1220–1228.

Kollins, S. (2003). Delay discounting is associated with substance use in college students. *Addictive Behaviors*, *28*(6), 1167–1173.

Koob, G., & Bloom, F. (1988). Cellular and molecular mechanisms of drug dependence. *Science*, *242*(4879), 715–723.

Koob, G., & Kreek, M. (2007). Stress, dysregulation of drug reward pathways, and the transition to drug dependence. *American Journal of Psychiatry*, *164*(8), 1149–1159.

Krishnan-Sarin, S., Reynolds, B., Duhig, A. M., Smith, A., Liss, T., McFetridge, A., et al. (2007). Behavioral impulsivity predicts treatment outcome in a smoking cessation program for adolescent smokers. *Drug and Alcohol Dependence*, *88*(1), 79–82.

Lim, A. C., Cservenka, A., & Ray, L. A. (2017). Effects of alcohol dependence severity on neural correlates of delay discounting. *Alcohol and Alcoholism*, *52*(4), 506–515.

Loxton, N. J., & Dawe, S. (2001). Alcohol abuse and dysfunctional eating in adolescent girls: The influence of individual differences in sensitivity to reward and punishment. *International Journal of Eating Disorders*, *29*(4), 455–462.

Lyvers, M., Hinton, R., Gotsis, S., Roddy, M., Edwards, M. S., & Thorberg, F. A. (2014). Traits linked to executive and reward systems functioning in clients undergoing residential treatment for substance dependence. *Personality and Individual Differences*, *70*, 194–199.

Mackillop, J. (2013). Integrating behavioral economics and behavioral genetics: Delayed reward discounting as an endophenotype for addictive disorders. *Journal of the Experimental Analysis of Behavior*, *99*(1), 14–31.

Mackillop, J., Amlung, M., Few, L., Ray, L., Sweet, L., & Munafò, M. (2011). Delayed reward discounting and addictive behavior: A meta-analysis. *Psychopharmacology*, *216*(3), 305–321.

Mackillop, J., & Kahler, C. W. (2009). Delayed reward discounting predicts treatment response for heavy drinkers receiving smoking cessation treatment. *Drug and Alcohol Dependence*, *104*(3), 197–203.

Madden, G., Begotka, A., Raiff, B., & Kastern, L. (2003). Delay discounting of real and hypothetical rewards. *Experimental and Clinical Psychopharmacology*, *11*(2), 139–145.

Madden, G. J., & Johnson, P. S. (2010). A delay-discounting primer. In G. J. Madden & W. K. Bickel (Eds.), *Impulsivity: The behavioral and neurological science of discounting*: American Psychological Association.

Madden, G. J., Petry, N. M., Badger, G. J., & Bickel, W. K. (1997). Impulsive and self-control choices in opioid-dependent patients and non-drug-using control participants: Drug and monetary rewards. *Experimental and Clinical Psychopharmacology*, *5*(3), 256–262.

Mahoney, A. E. J., & McEvoy, P. M. (2012). Changes in intolerance of uncertainty during cognitive behavior group therapy for social phobia. *Journal of Behavior Therapy and Experimental Psychiatry*, *43*(2), 849–854.

Malesza, M., & Ostaszewski, P. (2013). Relations between Cloninger's dimensions of temperament and steepness of delay and effort discounting of monetary rewards. *Psychological Reports*, *112*(3), 694–705.

Malesza, M., & Ostaszewski, P. (2016). Dark side of impulsivity — Associations between the dark triad, self-report and behavioral measures of impulsivity. *Personality and Individual Differences*, *88*, 197–201.

Manes, F., Sahakian, B., Clark, L., Rogers, R., Antoun, N., Aitken, M., et al. (2002). Decision making processes following damage to the prefrontal cortex. *Brain, 125*(3), 624–639.

Mantsch, J., Yuferov, V., Mathieu-Kia, A.-M., Ho, A., & Kreek, M. (2004). Effects of extended access to high versus low cocaine doses on self-administration, cocaine-induced reinstatement and brain mRNA levels in rats. *Psychopharmacology, 175*(1), 26–36.

Martinez, D., Saccone, P. A., Liu, F., Slifstein, M., Orlowska, D., Grassetti, A., et al. (2011). Deficits in dopamine D 2 receptors and presynaptic dopamine in heroin dependence: Commonalities and differences with other types of addiction. *Biological Psychiatry, 71*(3), 192–198.

McKerchar, T., & Renda, C. (2012). Delay and probability discounting in humans: An overview. *The Psychological Record, 62*(4), 817–834.

McKetin, R., Parasu, P., Cherbuin, N., Eramudugolla, R., & Anstey, K. (2016). Interrelationships between marijuana demand and discounting of delayed rewards: Convergence in behavioral economic methods. *Drug and Alcohol Dependence, 169*, 141–147.

Mejía-Cruz, D., Green, L., Myerson, J., Morales-Chainé, S., & Nieto, J. (2016). Delay and probability discounting by drug- dependent cocaine and marijuana users. *Psychopharmacology, 233*(14), 2705–2714.

Melis, M., Spiga, S., & Diana, M. (2005). The dopamine hypothesis of drug addiction: Hypodopaminergic state. *International Review of Neurobiology, 63*, 101–154.

Miller, M. A., Thome, A., & Cowen, S. (2013). Intersection of effort and risk: Ethological and neurobiological perspectives. *Frontiers in Neuroscience, 7*.

Mitchell, S. H. (1999). Measures of impulsivity in cigarette smokers and non-smokers. *Psychopharmacology, 146*(4), 455–464.

Mitchell, S. H. (2004). Effects of short-term nicotine deprivation on decision-making: Delay, uncertainty and effort discounting. *Nicotine & Tobacco Research, 6*(5), 819–828.

Mobini, S., Chiang, T. J., Ho, M. Y., Bradshaw, C. M., & Szabadi, E. (2000). Effects of central 5-hydroxytryptamine depletion on sensitivity to delayed and probabilistic reinforcement. *Psychopharmacology, 152*(4), 390–397.

Mobini, S., Kass, A. E., Yeomans, M. R., & Grant, A. (2007). Relationships between functional and dysfunctional impulsivity, delay discounting and cognitive distortions. *Personality and Individual Differences, 43*(6), 1517–1528.

Moody, L., Franck, C., & Bickel, W. K. (2016). Comorbid depression, antisocial personality, and substance dependence: Relationship with delay discounting. *Drug and Alcohol Dependence, 160*, 190–196.

Munafò, M. R., Zetteler, J. I., & Clark, T. G. (2007). Personality and smoking status: A meta-analysis. *Nicotine and Tobacco Research, 9*(3), 405–413.

Myerson, J., Green, L., Berk-Clark, C., & Grucza, R. (2015). Male, but not female, alcohol-dependent African Americans discount delayed gains more steeply than propensity-score matched controls. *Psychopharmacology, 232*(24), 4493–4503.

Myerson, J., Green, L., & Warusawitharana, M. (2001). Area under the curve as a measure of discounting. *Journal of the Experimental Analysis of Behavior, 76*(2), 235–243.

Nestler, E., & Aghajanian, G. (1997). Molecular and cellular basis of addiction. *Science, 278*(5335), 58–63.

Nishiyama, R. (2014). Response effort discounts the subjective value of rewards. *Behavioural Processes, 107*, 175–177.

Oglesby, M. E., Allan, N. P., & Schmidt, N. B. (2017). Randomized control trial investigating the efficacy of a computer-based intolerance of uncertainty intervention. *Behaviour Research and Therapy, 95*, 50–57.

Ohmura, Y., Takahashi, T., & Kitamura, N. (2005). Discounting delayed and probabilistic monetary gains and losses by smokers of cigarettes. *Psychopharmacology*, *182*(4), 508–515.

Ostaszewski, P., Bąbel, P., & Swebodziński, B. (2013). Physical and cognitive effort discounting of hypothetical monetary rewards. *Japanese Psychological Research*, *55*(4), 329–337.

Patak, M., & Reynolds, B. (2007). Question-based assessments of delay discounting: Do respondents spontaneously incorporate uncertainty into their valuations for delayed rewards? *Addictive Behaviors*, *32*(2), 351–357.

Patton, J. H., Stanford, M. S., & Barratt, E. S. (1995). Factor structure of the Barratt impulsiveness scale. *Journal of Clinical Psychology*, *51*(6), 768–774.

Perkins, F. N., & Freeman, K. B. (2018). Pharmacotherapies for decreasing maladaptive choice in drug addiction: Targeting the behavior and the drug. *Pharmacology, Biochemistry and Behavior*, *164*, 40–49.

Petry, N. M., Kirby, K. N., & Kranzler, H. R. (2002). Effects of gender and family history of alcohol dependence on a behavioral task of impulsivity in healthy subjects. *Journal of Studies on Alcohol*, *63*(1), 83–90.

Piazza, P., Deminiere, J.-M., Le Moal, M., & Simon, H. (1989). Factors that predict individual vulnerability to amphetamine self-administration. *Science*, *245*(4925), 1511.

Piazza, P., Deroche-Gamonent, V., Rouge-Pont, F., & Le Moal, M. (2000). Vertical shifts in self-administration dose-response functions predict a drug-vulnerable phenotype predisposed to addiction. *Journal of Neuroscience*, *20*(11), 4226–4232.

Portenoy, R., Farrar, J., Backonja, M. M., Cleeland, C. S., Yang, K., Friedman, M., et al. (2007). Long-term use of controlled-release oxycodone for noncancer pain: Results of a 3-year registry study. *Clinical Journal of Pain*, *23*(4), 287–299.

Prevost, C., Pessiglione, M., Metereau, E., Clery-Melin, M.-L., & Dreher, J.-C. (2010). Seperate valuation subsystems for delay and effort decision costs. *Journal of Neuroscience*, *30*(42), 14080–14090.

Rachlin, H., Raineri, A., & Cross, D. (1991). Subjective probability and delay. *Journal of the Experimental Analysis of Behavior*, *55*(2), 233–244.

Rahman, S., Sahakian, B. J., Hodges, J. R., Rogers, R. D., & Robbins, T. W. (1999). Specific cognitive deficits in mild frontal variant frontotemporal dementia. *Brain*, *122*(8), 1469.

Rasmussen, E. B., Lawyer, S. R., & Reilly, W. (2010). Percent body fat is related to delay and probability discounting for food in humans. *Behavioural Processes*, *83*(1), 23–30.

Reimers, S., Maylor, E. A., Stewart, N., & Chater, N. (2009). Associations between a one-shot delay discounting measure and age, income, education and real-world impulsive behavior. *Personality and Individual Differences*, *47*(8), 973–978.

Reynolds, B., Richards, J., Horn, K., & Karraker, K. (2004). Delay discounting and probability discounting as related to cigarette smoking status in adults. *Behavioural Processes*, *65*(1), 35–42.

Richards, J. B., Zhang, L., Mitchell, S. H., & Wit, H. (1999). Delay or probability discounting in a model of impulsive behavior: Effect of alcohol. *Journal of the Experimental Analysis of Behavior*, *71*(2), 121–143.

Robinson, T. E., & Berridge, K. C. (2001). Incentive-sensitization and addiction. *Addiction*, *96*(1), 103–114.

Robles, E., Huang, B., Simpson, P., & McMillan, D. (2011). Delay discounting, impulsiveness, and addiction severity in opioid-dependent patients. *Journal of Substance Abuse Treatment*, *41*(4), 354–362.

Rogers, R. D., Everitt, B. J., Baldacchino, A., Blackshaw, A. J., Swainson, R., Wynne, K., et al. (1999). Dissociable deficits in the decision-making cognition of chronic amphetamine abusers, opiate abusers, patients with focal damage to prefrontal cortex, and tryptophan-depleted normal volunteers: Evidence for monoaminergic mechanisms. *Neuropsychopharmacology*, *20*(4), 322–339.

Scherbaum, S., Haber, P., Morley, K., Underhill, D., & Moustafa, A. A. (2018). Biased and less sensitive: A gamified approach to delay discounting in heroin addiction. *Journal of Clinical and Experimental Neuropsychology*, *40*(2), 139–150.

Schultz, W. (2011). Potential vulnerabilities of neuronal reward, risk, and decision mechanisms to addictive drugs. *Neuron*, *69*(4), 603–617.

Sheffer, C. E., Christensen, D. R., Landes, R., Carter, L. P., Jackson, L., & Bickel, W. K. (2014). Delay discounting rates: A strong prognostic indicator of smoking relapse. *Addictive Behaviors*, *39*(11), 1682–1689.

Sheffer, C., Mackillop, J., McGeary, J., Landes, R., Carter, L., Yi, R., et al. (2012). Delay discounting, locus of control, and cognitive impulsiveness independently predict tobacco dependence treatment outcomes in a highly dependent, lower socioeconomic group of smokers. *American Journal on Addictions*, *21*(3), 221–232.

Sinha, R. (2008). Chronic stress, drug use, and vulnerability to addiction. *Annals of the New York Academy of Sciences*, *1141*(1), 105–130.

Smith, C., & Hantula, D. (2008). Methodological considerations in the study of delay discounting in intertemporal choice: A comparison of tasks and modes. *Behavior Research Methods*, *40*(4), 940–953.

Smith, C. T., Steel, E. A., Parrish, M. H., Kelm, M. K., & Boettiger, C. A. (2015). Intertemporal choice behavior in emerging adults and adults: Effects of age interact with alcohol use and family history status. *Frontiers in Human Neuroscience*, *9*.

Smyth, B., Barry, J., Keenan, E., & Ducray, K. (2010). Lapse and relapse following inpatient treatment for opiate dependence. *Irish Medical Journal*, *103*(6), 176–179.

Snider, S. E., Laconte, S. M., & Bickel, W. K. (2016). Episodic future thinking: Expansion of the temporal window in individuals with alcohol dependence. *Alcoholism: Clinical and Experimental Research*, *40*(7), 1558–1566.

Stanford, M. S., Mathias, C. W., Dougherty, D. M., Lake, S., Anderson, N., & Patton, J. (2009). Fifty years of the Barratt impulsiveness scale: An update and review. *Personality and Individual Differences*, *47*, 385–395.

Stanger, C., Elton, A., Ryan, S. R., James, G. A., Budney, A. J., & Kilts, C. D. (2013). Neuroeconomics and adolescent substance abuse: Individual differences in neural networks and delay discounting. *Journal of the American Academy of Child and Adolescent Psychiatry*, *52*(7), 747–755.

Stanger, C., Ryan, S., Fu, H., Landes, R. D., Jones, B. A., Bickel, W., et al. (2012). Delay discounting predicts adolescent substance abuse treatment outcome. *Experimental and Clinical Psychopharmacology*, *20*(3), 205–212.

Stein, J., Wilson, A., Koffarnus, M., Daniel, T., Epstein, L., & Bickel, W. (2016). Unstuck in time: Episodic future thinking reduces delay discounting and cigarette smoking. *Psychopharmacology*, *233*(21), 3771–3778.

Stevens, L., Verdejo-García, A., Roeyers, H., Goudriaan, A. E., & Vanderplasschen, W. (2015). Delay discounting, treatment motivation and treatment retention among substance-dependent individuals attending an in inpatient detoxification program. *Journal of Substance Abuse Treatment*, *49*(C), 58.

Sugiwaka, H., & Okouchi, H. (2004). Reformative self-control and discounting of reward value by delay or effort. *Japanese Psychological Research*, *46*(1), 1–9.

Takahashi, T., Ohmura, Y., Oono, H., & Radford, M. (2009). Alcohol use and discounting of delayed and probabilistic gain and loss. *Neuroendocrinology Letters*, *30*(6), 749–752.

Talkovsky, A. M., & Norton, P. J. (2016). Intolerance of uncertainty and transdiagnostic group cognitive behavioral therapy for anxiety. *Journal of Anxiety Disorders*, *41*, 108–114.

VanderBroek, L., Acker, J., Palmer, A., de Wit, H., & MacKillop, J. (2016). Interrelationships among parental family history of substance misuse, delay discounting, and personal substance use. *Psychopharmacology*, *233*(1), 39–48.

Volkow, N. D., Fowler, J. S., & Wang, G. J. (2003). The addicted human brain: Insights from imaging studies. *The Journal of Clinical Investigation*, *111*(10), 1444–1451.

Washio, Y., Higgins, S. T., Heil, S. H., McKerchar, T. L., Badger, G. J., Skelly, J. M., et al. (2011). Delay discounting is associated with treatment response among cocaine-dependent outpatients. *Experimental and Clinical Psychopharmacology*, *19*(3), 243–248.

Yi, R., Carter, A. E., & Landes, R. D. (2012). Restricted psychological horizon in active methamphetamine users: Future, past, probability, and social discounting. *Behavioral Pharmacology*, *23*(4), 358–366.

Yi, R., Chase, W. D., & Bickel, W. K. (2007). Probability discounting among cigarette smokers and nonsmokers: Molecular analysis discerns group differences. *Behavioral Pharmacology*, *18*(7), 633–639.

Yoon, J., Garza, R., Newton, T., Suchting, R., Weaver, M., Brown, G., et al. (2017). A comparison of Mazur's k and area under the curve for describing steep discounters. *The Psychological Record*, *67*(3), 355–363.

Yoon, J., Higgins, S., Heil, S., Sugarbaker, R., Thomas, C. S., & Badger, G. (2007). Delay discounting predicts postpartum relapse to cigarette smoking among pregnant women. *Experimental and Clinical Psychopharmacology*, *15*(2), 176–186.

Zhu, Y., Jiang, H., Zhong, N., Li, R., Li, X., Chen, T., et al. (2018). A newly designed mobile-based computerized cognitive addiction therapy app for the improvement of cognition impairments and risk decision making in methamphetamine use disorder: Randomized controlled trial. *JMIR mHealth and uHealth*, *6*(6).

Chapter 5

It's all about context: The environment and substance use

Justin Mahlberg[a], Ahmed A. Moustafa[b]
[a]*School of Social Sciences and Psychology, Western Sydney University, Sydney, NSW, Australia,* [b]*Marcs Institute for Brain, Behaviour, and Development and School of Psychology, Western Sydney University, Penrith, NSW, Australia*

Introduction

Drug and alcohol use is an important feature of human behavior; an activity that is associated with significant costs to the individual and to society (Collins & Lapsley, 2008; Teesson, Hall, Lynskey, & Degenhardt, 2000). We are not born with the desire to drink alcohol or take drugs, nor are we born with the behavioral repertoire necessary to successfully consume drugs. This means that drug use behaviors, like most human behaviors, are learned. The learning of drug use behaviors involves a complex intersection of a number of important associative processes. Firstly, substance use involves instrumental action; that is, learning to perform a voluntary action (the response [R]) that is executed because its consequence is positive (the outcome [O]). For example, the action of drinking a beer is a voluntary behavior that a person learns to execute more frequency because it results in the positive psychological effects of alcohol—this is clearly voluntary because people who do not like the effects of alcohol will not regularly partake in drinking it, but people who do enjoy the effects will drink regularly.

Environmental stimuli can trigger drug seeking

These drug use actions can also be modified by the context in which they are experienced. Over time, highly frequent actions are thought to become habitual, whereby the drug use behavior (R) can be elicited directly by the environment (e.g. by a stimulus [S]), regardless of how positive the consequences are when this behavior is enacted. Relatedly, stimuli in the environment can become associated with the effects of drugs, and through this process come to indirectly elicit greater instances of drug use behaviors compared to when the stimuli are not present. Therefore, in summary, drug use behaviors can be modified by stimulus (S) in the endogenous and exogenous environment, and the outcomes

Cognitive, Clinical, and Neural Aspects of Drug Addiction. https://doi.org/10.1016/B978-0-12-816979-7.00005-4

(O)—either positive or negative—that result from a behavioral response (R). The contribution of each of these are determined by the strength of each association but also by the particulars of the environment, which can determine how much each type of learning influences the observed behavior (e.g. see Holland, 2004; Holmes, Marchand, & Coutureau, 2010). Importantly, drug use and drug-seeking behavior can be influenced by discrete environmental stimuli, as well as being influenced by the diffuse environment as a whole (referred to as "context"). The first section of this chapter will focus on the various ways discrete stimuli in environment can influence drug use, and the second section focuses on context-drug use interactions.

Discrete stimuli can directly trigger drug seeking

It is hypothesized that stimuli can directly promote drug use behaviors through the development of a habit. This is where stimuli within the environment trigger a drug use response directly. This is thought to be an "automatic" response, such that the S can trigger a drug use R whether or not the outcome (the drug) is desired (as opposed to goal-directed control, where the value of the outcome is considered). The automaticity of drug use has been modeled in animals, where after extended experience initiating a response to retrieve an outcome, the behavior often becomes insensitive to changes in the value of the responses consequences (Dickinson, 1985). In drug use, this was illustrated by training rats to enact distinct responses for alcohol and food pellets (Dickinson, Wood, & Smith, 2002). After training, one group had the alcohol outcome devalued and another group had food devalued. Devaluation was achieved by pairing the target outcome with an injection of an aversive substance (lithium chloride). In an extinction test, it was found that, while food-seeking was reduced by devaluation and was lower than alcohol seeking, this effect was not present in the alcohol devaluation group: the rate of alcohol seeking was still statistically similarly to food seeking. This has also been demonstrated in animal models with other drugs, such as nicotine (Loughlin, Funk, Coen, & Lê, 2017) and cocaine (Miles, Everitt, & Dickinson, 2003).

Relevance of habit formation to addiction

A common theoretical framework for understanding why people become addicted to drugs or alcohol is the argument that addictive behavior is representative of a transition from a goal-directed to a general habitual response pattern. In humans, there is some weak evidence to support the notion people with drug dependence are more likely to use habit driven behavior over goal-directed decision-making. Sjoerds et al. (2013) showed that, in a food-seeking instrumental learning task, a group of people with alcohol dependence were less sensitive to outcome devaluation compared to a non-dependent comparison group. A weak group effect was found in Sebold et al. (2014), showing that people with alcohol dependence were less likely to use goal-directed behavioral strategies compared to healthy controls. Sebold et al. (2017), however, did not replicate this effect, as they did not observe behavioral differences between people with alcohol dependence and a non-dependent comparison group. Moreover, there was no association between drinking severity and the extent of goal-directed behavior observed in young social drinkers (Nebe et al., 2018).

There is some further evidence in humans that extended training of a reward-seeking behavior, including drug-seeking, might result in a transition to habitual responding. Tricomi, Balleine, and O'Doherty (2009) showed that, after learning to respond to two food rewards, devaluation of one of the rewards results in a reduction in responses relative to the still-valued reward. However, after 3 days of training, a second group did not show this devaluation effect. This supports the notion that reward-seeking can transition from goal-directed (i.e. driven by the value of the outcome) to a habitual behavioral pattern (i.e. behavior insensitive to reward value). Importantly, Ersche et al. (2016) found that people with a history of cocaine use, after extensive training, showed lower sensitivity to outcome devaluation compared to a comparison group with no history of substance misuse. These above-mentioned observations lend evidence to the notion that drug use—and drug addiction—is the development of a habitual behavioral pattern that is triggered directly from the environment, and thus drug use will persist despite a change in the value of a drug. However, the evidence discussed above is not consistent or conclusive. Indeed, an important study by de Wit et al. (2018) showed five failed attempts to induce habitual behavior in humans using overtraining in instrumental tasks. Thus, support for a habit theory of addiction using these paradigms are far from conclusive, and further work will be required to establish reliable observations of habit formation. However, the study of habit formation in drug use has also been studied by considering how environmental stimuli can also indirectly modify the expression of drug-seeking behavior in animals and humans. The next section considers how discrete stimuli can form associations with drug use, and subsequently increase the expression of drug-seeking.

Discrete stimuli can indirectly trigger drug seeking

Stimuli can also be indirectly involved in drug use through the development of associations with drug intake. In human drug use, this associative process is easily observable studying drug craving (i.e. the desire for drugs). There is an abundance of evidence that stimuli associated with drug use can promote an increase in desire for drugs (Carter & Tiffany, 1999). These stimuli are usually "proximal" to the drug use process, in that they are close to, and often predict the onset of, the drug use experience (Carter & Tiffany, 2001; Veilleux & Skinner, 2015). Therefore, it appears that stimuli in the environment can be directly involved in the fluctuation of desire for drugs.

This raises the important question of whether these stimuli that promote drug desire can subsequently contribute to drug use behaviors. This question can be probed using the increasingly popular Pavlovian-to-instrumental-transfer paradigm, which aims to evaluate the indirect effects of reward stimuli on drug-seeking behaviors, while limiting the possibility of observing direct S-R (habit) behavior. This is achieved by training stimulus-drug relationships (stimulus-outcome) and instrumental drug seeking responses (response-outcome) in separate stages, so that when the trained stimuli a presented during an extinction test where the instrumental behavior is assessed, any effect of stimuli must be the result of the interaction between these two discrete forms of learning and not a direct relationship between S and R (i.e. no direct "habit" learning), since they had never been experienced together until test.

The simplest way to observe a transfer effect between stimulus-outcome learning and drug-seeking is through a single-response PIT paradigm. This PIT effect was observed in animal models using ethanol in Krank (2003) and Krank, O'Neill, Squarey, and Jacob (2008). Rats were trained to self-administer ethanol using a lever press response (R-O). Subsequently, the rats were trained to learn a relationship between an initially neutral stimulus and ethanol administration (S-O). Upon test, exposure to the stimulus associated with ethanol administration resulted in an increase in lever presses for ethanol, relative to pre-stimulus exposure and compared to a group of rats that did not receive Pavlovian pairing between the S and ethanol administration. Overall, these experiments demonstrated that (1) stimuli never directly involved in learned drug use behavior can still modify drug-seeking; and (2) stimuli predicting the probability of drug intake (via a classically conditioned relationship) can promote drug self-administration. In humans, the single-response PIT effect has been observed using food, and similar psychological processes appear to mediate this effect (Colagiuri & Lovibond, 2015; Lovibond & Colagiuri, 2013; Lovibond, Satkunarajah, & Colagiuri, 2015). However, as far as we are aware the single-response paradigm has not been observed in a drug-use paradigm in humans.

A popular approach to studying the effects of drug stimuli on drug-seeking is using a choice PIT paradigm, where drug-seeking behavior can be observed relative to other appetitive behaviors (Cartoni, Balleine, & Baldassarre, 2016; Lamb, Schindler, & Pinkston, 2016). This choice paradigm has allowed observations of two distinct forms of Pavlovian transfer: specific and general PIT. The specific transfer effects of drug-stimuli on drug use behaviors have been thoroughly investigated with humans using the PIT paradigm. For example, Experiment 2 in Hogarth and Chase (2012) utilized a similar PIT design where participants were trained to press D and H to win chocolate and tobacco. The transfer test involved the introduction of images of chocolate or cigarettes, which are assumed to already have associative (S-O) histories with their respective outcomes. These are presented while participants have the opportunity to press for tobacco or chocolate during a test phase (under nominal extinction, so no new learning occurred). The results showed the percentage of key presses for tobacco was higher when people were presented with tobacco-related stimuli. Moreover, there have been a number of other investigations demonstrating a transfer effect for cigarettes across differing methodologies (e.g. Hogarth et al., 2014; Hogarth, Dickinson, Wright, Kouvaraki, & Duka, 2007; Hogarth, Field, & Rose, 2013; Rose, Brown, MacKillop, Field, & Hogarth, 2018), and has also been observed in human alcohol-seeking behavior (Hardy, Mitchell, Seabrooke, & Hogarth, 2017; Mahlberg, Weidemann, Hogarth, & Moustafa, 2019; Martinovic et al., 2014; Rose et al., 2018) indicating the reliability of this effect.

A general transfer effect was observed in Corbit and Janak (2007) using ethanol in an animal model. Ethanol and sucrose-seeking behaviors (R1-O1 and R2-O2, respectively) were trained separately from stimuli that were associated with the delivery of each of these rewards (S1-O1 and S2-O2, respectively). In a subsequent extinction test, both specific and general forms of transfer were observed for the ethanol seeking behavior: when the ethanol-paired stimulus was presented, the frequency of ethanol seeking increased compared to baseline and compared to when the sucrose stimulus was presented. However, the ethanol stimulus also promoted sucrose seeking above baseline levels in a similar manner to the sucrose stimulus. These observations provide evidence of a general transfer effect, as it indicated that alcohol-paired stimuli can also promote non-specific excitation of all other available appetitive behaviors. Overall, this shows that the context surrounding us can have both targeted and diffuse effects on the extent to which we engage in the use of drugs.

Psychological processes in stimulus elicited drug seeking

There were two main theories that aimed to account for these transfer effects. A motivational hypothesis explains the transfer effect by arguing that the conditional stimulus elicits a conditioned motivation to consume tobacco, which in turn increases instrumental action for drugs (Hogarth, Balleine, Corbit, & Killcross, 2013; Rescorla & Solomon, 1967). An alternative account, the "expectancy" hypothesis, states that a conditional stimulus elicits a conditioned expectation that drug is available (Trapold & Overmier, 1972). This, in turn increases the probability of instrumental action for the drug.

Importantly Hogarth and Chase (2011a) showed, using a similar paradigm, that the specific transfer effect was insensitive to changes in drug value. After training, participants were given a session prior to the transfer test to smoke as much as they desired. This was implemented to significantly reduce the desire for nicotine and thus reduce the value of cigarettes as an outcome. It was found that, despite a significantly reduced motivation for cigarettes, upon transfer test participants still demonstrated a magnified response for nicotine when confronted with cigarette-paired stimuli. These results were replicated with a nicotine nasal spray used for devaluation (Hogarth, 2012), and also persisted in experiment 2 of Hogarth and Chase (2011a) when the devaluation procedure reduced motivation for cigarettes by requiring participants to view and rate the impact of health warnings that highlighted the negative outcomes of cigarette smoking. Moreover, Rose et al. (2018) also replicated this result in alcohol seeking, showing that alcohol-paired stimuli could increase choice for alcohol above baseline, even after alcohol was devaluated. Overall, this evidence demonstrates that (a) drug stimulus can increase drug seeking in humans, (b) that this transfer effect can be stimulus specific (i.e. a specific transfer effect), (c) and that the transfer effect is resistant to changes in incentive value of the drug. This body of work is supportive of the expectancy account of Pavlovian transfer (Trapold & Overmier, 1972), in that the drug stimuli appears to promote drug-seeking regardless of the current value of the drug, by triggering a learned expectancy for the onset of the drug outcome (an S-O expectancy).

One problem with the theory that expectancy for the outcome (i.e. a S-O expectancy) is the process underlying specific transfer is the observation that specific PIT is resistant to both Pavlovian and instrumental extinction (Hogarth et al., 2014; Hogarth & Chase, 2012). That is, extinction of the relationship between S and O, or the relationship between R and O does not appear to remove the transfer effect on drug-seeking. If a simple S-O signal was responsible for exciting a subsequent R-O (instrumental) drug behavior, then reducing the relationship between the S and the O or the R and the O should weaken the effect drug stimuli has on drug seeking. However, in human drug choice there have been two effective methods of rendering stimuli ineffectual on drug seeking: discriminative extinction, where the drug stimulus is presented while drug-seeking is continued, but no drug outcomes are delivered (Hogarth et al., 2014; Rosas, Paredes-Olay, García-Gutiérrez, Espinosa, & Abad, 2010); or instructive extinction, where participants are explicitly instructed that the stimulus did not signal a more effective drug-seeking response (Hogarth et al., 2014). Indeed, some experimental observations have also indicated that the excitation of drug seeking responses might rely on a more complicated hierarchical expectancy, where the S provokes an expectation that the drug-seeking behavior is more likely to be effective (Colwill & Rescorla, 1990; Hardy et al., 2017), rather than a simple binary S-O-R associative chain.

However, there is also evidence that is supportive of the motivational hypothesis. In the single response PIT tasks use in Krank (2003) and Krank et al. (2008), the alcohol cueing effect in rats was strongest when the cue was near the self-administration apparatus. This is evidence of sign-tracking (or autoshaping) approach behaviors. In animals, rats that showed autoshaping behaviors toward reward stimuli subsequently showed that new instrumental behaviors could be trained effectively using these stimuli as a reward (Robinson & Flagel, 2009). Thus, through Pavlovian learning drug stimuli might acquire incentive salience that results in animals and humans increasing their motivation to seek the drug-stimuli themselves. In humans, there is some preliminary evidence of this process in a PIT task using eye tracking to distinguish sign-trackers and goal-trackers. In a single-response PIT task where participants learned to win money, and learned to associate arbitrary stimuli with either the delivery of money or the delivery of nothing, a transfer test revealed that the strength of the transfer effect was larger in people that were more likely to look toward the area of the screen where the reward stimuli would appear compared to people who looked toward where the rewarding outcome (the money) would appear. This is preliminary evidence for sign-tracking behavior in humans modulating the effect rewarding stimuli has on reward-seeking behavior. However, as far as we are aware there is currently no demonstration of drug stimuli sign-tracking in humans.

The general transfer effect is considered a demonstration of drug-stimuli modifying behavior through acquired incentive motivation. Drug cues, in conjunction with having effects on behaviors specific to their associated reward (e.g. alcohol-related stimuli increasing alcohol-seeking), might also have generally motivating effects that can elicit increases in other drug and non-drug reward-seeking. Animal studies have shown that rats will respond to the presence of ethanol-associated cues by increasing both ethanol-seeking and food seeking relative to baseline levels (Corbit & Janak, 2007; Glasner, Overmier, & Balleine, 2005). The dominant understanding of this effect is that it is a *general* motivational effect. General, because the reward stimulus appeared to elicit increases in both available behaviors. The general transfer effect appears to be a motivational effect. In Corbit, Janak, and Balleine (2007), a general transfer effect was demonstrated with food, but this cueing effect was abolished when rats were shifted from a hungry state to a satiated state (i.e. they were not hungry). This seems to suggest that the motivational state of the subject is involved in the general transfer process.

There is currently no available human evidence that examined the general transfer effect with drug-rewards. However, human studies examining the effect of non-drug reward cues on appetitive behavior has supported the idea that reward cues can elicit increases in other reward-seeking behavior (e.g. Nadler, Delgado, & Delamater, 2011; Prévost, Liljeholm, Tyszka, & O'Doherty, 2012; Watson, Wiers, Hommel, & de Wit, 2014). Correlational evidence has also suggested that these cueing effects are related to hunger (Watson et al., 2014) and that high levels of stress results in strong general transfer (Quail, Morris, & Balleine, 2017). In sum, these studies evidence the notion that the general transfer effect appears to operate through as a motivational effect that elicits a general enhancement of available reward-seeking behaviors.

There is broad evidence for the notion that drug cues can elicit an increase in desire for other drugs (Cooney et al., 2007; Drobes, 2002; Sayette, Martin, Wertz, Perrott, & Peters, 2005; Taylor, Harris, Singleton, Moolchan, & Heishman, 2000). However, it is unclear what associative mechanisms are in operation in these drug craving effects. In particular, with general PIT effects, there is currently very little research that has explored this effect in humans using drug stimuli, and no published research that has studied the effect using drug rewards. Sommer et al. (2017) used a variation of the PIT paradigm to examine the effect of alcohol stimuli on money-seeking behavior in people who were in treatment for alcohol dependence. Contrary to expectations, alcohol stimuli reduced money-seeking behavior relative to water stimuli exposure. Furthermore, we recently examined the effect of alcohol stimuli on food-seeking behavior in non-dependent students who drank alcohol (Mahlberg, Weidemann, Hogarth, & Moustafa, n.d.). We found that presses for food points increased in the presence of alcohol stimuli relative to the no stimulus condition. However, we found that the alcohol cueing effect was reward specific: presses for potato chip points increased above baseline during alcohol stimulus exposure, but presses for chocolate points did not. Overall, the current body of research that has studied the general PIT effect in humans with drug stimuli is mixed, and only partially supports the notion that (a) the effect is non-specific, and (b) the effect increases reward-seeking through motivation for drugs. Since there is currently only limited evidence examining the psychological processes involved, more research is necessary to understand how drug stimuli can change other reward-seeking behaviors.

Relevance of the PIT model to addiction

Modeling drug stimulus effects on drug-seeking behavior using the PIT paradigm has utility in helping understand the psychological mechanisms in action that promote stimulus-driven drug seeking. This is important, as understanding how stimulus driven drug-seeking works can help develop our understanding of why people might use drugs in different situations. For example, the demonstration that specific transfer effects appear to be insensitive to drug devaluation is important as it is taken as evidence of the habit theory of addiction. This is one broad perspective that suggests that exposure to drugs results in drug-seeking behavior that is increasingly automatic or habitual (Hogarth, Balleine, et al., 2013), resulting in the development of what is described as addiction to drugs. The habit hypothesis of drug addiction has strong face validity (Hogarth & Chase, 2011b). Clinicians generally consider loss of control over drug use, and persistent drug use despite negative consequences, an important feature of substance use disorders (American Psychiatric Association, 2013). Moreover, a popular cognitive theory of addiction has described drug use behavior as set of automatic behaviors (i.e. a habit) to account for the observation that drug desire and drug-seeking often appear to be independent of one another (Tiffany, 1990). Thus, the specific PIT paradigm has utility in observing drug seeking behavior that appears to correspond to this habit model of addiction, making the case that discrete stimuli within the environment can trigger habitual drug seeking (Watson & de Wit, 2018).

Using the PIT paradigm, we can also observe the effects drug use has on how humans and animals respond to the environment. It has been shown that amphetamine exposure after training can abolish outcome selective responding to food stimuli in rats (Hall & Gulley, 2011; Shiflett, 2012). Moreover, humans exposed to alcohol stimuli can display stronger stimulus-directed behavior for cigarettes despite devaluation (Hogarth, Field, et al., 2013). Therefore, this PIT model also has utility in examining how the environment affects drug-use behavior under conditions of acute drug intake or when other drug stimuli are present in the environment.

Importantly, though, there is little evidence that the drug stimulus effects modeled in this paradigm correlate with drug dependence (Glasner et al., 2005; Hardy et al., 2017; Hogarth & Chase, 2011a, 2012) or with drug stimulus-elicited craving (Mahlberg et al., 2019). Moreover, it appears as though people with drug dependence do not display different stimulus-elicited behavior generally compared to a non-dependent comparison group (Hogarth et al., 2018). Therefore, the utility of the PIT paradigm is that it has shown that people are vulnerable to stimulus-induced changes in drug seeking, but these stimulus effects are not necessarily a central feature of drug dependence.

There are also important issues with external validity of the PIT task in generalizing how drug-seeking behavior works in laboratory experiments to real drug seeking patterns. The PIT task, in an attempt to isolate specific mediating roles of stimuli-outcome associations on drug-seeking, provides an impoverished empirical scenario that might not be representative of how stimulus-elicited drug-seeking operates in real life. That is, a situation where there is only one stimulus presented that signals information about only one rewarding outcome, and where there is very little cost to responding (Lovibond & Colagiuri, 2013), is highly unlikely to be representative of a real drug use situation. While under this empirical situation, humans and animals alike appear to have elevated drug seeking regardless of their motivation to consume drugs, which corresponds to a habit theory of addiction. However, more recent research has shown in situations with more than one stimulus presented, decision-making appears to be sensitive to outcome desire (i.e. goal-directed, Seabrooke, Hogarth, Edmunds, & Mitchell, 2019; Seabrooke, Le Pelley, Hogarth, & Mitchell, 2017). Future research should consider these alternative designs with greater stimulus information and choice in drug-seeking PIT paradigms, to understand how drug-seeking in modified by the environment in complex scenarios that are more representative of the decision-making process in human drug users.

Finally, there is currently very little empirical evidence regarding the general transfer effect in humans. Some research supports the animal models (e.g. Quail et al., 2017; Watson et al., 2014), but other research has found evidence that forces us to question the specificity (Mahlberg et al., n.d.) and the automaticity (Sommer et al., 2017) of the effect of alcohol cues on other reward-seeking behaviors. Thus, this motivational form of the transfer effect needs more research in humans, to develop a more sophisticated knowledge of how human drug users are motivated by different environmental stimuli.

Drug and context associations

As highlighted above, drug addiction constitutes a repertoire of learned behavior involving the use of drugs. These include instrumental behaviors that procure highly rewarding outcomes (drugs) and learning to respond to stimuli within our environment that become predictive of the onset of drugs. But the learning of these processes do not occur in isolation (Rosas, Todd, & Bouton, 2013). Learning of discrete stimulus-drug and response-drug behaviors occur within a wider environment or context. "Context" is a difficult to define concept, because it can involve a wide variety of varying characteristics. However, in general, a context is the constellation of stimuli that surround or envelop the learning environment, without necessarily being an active component of the specific learning situation. For example, a context for someone learning about alcohol use behaviors can be the room where the learning takes place, the smells of the room, the surrounding sounds, et cetera.

Notice that none of these individual characteristics described are necessarily directly involved in the process of learning about drinking beer. But despite this, context can have quite important effects on drug use and drug related behavior. This section will describe evidence that has shown drug contexts can modulate the sensitivity to drugs as well as develop motivational properties from historical drug experience. The most important aspect of this section will be the importance of context in the extinction of drug-related learning. It will be shown in this chapter that context is inherently involved in the acquisition, extinction and reinstatement (renewal) of drug-seeking behaviors. We will discuss various forms of "context", including contexts involving exogenous stimuli (e.g. place preference, ABA renewal paradigms) and endogenous contexts (e.g. internal states like drug-affected or stressed), and their ability to trigger a return of drug-seeking responses drug-seeking behaviors were extinguished. Next, we will discuss theoretical and empirical models that demonstrate the various psychological processes involved, as well as examining the neuropsychological mechanisms in operation in these context effects. Finally, we will discuss the validity of the popular empirical models used to demonstrate context effects on drug seeking, with a particular focus on their external validity as models for clinical drug use.

Contexts can trigger drug craving

The context can promote drug craving in humans. Thewissen, Van Den Hout, Havermans, and Jansen (2005) showed that a context that signaled the availability of cigarettes could provoke higher cigarette craving compared to a context that did not signal cigarette availability in cigarette smokers. Moreover, Thewissen et al. (2005) showed that a change from an acquisition context to a novel context in test resulted in *lower* cigarette craving overall compared to a group that did not experience a context change. This, in sum, suggested that the acquisition context developed an association that supported the strength of cigarette craving.

Context induced renewal has also been shown in craving studies. Thewissen, Snijders, Havermans, van den Hout, and Jansen (2006) showed that, after acquisition of an association between a neutral cue and cigarette smoking in one context (context A), and extinction in a different context (context B), the strength of cigarette craving was dependent on the test context. Testing the level of participants craving in context A showed a large increase in craving from extinction to test, while testing in context B showed little change in craving.

Collins and Brandon (2002) tested the effect of context on alcohol craving in non-dependent alcohol drinkers. They first examined the extent of baseline alcohol craving in one context (context A). Then in an extinction phase, alcohol craving was extinguished by presenting a novel cue (a discrete cue that was used as an extinction-associated cue). For group one, the novel cue was presented in context A (the baseline measurement context). For groups two and three, the novel cue was presented during extinction in a different context (context B). After extinction, the three groups experienced a renewal test, where their craving levels were tested again. All groups were tested in context A. Group two also had the extinguished cue presented during the renewal test. It was found that the groups exposed to extinction in context B had greater alcohol craving and salivation when tested in context A, relative to the group that received extinction in the same context. Furthermore, the presentation of the extinction-associated cue in group two attenuated this context renewal effect, such that alcohol craving and salivation levels were lower than group three (although craving was still higher than group one). Overall this study showed that (a) alcohol craving was vulnerable to context change effects after extinction, and (b) discrete cues associated with extinction learning can attenuate but not totally eradicate this effect.

However, Stasiewicz, Brandon, and Bradizza (2007) used an extinction protocol do reduce the extent of alcohol craving to alcohol stimuli in a group of alcohol-dependent outpatients. Then, in a test phase, participants were tested for their cue-elicited alcohol craving in either the same context (the extinction context) or in a different context. They also tested whether the presence of a discrete extinction cue modified the context effect. However, they found no renewal effect in the different context, nor did they find any attenuation effects of discrete extinction cues on renewal. A similar null effect of context was found by MacKillop and Lisman (2008), where a context shift after extinction did not result in noteworthy elevation of craving in people dependent on alcohol. These null results could be due to the particular contexts not being discriminable enough to provoke large effects on craving (Stasiewicz et al., 2007). Thus, overall, there is some preliminary evidence that contexts associated with drug use can provoke drug desire, although more work should be done in the future to clarify under what circumstances these associations provide the greatest effects on drug desire.

Context can acquire drug-like motivational properties

The context in which drug use takes place can acquire motivational properties, which can be shown in a conditioned placed preference paradigm. Conditioned placed preference (CPP) is shown in animals when the rewarding effects of drugs (e.g. received intravenously) are reliably paired with a particular context. Place preference is shown in a test where the animal is allowed to access both contexts, and the time spent in the drug-administration context is measured as an index of preference for the drug-associated context.

An early demonstration by Beach (1957) showed that the context surrounding drug experiences can become important. This study aimed to observe whether rats would learn to prefer the place where they experienced injections of morphine (drug context) compared to a context that they did not receive morphine. After a period of daily morphine injections, across a series of experiments rats demonstrated a "preference" for the morphine context, in that they moved toward the drug context more after training compared to prior training figures. Subsequent to this demonstration of a preference for the drug-paired place, a multitude of experiments have been conducted examine the conditioned place preference (CPP) effect, using a wide variety of drug types and administration methods (for an exhaustive list, see Tzschentke, 1998, 2007).

There have been two experimental studies that have examined the CPP effect in humans that provide positive evidence for the current conceptualization of this paradigm as a measure of preference. In the first series of studies to demonstrate the CPP effect in humans, Childs and de Wit (2009) used an unbiased design, where two groups received four experimental sessions where they experienced amphetamine administration or placebo, and one final test experimental session. In one group (the "paired" group), participants always received the drug in the same room (context A) and placebo in a different room (context B); while the other group (the "unpaired" group) received the drug once in each room. On test, participants in the paired group showed subjective liking ratings for the drug room that were marginally higher than the placebo room, while the unpaired group showed no difference between groups. A subsequent study provided a stronger replication of this trend by using a biased CPP design (Childs & de Wit, 2013). This accounted for individual differences in room preference before training in the paired group, by pairing the drug with a participants least preferred room. In Childs and de Wit (2013), there was a significant increase in preference ratings for the initially non-preferred room, but only in the group of participants that had their drug administration paired. Moreover, subjective ratings for how much participants liked the drug effects uniquely correlated with the change in ratings for room liking. Moreover, Childs and de Wit (2016) recently replicated the above subjective ratings using an alcohol CPP procedure, but notably extended the evidence for CPP in humans by also measuring preference for the room by measuring time spent in each room. They found that people spent more time in the alcohol-paired context in comparison to the control group, replicating the behavioral measure of CPP taken in animals, as well as capturing the subjective change in liking of the context found their previous work.

This provided evidence in humans that (a) drug administrations can result in preferences for a particular context, and (b) this preference appears to be related to a person's positive experience of the drug. Moreover, it provides some convergent evidence for the speculation of the psychological processes that underpin place preference: the fact that preference only changed for those in paired group indicate that there is indeed a learned associative component to the CPP effect, and that it might be driven by a predictive relationship between the place and the drug effects (see Rescorla, 1967).

The psychological processes of conditioned place preference

While the CPP task appears quite conceptually intuitive, it is difficult to elucidate the psychological processes that underpin the development of a "preference" (Stephens, Crombag, & Duka, 2013). This is because there is likely a number of mechanisms working in parallel that result in a preference for a particular context, and the relative balance of these mechanisms would be dependent on the circumstances that produce a place preference.

First and foremost, the CPP effect follows principles of classical conditioning (Tzschentke, 1998). That is conditioned preference for a context requires a learned association between a context and the drug experience. For example, Brabant, Quertemont, and Tirelli (2005) showed that one session of pairing cocaine injections did not establish an observable place preference. However, two and four cocaine paring sessions established a place preference that endured for less than 14 and for 28 days, respectively. This illustrated that the CPP effect is associative, and the strength of the preference is relative to the extent of the associative history. In CPP experiments, there is a perfect predictive relationship between the context and the drug effects, where drug effects only happen when in the context. When the context does not predict the drug effects, no place preference develops (e.g. Childs & de Wit, 2013). Thus, a place preference requires a contingent relationship between context and drug delivery (see Rescorla, 1967), where the context becomes a predictive signal about the impending experience of drug effects.

Further evidence for the associative nature of the CPP effect is that an extinction protocol can reduce the strength of the place preference. Extinction in this situation is the process of either exposing the animal to a drug-paired context but now receiving saline instead of the drug, or experiencing the context without receiving any drug or vehicle administrations. The purpose of an extinction phase is to observe a reduction in time spent in the once-preferred context, compared to the other non-drug context and/or to a control group (Aguilar, Rodríguez-Arias, & Miñarro, 2009). This shows that once learning that the context is no longer associated with the experience of the drug then the preference decreases, indicating that the association with the drug effect is important.

Furthermore, the stimulus conditions produced by the drug (i.e. the drug effects) become associated with the context (see Eikelboom & Stewart, 1982). Because of this, CPP is thought to be evidence of the context acquiring secondary "drug like" reinforcing properties (Bardo, Rowlett, & Harris, 1995), where the context comes to acquire motivational properties that results in approach behaviors and a "preference" for that context. For example, Spiteri, Le Pape, and Ågmo (2000) examined the behavior of mice in the test session of a conditioned place preference paradigm, conditioned with either food or morphine. In the morphine condition, mice showed different behavior compared to the food condition. While both groups showed a place preference for the reward delivery context, the morphine group remained in close contact with stimuli within the context that was associated with morphine delivery by spending more time (on average) in the preferred context, whereas the mice in the food condition showed greater exploratory behavior by increasing their number of visits but not remaining in the context for longer. This suggested that place preference conditioned with food results in exploratory behaviors where mice increased their checking for food, the morphine place preference group showed stronger affective behavior toward the morphine context, potentially indicating a conditioned motivation response to "prefer" the context.

Further evidence for this motivational effect is the observation that there is temporal moderator involved in the CPP effect: the time taken between administering the drug in the rat and when the rat experiences the context can change the effect from a place preference to a place aversion, where instead of seeking the context the animal avoids the context (Bardo & Bevins, 2000). Ettenberg, Raven, Danluck, and Necessary (1999) showed this with cocaine. In their conditioned place preference procedure, they paired the context with the effects of cocaine either immediately after injections, 5 min after injections, or 15 min after injections. The immediate group and the 5 min group both displayed a place preference for the drug context. However, the grouped paired with the delayed cocaine effects (the 15 min group) displayed the opposite effect: they showed an avoidance of the drug context. This indicated that the motivational qualities of the cocaine injection were time sensitive, in that it is positive immediately but becomes aversive over time. This provided evidence for the fact that the CPP effect is a demonstration of the positive motivation for the drug that comes to pair with the context. Therefore, the CPP paradigm can be useful for studying the motivational effects of drugs. In this instance, the motivational effects of cocaine were shown to be biphasic: initially positive, but the experience changes to a negative or aversive experience after a certain period.

Reinstatement effects in the conditioned place preference paradigm

While extinguishing the relationship between the context and the experience of drug effects can reduce the strength of the place preference, there are situations that can result in a place preference returning, despite no further pairings between the context and the drug effects.

A non-contingent presentation of the drug experience can result in reinstatement of the place preference. This was demonstrated in Mueller and Stewart (2000), where conditioned place preference was established using cocaine injections. This preference was then extinguished so that time spent in the drug context no longer differed from other contexts. However, animals that received a priming injection of cocaine before test showed a reinstatement effect: they showed a conditioned place preference for the cocaine context, even after this preference was extinguished. This reinstatement effect has also been shown using morphine (Mueller, Perdikaris, & Stewart, 2002), and ethanol (Kuzmin, Sandin, Terenius, & Ögren, 2003). This was taken as evidence that the drug prime "reminded" the animals of the incentive value of the drug, in turn motivating the animal to renew their preference for the context that also has come to signal the drugs incentive value (Aguilar et al., 2009).

Moreover, these reinstatement effects are not specific effects created by the unique pharmacological effects of the target drug. Other drugs (i.e. drugs not experienced during CPP learning) can result in reinstatement of place preference (Itzhak & Martin, 2002). Moreover, exposure to stress can also prime a return to conditioned place preference (Wang, Luo, Zhang, & Han, 2000). Thus, reinstatement of conditioned place preference effects appears to be the result of elevating the incentive motivation for drugs.

Relevance of conditioned place preference to the study of addiction

The CPP effect is not a phenomenon unique to drugs. Animals can also show a place preference to contexts that is paired with social interactions (Trezza, Damsteegt, & Vanderschuren, 2009). Humans can also develop place preferences for food rewards, such as chocolate (e.g. Astur, Carew, & Deaton, 2014). Importantly, the CPP effect does not appear to differ in people with a history of drug dependence. Radell et al. (2018) used a modified conditioned place preference task to investigate preference for a virtual room with high probability of rewards available compared to a different virtual room with a low probability of rewards available. This task was used to examine group differences between people with opiate dependence and a non-dependent comparison group. It was found that both groups developed a place preference for the reward rich context, such that they spent more time in the reward rich context compared to the reward poor context. Importantly, this CPP effect did not differ by drug dependence status. While there are important differences between this task and the animal and translational models of CPP—in this task, instrumental reward seeking was used to learn the context-reward associations instead of Pavlovian associations—it illustrated that the development of context-reward associations does not appear to differ in people with a history of drug dependence. Thus, the context effects observed in conditioned place preference studies are not evidence of unique conditioned motivational effects, and is unlikely to be useful as a behavioral marker or risk assessment for drug dependence.

Despite this, the strength of the CPP model is that it shows that the drug context can enter into associations with the effects of drugs that can result in changes to behavior, and thus provides a clear demonstration of how the wider environment can become implicated in drug use. Because of this, the CPP paradigm has been useful to study the psychological and neurobiological processes of incentive learning in drugs. Indeed, the CPP paradigm can be useful to study the utility of drugs to assist in the treatment of substance use disorders, such as pharmaceutical interventions that intervene at the level of incentive motivation for drug use. Thus, the CPP paradigm, as a measure of incentive motivation for a drug, can be useful to study the therapeutic effects of intervention drugs, by considering how organisms respond to conditioned drug effects and whether this can be reduced or eliminated with pharmacological or psychological interventions (Napier, Herrold, & de Wit, 2013).

Limitations of CPP paradigm

The idea that the context effects in animals indeed represent an expression of positive motivation for the conditioned place (i.e. a "preference") is difficult to elucidate from the animal CPP model, because there could be a number of reasons why animals spend more time in the drug-paired context that are not indicative of "preference". During acquisition of the place preference, the drug-state enhances instrumental behavior that results in a context becoming a discriminative stimulus for those instrumental behaviors, or the context elicits drug-like behavior learned during conditioning. Either way, the animal in test might spend longer in the context not because it is demonstrating a motivated "preference", but because the context is directly triggering learned behaviors and the animal is not necessarily remaining in the context longer because it prefers it or finds the consequence of this behavior to be positive (Huston, de Souza Silva, & Müller, 2013).

However, the main limitation of this model of context effects in drug addiction is its inability to translate ecologically to a human population, since it is a purely Pavlovian design that involved the experimental administration of drugs to subjects (Bardo & Bevins, 2000), and this kind of drug delivery is not how people who use drugs come to acquire drug use behaviors. Other experimental models, such as the context reinstatement procedure, have been developed to capture the effects of context on drug-seeking directly.

Context can trigger an increase in drug use behavior

A popular approach to studying the effects of context on drug use behaviors is the context reinstatement paradigm. This model studies the effects of context on drug self-administration using a context renewal procedure, a paradigm that was modeled off context reinstatement of fear learned fear responses (e.g. Bouton & Bolles, 1979). In this three stage paradigm first reported by Crombag and Shaham (2002), a sample of rats learned to self-administer a drug cocktail ("speedball", a mix of heroin and speed) in an operant chamber (context A). Next, in an extinction phase, one group remained in context A, while the other animals were transferred to a different operant chamber with distinct changes in tactile, visual, auditory, and olfactory properties (context B). All animals underwent extinction of this self-administration behavior, where the animals learned over many sessions that the operant behavior no longer acquired the drug and thus reduced their operant response frequency. The final test phase involved an assessment of the ability of context to reinstate drug-seeking behavior by increasing the frequency of an extinguished operant response. This was done by comparing three groups. A control group, that consisted of both rats that remained in context A for all three stages of the experiment (AAA), and rats that acquired drug self-administration in context A but experienced extinction and tested for renewal in context B (ABB). This was compared to the renewal group, which experienced acquisition of the drug self-administration behavior in context A, extinction of this learned behavior in context B, but crucially was returned to the acquisition context for test (i.e. context A). It was observed that the renewal group (ABA) demonstrated more frequent drug-seeking behaviors in test compared to the controls (i.e. both the AAA and ABA groups). This was a clear demonstration that a return to the context where acquisition of drug self-administration behaviors was experienced resulted in a return of drug-seeking behaviors, despite successful extinction of this behavior in a different context. In test, the ABA group had more frequent drug-seeking responses compared to the other conditions. Therefore, the learning that resulted in extinction of the drug-seeking behaviors was context specific, and evidences the notion that extinguishing a learned behavior does not remove this learning but merely masks or moderates the learned behaviors in some way (Bouton, Winterbauer, & Todd, 2012).

Psychological processes in context reinstatement

Initial research suggested an excitatory process learned during the acquisition context (i.e. context A) was the crucial mechanism that elicited reinstatement when subjects were returned to this context. A non-significant difference between AAA and ABB was found by Crombag and Shaham (2002), which suggested that it something important about the shift to a novel context in extinction and then the reintroduction of the acquisition context that results in drug-seeking reinstatement. That is, the contextual learning in acquisition appears to be central to the renewal effect. Fuchs et al. (2005) also assessed this in a drug self-administration paradigm by comparing an ABA renewal group with an AAB group. Context reinstatement was observed in the ABA group as responding for the drug increased in test compared to the extinction phase. However, there was no observed increase in responding for the drug in the AAB group, suggesting that a transition to a novel context might not be enough to induce drug reinstatement. This supports the hypothesis that there was an excitatory association developed between the acquisition context and the drug self-administration behaviors. It has been argued that this excitatory association might act on drug seeking through PIT mechanisms (Crombag, Bossert, Koya, & Shaham, 2008; Khoo, Gibson, Prasad, & McNally, 2017; see above for a discussion of PIT mechanisms).

Another view is that it is in fact something learned in the extinction context that is important for the reinstatement effect. If there was something specific learned during extinction, then a shift from the extinction context to a novel context should also result in reinstatement, since the novel context will not signal anything about the extinction learning or the acquisition learning. Currently, drug self-administration research has found null results for this effect (Crombag & Shaham, 2002; Fuchs et al., 2005; described above), suggesting the learning during extinction is not sufficient to promote context induced reinstatement in drug self-administration.

However, context reinstatement may have been impeded because, in the above paradigm, extinction may have been more effective since it occurred within the original acquisition phase. Thus, novel context reinstatement can also be tested in an ABC paradigm, where there is a novel context for acquisition, extinction, and test. ABC reinstatement is difficult to show in drugs (Crombag et al., 2008). This is suggested by Crombag et al. (2008) as evidence against an inhibitory model of context reinstatement. However, subsequent demonstrations of novel context renewal by using food rewards have strengthened the evidence for this process (Bouton & Todd, 2014; Todd, 2013). The null results in drug studies using a novel context reinstatement procedure could have occurred because the novel context needs to have prior experience with instrumental conditioning for it to renew drug seeking (Khoo et al., 2017). This hypothesis is yet to be tested. However, this new evidence suggests that it is likely there is an inhibitory process learned during the extinction phase whereby the context comes to either directly or indirectly reduce the activation of drug use behaviors (see Todd, Vurbic, & Bouton, 2014).

Despite this, overall the context reinstatement effects in novel contexts were weaker than they are in the classic ABA paradigm (Podlesnik, Kelley, Jimenez-Gomez, & Bouton, 2017). This could be because transitioning back to the acquisition context (context A) results in two changes: (1) the removal of the extinction context inhibitory effects (Bouton & Todd, 2014) and (2) the reintroduction of the acquisition contexts excitatory effects.

Relevance of context reinstatement to drug addiction

This paradigm has an advantage over other procedures like the CPP paradigm because it observes patterns of self-administration, which is less ambiguous than a "preference" test, but also has more face validity than the CPP procedure because context effects in drug users involve effects on drug-use, not just a preference for a particular context. Because of this, context reinstatement paradigms have been useful in studying the effects of variables such as drug intake, withdrawal, and stress on context-induced drug-seeking (Shaham, Shalev, Lu, de Wit, & Stewart, 2003). Context reinstatement paradigms have also been useful for assessing the utility of pharmaceutical interventions in context induced drug-seeking. For example, Torregrossa, Sanchez, and Taylor (2010) showed that D-cycloserine administration, by activating NMDA receptors in the Nucleus Accumbens, reduced the context dependency of extinction. That is, self-administration of cocaine remained at similar levels for those animals that experienced DCS regardless of the context. Therefore, context reinstatement paradigms have utility in developing a greater understanding of how context affects drug use behaviors, but also can be used as conceptual models for the development of novel pharmacological interventions for drug addiction (Torregrossa, Corlett, & Taylor, 2011).

One problem with the ecological validity of the context reinstatement paradigm is that the extinction component of the procedure does not model a clinical reality—it is unlikely that a person becomes abstinent from drugs because they experience a situation where a previously learned drug-seeking response is still available to enact but no longer retrieves the drug outcome. A more likely scenario is a person might undergo a period of voluntary abstinence when the balance between drug reward and the negative effects of drug use shift in their balance toward the latter. That is, a person might become abstinent for a period of time because their drug use resulted in increasingly aversive effects on their health, social, or financial situation. Therefore, a modified version of the classic context reinstatement paradigm was developed to try and model this form of abstinence (Marchant, Khuc, Pickens, Bonci, & Shaham, 2013). This procedure follows closely to the original ABA design. However, instead of extinction experienced in context B, drug use behaviors are still accessible in Context B but the frequency of drug-seeking is reduced by introducing a contingent punisher (e.g. a footshock) which results in punishment-induced abstinence. Marchant et al. (2013) found that after punishment induced abstinence renewal still occurred after transitioning to the acquisition context (context A), showing a context reinstatement effect can be observed in a paradigm that is more generalizable for human drug addiction. However, this procedure still has issues with external validity, since the response contingent punishment used to model punishment-induced abstinence is not very representative of the more diffuse general negative effects that occur from human drug use (Marchant et al., 2013). There are also potentially different mechanisms of action in this paradigm compared to the original context reinstatement paradigm because the punishment learning introduces additional Pavlovian/instrumental interactions.

There is currently no demonstration of context renewal effects on drug-seeking in humans. There is one recent human study that captured this effect in chocolate seeking. Bezzina, Lee, Lovibond, and Colagiuri (2016) used a modified PIT task to study the differential effects of Pavlovian extinction on the ability of chocolate-associated cues to elicit increases in chocolate-seeking. In addition, they also examined the effect of context: one group received acquisition, extinction, and the PIT test in the same context (an AAA group); whereas the other group received acquisition and test in the same context, but extinction in a different context (an ABA group). As expected, they found significant cueing effects in both groups, as Pavlovian extinction did not suppress the PIT effect. The novel finding, though, was that there was an effect of context: the PIT effect was larger in the ABA group, suggesting that a context switch resulted in an increase in the strength of the reward cue to facilitate food seeking. This result is important, as it demonstrates that both discrete and context cues work in parallel to modify reward seeking in humans. Future research needs to be done to examine how these processes work in human drug seeking.

Summary and conclusion

This chapter shows that the environment where drug use takes place is important to consider when studying drug use and addiction. Discrete stimuli can form associative relationships with the use of drugs, and through this association, environmental stimuli can provoke an increase in drug desire and use in animals and humans in a variety of ways. This has been modeled experimentally using a combination of instrumental learning and cue exposure paradigms, as well as studying the interaction between these using the Pavlovian-instrumental-transfer paradigm. Moreover, the broader environment that envelops the learning of these discrete stimuli and response associations (i.e. the context) has also been shown to be important when studying drug use and addiction. Using conditioned place preference and context reinstatement paradigms, research has shown that the context can enter into relationships with drug use. These context-drug interactions result in the context mediating the expression of drug motivation and drug-seeking.

Across the paradigms discussed in this chapter, we have shown that much of this research is useful in the study of drug addiction, as it helps to elucidate the psychological mechanisms that are in operation in these environment-drug use interactions. Moreover, these paradigms have been useful in providing experimental models to test the utility of pharmacological interventions to help assist in therapeutic situations for people who aim to reduce or cease their drug use. However, there are some caveats in this body of literature, as there is a general dearth of translational research that has studied systematically some of the effects discussed above in humans. Moreover, the experimental models discussed above all share a caveat of ecological validity, where their designs limit the generalizability of their findings to real life drug use experiences. Therefore, necessary future research includes continuing to attempt translations of animal modeling to human populations, as well as improving these experimental designs so the effects observed have greater translational value to clinical drug addiction populations.

References

Aguilar, M. A., Rodríguez-Arias, M., & Miñarro, J. (2009). Neurobiological mechanisms of the reinstatement of drug-conditioned place preference. *Brain Research Reviews*, *59*(2), 253–277. https://doi.org/10.1016/j.brainresrev.2008.08.002.

American Psychiatric Association. (2013). *Diagnostic and statistical manual of mental disorders*. Vol. 5. American Psychiatric Pub.

Astur, R. S., Carew, A. W., & Deaton, B. E. (2014). Conditioned place preferences in humans using virtual reality. *Behavioural Brain Research*, *267*, 173–177. https://doi.org/10.1016/j.bbr.2014.03.018.

Bardo, M. T., & Bevins, R. A. (2000). Conditioned place preference: What does it add to our preclinical understanding of drug reward? *Psychopharmacology*, *153*(1), 31–43. https://doi.org/10.1007/s002130000569.

Bardo, M. T., Rowlett, J. K., & Harris, M. J. (1995). Conditioned place preference using opiate and stimulant drugs: A meta-analysis. *Neuroscience & Biobehavioral Reviews*, *19*(1), 39–51. https://doi.org/10.1016/0149-7634(94)00021-R.

Beach, H. D. (1957). Morphine addiction in rats. *Canadian Journal of Psychology/Revue canadienne de psychologie*, *11*(2), 104. https://doi.org/10.1037/h0083703.

Bezzina, L., Lee, J. C., Lovibond, P. F., & Colagiuri, B. (2016). Extinction and renewal of cue-elicited reward-seeking. *Behaviour Research and Therapy*, *87*, 162–169. https://doi.org/10.1016/j.brat.2016.09.009.

Bouton, M. E., & Bolles, R. C. (1979). Contextual control of the extinction of conditioned fear. *Learning and Motivation*, *10*(4), 445–466. https://doi.org/10.1016/0023-9690(79)90057-2.

Bouton, M. E., & Todd, T. P. (2014). A fundamental role for context in instrumental learning and extinction. *Behavioural Processes*, *104*, 13–19. https://doi.org/10.1016/j.beproc.2014.02.012.

Bouton, M. E., Winterbauer, N. E., & Todd, T. P. (2012). Relapse processes after the extinction of instrumental learning: Renewal, resurgence, and reacquisition. *Behavioural Processes*, *90*(1), 130–141. https://doi.org/10.1016/j.beproc.2012.03.004.

Brabant, C., Quertemont, E., & Tirelli, E. (2005). Influence of the dose and the number of drug-context pairings on the magnitude and the long-lasting retention of cocaine-induced conditioned place preference in C57BL/6J mice. *Psychopharmacology*, *180*(1), 33–40. https://doi.org/10.1007/s00213-004-2138-6.

Carter, B. L., & Tiffany, S. T. (1999). Meta-analysis of cue-reactivity in addiction research. *Addiction*, *94*(3), 327–340. https://doi.org/10.1046/j.1360-0443.1999.9433273.x.

Carter, B. L., & Tiffany, S. T. (2001). The cue-availability paradigm: The effects of cigarette availability on cue reactivity in smokers. *Experimental and Clinical Psychopharmacology*, *9*(2), 183–190. https://doi.org/10.1037/1064-1297.9.2.183.

Cartoni, E., Balleine, B., & Baldassarre, G. (2016). Appetitive Pavlovian-instrumental transfer: A review. *Neuroscience & Biobehavioral Reviews*, *71*. https://doi.org/10.1016/j.neubiorev.2016.09.020.

Childs, E., & de Wit, H. (2009). Amphetamine-induced place preference in humans. *Biological Psychiatry*, *65*(10), 900–904. https://doi.org/10.1016/j.biopsych.2008.11.016.

Childs, E., & de Wit, H. (2013). Contextual conditioning enhances the psychostimulant and incentive properties of D-amphetamine in humans. *Addiction Biology*, *18*(6), 985–992. https://doi.org/10.1111/j.1369-1600.2011.00416.x.

Childs, E., & de Wit, H. (2016). Alcohol-induced place conditioning in moderate social drinkers. *Addiction*, *111*(12), 2157–2165. https://doi.org/10.1111/add.13540.

Colagiuri, B., & Lovibond, P. F. (2015). How food cues can enhance and inhibit motivation to obtain and consume food. *Appetite*, *84*, 79–87. https://doi.org/10.1016/j.appet.2014.09.023.

Collins, B. N., & Brandon, T. H. (2002). Effects of extinction context and retrieval cues on alcohol cue reactivity among nonalcoholic drinkers. *Journal of Consulting and Clinical Psychology*, *70*(2), 390. https://doi.org/10.1037/0022-006x.70.2.390.

Collins, D., & Lapsley, H. M. (2008). *The costs of tobacco, alcohol and illicit drug abuse to Australian society in 2004/05*. Department of Health and Ageing Canberra.

Colwill, R. M., & Rescorla, R. A. (1990). Evidence for the hierarchical structure of instrumental learning. *Animal Learning & Behavior*, *18*(1), 71–82. https://doi.org/10.3758/bf03205241.

Cooney, N. L., Litt, M. D., Cooney, J. L., Pilkey, D. T., Steinburg, H. R., & Oncken, C. A. (2007). Alcohol and tobacco cessation in alcohol-dependent smokers: Analysis of real-time reports. *Psychology of Addictive Behaviors*, *21*(3), 277–286. https://doi.org/10.1037/0893-164X.21.3.277.

Corbit, L. H., & Janak, P. H. (2007). Ethanol-associated cues produce general Pavlovian-instrumental transfer. *Alcoholism: Clinical and Experimental Research*, *31*(5), 766–774. https://doi.org/10.1111/j.1530-0277.2007.00359.x.

Corbit, L. H., Janak, P. H., & Balleine, B. W. (2007). General and outcome-specific forms of Pavlovian-instrumental transfer: The effect of shifts in motivational state and inactivation of the ventral tegmental area. *European Journal of Neuroscience*, *26*(11), 3141–3149. https://doi.org/10.1111/j.1460-9568.2007.05934.x.

Crombag, H. S., Bossert, J. M., Koya, E., & Shaham, Y. (2008). Context-induced relapse to drug seeking: A review. *Philosophical Transactions of the Royal Society B: Biological Sciences*, *363*(1507), 3233–3243. https://doi.org/10.1098/rstb.2008.0090.

Crombag, H. S., & Shaham, Y. (2002). Renewal of drug seeking by contextual cues after prolonged extinction in rats. *Behavioral Neuroscience*, *116*(1), 169. https://doi.org/10.1037/0735-7044.116.1.169.

de Wit, S., Kindt, M., Knot, S. L., Verhoeven, A. A. C., Robbins, T. W., Gasull-Camos, J., … Gillan, C. M. (2018). Shifting the balance between goals and habits: Five failures in experimental habit induction. *Journal of Experimental Psychology: General*, *147*(7), 1043–1065. https://doi.org/10.1037/xge0000402.

Dickinson, A. (1985). Actions and habits: The development of behavioural autonomy. *Philosophical Transactions of the Royal Society B*, *308*(1135), 67–78. https://doi.org/10.1098/rstb.1985.0010.

Dickinson, A., Wood, N., & Smith, J. W. (2002). Alcohol seeking by rats: Action or habit? *The Quarterly Journal of Experimental Psychology B: Comparative and Physiological Psychology*, *55B*(4), 331–348. https://doi.org/10.1080/0272499024400016.

Drobes, D. J. (2002). Cue reactivity in alcohol and tobacco dependence. *Alcoholism: Clinical and Experimental Research*, *26*(12), 1928–1929. https://doi.org/10.1111/j.1530-0277.2002.tb02506.x.

Eikelboom, R., & Stewart, J. (1982). Conditioning of drug-induced physiological responses. *Psychological Review*, *89*(5), 507. https://doi.org/10.1037/0033-295X.89.5.507.

Ersche, K. D., Gillan, C. M., Jones, P. S., Williams, G. B., Ward, L. H. E., Luijten, M., … Robbins, T. W. (2016). Carrots and sticks fail to change behavior in cocaine addiction. *Science*, *352*(6292), 1468–1471. https://doi.org/10.1126/science.aaf3700.

Ettenberg, A., Raven, M. A., Danluck, D. A., & Necessary, B. D. (1999). Evidence for opponent-process actions of intravenous cocaine. *Pharmacology Biochemistry and Behavior*, *64*(3), 507–512. https://doi.org/10.1016/S0091-3057(99)00109-4.

Fuchs, R. A., Evans, K. A., Ledford, C. C., Parker, M. P., Case, J. M., Mehta, R. H., & See, R. E. (2005). The role of the dorsomedial prefrontal cortex, basolateral amygdala, and dorsal hippocampus in contextual reinstatement of cocaine seeking in rats. *Neuropsychopharmacology, 30*(2), 296. https://doi.org/10.1038/sj.npp.1300579.

Glasner, S. V., Overmier, J. B., & Balleine, B. W. (2005). The role of Pavlovian cues in alcohol seeking in dependent and nondependent rats. *Journal of Studies on Alcohol, 66*(1), 53–61. https://doi.org/10.15288/jsa.2005.66.53.

Hall, D. A., & Gulley, J. M. (2011). Disruptive effect of amphetamines on Pavlovian to instrumental transfer. *Behavioural Brain Research, 216*(1), 440–445. https://doi.org/10.1016/j.bbr.2010.08.040.

Hardy, L., Mitchell, C., Seabrooke, T., & Hogarth, L. (2017). Drug cue reactivity involves hierarchical instrumental learning: Evidence from a biconditional Pavlovian to instrumental transfer task. *Psychopharmacology, 234*(13), 1977–1984. https://doi.org/10.1007/s00213-017-4605-x.

Hogarth, L. (2012). Goal-directed and transfer-cue-elicited drug-seeking are dissociated by pharmacotherapy: Evidence for independent additive controllers. *Journal of Experimental Psychology: Animal Behavior Processes, 38*(3), 266–278. https://doi.org/10.1037/a0028914.

Hogarth, L., Balleine, B. W., Corbit, L. H., & Killcross, S. (2013). Associative learning mechanisms underpinning the transition from recreational drug use to addiction. *Annals of the New York Academy of Sciences, 1282*(1), 12–24. https://doi.org/10.1111/j.1749-6632.2012.06768.x.

Hogarth, L., & Chase, H. W. (2011a). Parallel goal-directed and habitual control of human drug-seeking: Implications for dependence vulnerability. *Journal of Experimental Psychology: Animal Behavior Processes, 37*(3), 261–276. https://doi.org/10.1037/a0022913.

Hogarth, L., & Chase, H. W. (2011b). Vulnerabilities underlying human drug dependence: Goal valuation versus habit learning. In M. Haselgrove & L. Hogarth (Eds.), *Clinical applications of learning theory* (pp. 75–102). Hove, United Kingdom: Taylor & Francis Group.

Hogarth, L., & Chase, H. W. (2012). Evaluating psychological markers for human nicotine dependence: Tobacco choice, extinction, and Pavlovian-to-instrumental transfer. *Experimental and Clinical Psychopharmacology, 20*(3), 213–224. https://doi.org/10.1037/a0027203.

Hogarth, L., Dickinson, A., Wright, A., Kouvaraki, M., & Duka, T. (2007). The role of drug expectancy in the control of human drug seeking. *Journal of Experimental Psychology: Animal Behavior Processes, 33*(4), 484–496. https://doi.org/10.1037/0097-7403.33.4.484.

Hogarth, L., Field, M., & Rose, A. K. (2013). Phasic transition from goal-directed to habitual control over drug-seeking produced by conflicting reinforcer expectancy. *Addiction Biology, 18*(1), 88–97. https://doi.org/10.1111/adb.12009.

Hogarth, L., Lam-Cassettari, C., Pacitti, H., Currah, T., Mahlberg, J., Hartley, L., & Moustafa, A. (2018). Intact goal-directed control in treatment-seeking drug users indexed by outcome-devaluation and Pavlovian to instrumental transfer: Critique of habit theory. *European Journal of Neuroscience*, https://doi.org/10.1111/ejn.13961.

Hogarth, L., Retzler, C., Munafò, M. R., Tran, D. M. D., Troisi, J. R., II, Rose, A. K., … Field, M. (2014). Extinction of cue-evoked drug-seeking relies on degrading hierarchical instrumental expectancies. *Behaviour Research and Therapy, 59*, 61–70. https://doi.org/10.1016/j.brat.2014.06.001.

Holland, P. C. (2004). Relations between Pavlovian-instrumental transfer and reinforcer devaluation. *Journal of Experimental Psychology: Animal Behavior Processes, 30*(2), 104–117. https://doi.org/10.1037/0097-7403.30.2.104.

Holmes, N. M., Marchand, A. R., & Coutureau, E. (2010). Pavlovian to instrumental transfer: A neurobehavioural perspective. *Neuroscience and Biobehavioral Reviews, 34*(8), 1277–1295. https://doi.org/10.1016/j.neubiorev.2010.03.007.

Huston, J. P., de Souza Silva, M. A., & Müller, C. P. (2013). What's conditioned in conditioned place preference? *Trends in Pharmacological Sciences*, *34*(3), 162–166. https://doi.org/10.1016/j.tips.2013.01.004.

Itzhak, Y., & Martin, J. L. (2002). Cocaine-induced conditioned place preference in mice: Induction, extinction and reinstatement by related psychostimulants. *Neuropsychopharmacology*, *26*(1), 130. https://doi.org/10.1016/s0893-133x(01)00303-7.

Khoo, S. S., Gibson, G., Prasad, A., & McNally, G. (2017). How contexts promote and prevent relapse to drug seeking. *Genes, Brain and Behavior*, *16*(1), 185–204. https://doi.org/10.1111/gbb.12328.

Krank, M. D. (2003). Pavlovian conditioning with ethanol: Sign-tracking (autoshaping), conditioned incentive, and ethanol self-administration. *Alcoholism: Clinical and Experimental Research*, *27*(10), 1592–1598. https://doi.org/10.1097/01.alc.0000092060.09228.de.

Krank, M. D., O'Neill, S., Squarey, K., & Jacob, J. (2008). Goal-and signal-directed incentive: Conditioned approach, seeking, and consumption established with unsweetened alcohol in rats. *Psychopharmacology*, *196*(3), 397–405. https://doi.org/10.1007/s00213-007-0971-0.

Kuzmin, A., Sandin, J., Terenius, L., & Ögren, S. O. (2003). Acquisition, expression, and reinstatement of ethanol-induced conditioned place preference in mice: Effects of opioid receptor-like 1 receptor agonists and naloxone. *Journal of Pharmacology and Experimental Therapeutics*, *304*(1), 310–318. https://doi.org/10.1124/jpet.102.041350.

Lamb, R. J., Schindler, C. W., & Pinkston, J. W. (2016). Conditioned stimuli's role in relapse: Preclinical research on pavlovian-instrumental-transfer. *Psychopharmacology*, https://doi.org/10.1007/s00213-016-4216-y.

Loughlin, A., Funk, D., Coen, K., & Lê, A. D. (2017). Habitual nicotine-seeking in rats following limited training. *Psychopharmacology*, *234*(17), 2619–2629. https://doi.org/10.1007/s00213-017-4655-0.

Lovibond, P. F., & Colagiuri, B. (2013). Facilitation of voluntary goal-directed action by reward cues. *Psychological Science*, *24*(10), 2030–2037. https://doi.org/10.1177/0956797613484043.

Lovibond, P. F., Satkunarajah, M., & Colagiuri, B. (2015). Extinction can reduce the impact of reward cues on reward-seeking behavior. *Behavior Therapy*, *46*(4), 432–438. https://doi.org/10.1016/j.beth.2015.03.005.

MacKillop, J., & Lisman, S. A. (2008). Effects of a context shift and multiple context extinction on reactivity to alcohol cues. *Experimental and Clinical Psychopharmacology*, *16*(4), 322–331. https://doi.org/10.1037/a0012686.

Mahlberg, J., Weidemann, G., Hogarth, L., & Moustafa, A. A. (n.d.). Do alcohol stimuli produce general transfer effects in human food-seeking? *Experimental Brain Research*, Under Review.

Mahlberg, J., Weidemann, G., Hogarth, L., & Moustafa, A. A. (2019). Cue-elicited craving and human Pavlovian-to-instrumental transfer. *Addiction Research & Theory*, 1–7. https://doi.org/10.1080/16066359.2018.1544625.

Marchant, N. J., Khuc, T. N., Pickens, C. L., Bonci, A., & Shaham, Y. (2013). Context-induced relapse to alcohol seeking after punishment in a rat model. *Biological Psychiatry*, *73*(3), 256–262. https://doi.org/10.1016/j.biopsych.2012.07.007.

Martinovic, J., Jones, A., Christiansen, P., Rose, A. K., Hogarth, L., & Field, M. (2014). Electrophysiological responses to alcohol cues are not associated with Pavlovian-to-instrumental transfer in social drinkers. *PLoS ONE*, *9*(4), e94605. https://doi.org/10.1371/journal.pone.0094605.

Miles, F. J., Everitt, B. J., & Dickinson, A. (2003). Oral cocaine seeking by rats: Action or habit? *Behavioral Neuroscience*, *117*(5), 927–938. https://doi.org/10.1037/0735-7044.117.5.927.

Mueller, D., Perdikaris, D., & Stewart, J. (2002). Persistence and drug-induced reinstatement of a morphine-induced conditioned place preference. *Behavioural Brain Research*, *136*(2), 389–397. https://doi.org/10.1016/s0166-4328(02)00297-8.

Mueller, D., & Stewart, J. (2000). Cocaine-induced conditioned place preference: Reinstatement by priming injections of cocaine after extinction. *Behavioural Brain Research*, *115*(1), 39–47. https://doi.org/10.1016/S0166-4328(00)00239-4.

Nadler, N., Delgado, M. R., & Delamater, A. R. (2011). Pavlovian to instrumental transfer of control in a human learning task. *Emotion*, *11*(5), 1112–1123. https://doi.org/10.1037/a0022760.

Napier, T. C., Herrold, A. A., & de Wit, H. (2013). Using conditioned place preference to identify relapse prevention medications. *Neuroscience & Biobehavioral Reviews*, *37*(9 (Part A)), 2081–2086. https://doi.org/10.1016/j.neubiorev.2013.05.002.

Nebe, S., Kroemer, N. B., Schad, D. J., Bernhardt, N., Sebold, M., Müller, D. K., … Rapp, M. A. (2018). No association of goal-directed and habitual control with alcohol consumption in young adults. *Addiction Biology*, *23*(1), 379–393.

Podlesnik, C. A., Kelley, M. E., Jimenez-Gomez, C., & Bouton, M. E. (2017). Renewed behavior produced by context change and its implications for treatment maintenance: A review. *Journal of Applied Behavior Analysis*, *50*(3), 675–697. https://doi.org/10.1002/jaba.400.

Prévost, C., Liljeholm, M., Tyszka, J. M., & O'Doherty, J. P. (2012). Neural correlates of specific and general Pavlovian-to-instrumental transfer within human amygdalar subregions: A high-resolution fMRI study. *Journal of Neuroscience*, *32*(24), 8383–8390. https://doi.org/10.1523/JNEUROSCI.6237-11.2012.

Quail, S. L., Morris, R. W., & Balleine, B. W. (2017). Stress associated changes in Pavlovian-instrumental transfer in humans. *The Quarterly Journal of Experimental Psychology*, *70*(4), 675–685. https://doi.org/10.1080/17470218.2016.1149198.

Radell, M. L., Allen, M. T., Favaloro, B., Myers, C. E., Haber, P., Morley, K., & Moustafa, A. A. (2018). Intolerance of uncertainty and conditioned place preference in opioid addiction. *PeerJ*, *6*, e4775, https://doi.org/10.7717/peerj.4775.

Rescorla, R. A. (1967). Pavlovian conditioning and its proper control procedures. *Psychological Review*, *74*(1), 71. https://doi.org/10.1037/h0024109.

Rescorla, R. A., & Solomon, R. L. (1967). Two-process learning theory: Relationships between pavlovian conditioning and instrumental learning. *Psychological Review*, *74*(3), 151–182. https://doi.org/10.1037/h0024475.

Robinson, T. E., & Flagel, S. B. (2009). Dissociating the predictive and incentive motivational properties of reward-related cues through the study of individual differences. *Biological Psychiatry*, *65*(10), 869–873. https://doi.org/10.1016/j.biopsych.2008.09.006.

Rosas, J. M., Paredes-Olay, M. C., García-Gutiérrez, A., Espinosa, J. J., & Abad, M. J. F. (2010). Outcome-specific transfer between predictive and instrumental learning is unaffected by extinction but reversed by counterconditioning in human participants. *Learning and Motivation*, *41*(1), 48–66. https://doi.org/10.1016/j.lmot.2009.09.002.

Rosas, J. M., Todd, T. P., & Bouton, M. E. (2013). Context change and associative learning. *Wiley Interdisciplinary Reviews: Cognitive Science*, *4*(3), 237–244. https://doi.org/10.1002/wcs.1225.

Rose, A. K., Brown, K., MacKillop, J., Field, M., & Hogarth, L. (2018). Alcohol devaluation has dissociable effects on distinct components of alcohol behaviour. *Psychopharmacology*, *235*(4). https://doi.org/10.1007/s00213-018-4839-2.

Sayette, M. A., Martin, C. S., Wertz, J. M., Perrott, M. A., & Peters, A. R. (2005). The effects of alcohol on cigarette craving in heavy smokers and tobacco chippers. *Psychology of Addictive Behaviors*, *19*(3), 263. https://doi.org/10.1037/0893-164x.19.3.263.

Seabrooke, T., Hogarth, L., Edmunds, C. E. R., & Mitchell, C. J. (2019). Goal-directed control in Pavlovian-instrumental transfer. *Journal of Experimental Psychology: Animal Learning and Cognition*, *45*(1), 95. https://doi.org/10.1037/xan0000191.

Seabrooke, T., Le Pelley, M. E., Hogarth, L., & Mitchell, C. J. (2017). Evidence of a goal-directed process in human Pavlovian-instrumental transfer. *Journal of Experimental Psychology: Animal Learning and Cognition*, *43*(4). https://doi.org/10.1037/xan0000147.

Sebold, M., Deserno, L., Nebe, S., Schad, D. J., Garbusow, M., Hägele, C., et al.Huys, Q. J. M. (2014). Model-based and model-free decisions in alcohol dependence. *Neuropsychobiology*, *70*(2), 122–131.

Sebold, M., Nebe, S., Garbusow, M., Guggenmos, M., Schad, D. J., Beck, A., ... Neu, P. (2017). When habits are dangerous: Alcohol expectancies and habitual decision making predict relapse in alcohol dependence. *Biological Psychiatry*, *82*(11), 847–856.

Shaham, Y., Shalev, U., Lu, L., de Wit, H., & Stewart, J. (2003). The reinstatement model of drug relapse: History, methodology and major findings. *Psychopharmacology*, *168*(1), 3–20. https://doi.org/10.1007/s00213-002-1224-x.

Shiflett, M. W. (2012). The effects of amphetamine exposure on outcome-selective Pavlovian-instrumental transfer in rats. *Psychopharmacology*, *223*(3), 361–370. https://doi.org/10.1007/s00213-012-2724-y.

Sjoerds, Z., De Wit, S., Van Den Brink, W., Robbins, T. W., Beekman, A. T. F., Penninx, B.W.J.H., & Veltman, D. J. (2013). Behavioral and neuroimaging evidence for overreliance on habit learning in alcohol-dependent patients. *Translational Psychiatry*, *3*(12), e337.

Sommer, C., Garbusow, M., Jünger, E., Pooseh, S., Bernhardt, N., Birkenstock, J., ... Huys, Q. (2017). Strong seduction: Impulsivity and the impact of contextual cues on instrumental behavior in alcohol dependence. *Translational Psychiatry*, *7*(8), https://doi.org/10.1038/tp.2017.158.

Spiteri, T., Le Pape, G., & Ågmo, A. (2000). What is learned during place preference conditioning? A comparison of food- and morphine-induced reward. *Psychobiology*, *28*(3), 367–382. https://doi.org/10.3758/bf03331994.

Stasiewicz, P. R., Brandon, T. H., & Bradizza, C. M. (2007). Effects of extinction context and retrieval cues on renewal of alcohol-cue reactivity among alcohol-dependent outpatients. *Psychology of Addictive Behaviors*, *21*(2), https://doi.org/10.1037/0893-164X.21.2.244.

Stephens, D. N., Crombag, H. S., & Duka, T. (2013). The challenge of studying parallel behaviors in humans and animal models. In W. H. Sommer & R. Spanagel (Eds.), *Behavioral neurobiology of alcohol addiction* (pp. 611–645). Berlin, Heidelberg: Springer Berlin Heidelberg.

Taylor, R. C., Harris, N. A., Singleton, E. G., Moolchan, E. T., & Heishman, S. J. (2000). Tobacco craving: Intensity-related effects of imagery scripts in drug abusers. *Experimental and Clinical Psychopharmacology*, *8*(1), 75. https://doi.org/10.1037/1064-1297.8.1.75.

Teesson, M., Hall, W., Lynskey, M., & Degenhardt, L. (2000). Alcohol-and drug-use disorders in Australia: Implications of the national survey of mental health and wellbeing. *Australian and New Zealand Journal of Psychiatry*, *34*(2), 206–213. https://doi.org/10.1080/j.1440-1614.2000.00715.x.

Thewissen, R., Snijders, S.J.B.D., Havermans, R. C., van den Hout, M., & Jansen, A. (2006). Renewal of cue-elicited urge to smoke: Implications for cue exposure treatment. *Behaviour Research and Therapy*, *44*(10), 1441–1449. https://doi.org/10.1016/j.brat.2005.10.010.

Thewissen, R., Van Den Hout, M., Havermans, R. C., & Jansen, A. (2005). Context-dependency of cue-elicited urge to smoke. *Addiction*, *100*(3), 387–396. https://doi.org/10.1111/j.1360-0443.2005.00996.x.

Tiffany, S. T. (1990). A cognitive model of drug urges and drug-use behavior: Role of automatic and nonautomatic processes. *Psychological Review*, *97*(2), 147. https://doi.org/10.1037/0033-295x.97.2.147.

Todd, T. P. (2013). Mechanisms of renewal after the extinction of instrumental behavior. *Journal of Experimental Psychology: Animal Behavior Processes*, *39*(3), 193–207. https://doi.org/10.1037/a0032236.

Todd, T. P., Vurbic, D., & Bouton, M. E. (2014). Mechanisms of renewal after the extinction of discriminated operant behavior. *Journal of Experimental Psychology: Animal Learning and Cognition*, *40*(3), 355. https://doi.org/10.1037/xan0000021.

Torregrossa, M. M., Corlett, P. R., & Taylor, J. R. (2011). Aberrant learning and memory in addiction. *Neurobiology of Learning and Memory*, *96*(4), 609–623. https://doi.org/10.1016/j.nlm.2011.02.014.

Torregrossa, M. M., Sanchez, H., & Taylor, J. R. (2010). Cycloserine reduces the context specificity of Pavlovian extinction of cocaine cues through actions in the nucleus accumbens. *Journal of Neuroscience*, *30*(31), 10526–10533. https://doi.org/10.1523/jneurosci.2523-10.2010.

Trapold, M. A., & Overmier, J. B. (1972). The second learning process in instrumental conditioning. In A. A. Black & W. F. Prokasy (Eds.), *Classical conditioning II: Current theory and research.* (1st ed.)(pp. 427–452). New York: Appelton-Century-Crofts.

Trezza, V., Damsteegt, R., & Vanderschuren, L.J.M.J. (2009). Conditioned place preference induced by social play behavior: Parametrics, extinction, reinstatement and disruption by methylphenidate. *European Neuropsychopharmacology*, *19*(9), 659–669. https://doi.org/10.1016/j.euroneuro.2009.03.006.

Tricomi, E., Balleine, B. W., & O'Doherty, J. P. (2009). A specific role for posterior dorsolateral striatum in human habit learning. *European Journal of Neuroscience*, *29*(11), 2225–2232. https://doi.org/10.1111/j.1460-9568.2009.06796.x.

Tzschentke, T. M. (1998). Measuring reward with the conditioned place preference paradigm: A comprehensive review of drug effects, recent progress and new issues. *Progress in Neurobiology*, *56*(6), 613–672. https://doi.org/10.1016/S0301-0082(98)00060-4.

Tzschentke, T. M. (2007). Review on CPP: Measuring reward with the conditioned place preference (CPP) paradigm: Update of the last decade. *Addiction Biology*, *12*(3–4), 227–462. https://doi.org/10.1111/j.1369-1600.2007.00070.x.

Veilleux, J. C., & Skinner, K. D. (2015). Smoking, food, and alcohol cues on subsequent behavior: A qualitative systematic review. *Clinical Psychology Review*, *36*, 13–27. https://doi.org/10.1016/j.cpr.2015.01.001.

Wang, B., Luo, F., Zhang, W.-T., & Han, J.-S. (2000). Stress or drug priming induces reinstatement of extinguished conditioned place preference. *Neuroreport*, *11*(12), 2781–2784. https://doi.org/10.1097/00001756-200008210-00034.

Watson, P., & de Wit, S. (2018). Current limits of experimental research into habits and future directions. *Current Opinion in Behavioral Sciences*, *20*, 33–39. https://doi.org/10.1016/j.cobeha.2017.09.012.

Watson, P., Wiers, R. W., Hommel, B., & de Wit, S. (2014). Working for food you don't desire. Cues interfere with goal-directed food-seeking. *Appetite*, *79*, 139–148. https://doi.org/10.1016/j.appet.2014.04.005.

Chapter 6

Avoidance learning and behavior in patients with addiction

Milen L. Radell[a], Farahnaz Ghafar[b], Peter Casbolt[b], Ahmed A. Moustafa[c]

[a]*Department of Psychology, Niagara University, Lewiston, NY, United States,* [b]*School of Social Sciences and Psychology, Western Sydney University, Sydney, NSW, Australia,* [c]*Marcs Institute for Brain, Behaviour, and Development and School of Psychology, Western Sydney University, Penrith, NSW, Australia*

Introduction

Avoidance of particular stimuli or situations associated with aversive outcomes is a common behavior found in daily life. For instance, most humans will avoid a stray dog for fear of being bitten and stop at a red traffic light to avoid an accident. Avoidance behavior can be defined as either the withholding or the performance of a particular response in order to prevent an imminent aversive event, and includes both avoidance and escape in all their forms (Hayes, Wilson, Gifford, Follette, & Strosahl, 1996). While this behavior is adaptive, excessive avoidance in the absence of actual danger can contribute to psychopathology. This type of pathological avoidance is a common feature of a variety of mental disorders, including most anxiety disorders, post-traumatic stress disorder, obsessive compulsive disorder, and substance use disorders (Hayes et al., 1996).

Addiction has been described as a "behavioral pattern of compulsive drug use, characterized by overwhelming involvement with the use of a drug, the securing of its supply, and a high tendency to relapse" (Jaffe, 1965, p. 286) after the drug is withdrawn. This definition underscores the excessive preoccupation with the drug, at the expense of other activities and irrespective of the associated negative consequences (e.g., job or relationship loss) that form essential features of substance use disorders as defined by the current 5th edition of the Diagnostic and Statistical Manual of Mental Disorders (DSM-5; American Psychiatric Association, 2013). It also highlights that the drug-related behavior tends to be compulsive, as well as recurrent, which poses a major hurdle for successful long-term treatment. Current models of addiction recognize that drug use may begin for some combination of positive and negative reinforcement (Hayes et al., 1996). From the perspective of negative reinforcement models

Cognitive, Clinical, and Neural Aspects of Drug Addiction. https://doi.org/10.1016/B978-0-12-816979-7.00006-6

(e.g., self-medication, tension-reduction, emotion regulation), drug use may represent a form of avoidance behavior that helps reduce or prevent aversive cognitive and emotional states (e.g., anger, sadness, grief, or craving). However, in doing so, the drug-related behavior is strengthened, ultimately contributing to the transition from abuse to addiction.

This chapter will review research on avoidance in the context of substance use disorders, with a primary focus on the human literature. However, comparisons will also be made to relevant animal work, which has greatly contributed to understanding how avoidance might be related to drug use. Since the translation of animal data to human populations has been limited, we will also discuss the use of computer-based tasks that mimic animal avoidance learning paradigms as a way to bridge the two literatures.

Avoidance of physical withdrawal

Early negative reinforcement models of addiction focused on the avoidance of physical withdrawal symptoms in order to explain the pervasive nature of substance use that occurs in addiction (Baker, Piper, McCarthy, Majeskie, & Fiore, 2004; Hutcheson, Everitt, Robbins, & Dickinson, 2001). Although the specific withdrawal symptoms and their intensity can vary greatly depending on the drug, according to this model, withdrawal represents an aversive state that is common to most drugs of abuse. Taking the drug provides immediate relief of this state, and thus reinforces drug-related behavior via a negative reinforcement mechanism. This view has also been presented in the opponent process theory (Solomon & Corbit, 1974), according to which drug effects disrupt homeostasis, leading to the activation of an opponent process that counteracts these effects in order to restore balance. Furthermore, this opponent process is proposed to underlie tolerance, and is slow to dissipate after drug use is discontinued, leading to withdrawal (Solomon & Corbit, 1974). Thus, according to both of these models, negative reinforcement provides the basis for the maintenance of drug use. However, a limitation of this idea is that withdrawal appears to be neither a necessary nor sufficient condition for addiction (Edwards, 1990). For example, many drugs lead to dependence and withdrawal (e.g., haloperidol) when use is discontinued yet these drugs are seldom, if ever, abused. In fact, in the case of haloperidol, or other antipsychotic medications, patient compliance can be an issue. Additionally, even for some highly abused drugs, such as cocaine, withdrawal symptoms can be mild and disappear quickly after use is discontinued, which fails to explain the high propensity for relapse observed in addiction (Jaffe, 1992).

Avoidance of negative affect

More recent learning-based models of addiction combine both positive and negative reinforcement in order to explain drug-related behavior, recognize that each represents only a single factor that contributes to this behavior, and place more emphasis on negative affect. For example, Baker et al. (2004) reframed the negative reinforcement model within an emotion regulation framework, and like others, recognized that avoidance of aversive states is not the sole factor involved in substance abuse. Nonetheless, these authors argue that this mechanism has been unfairly undervalued, or even ignored, in recent models of addiction, which have focused on appetitive processes (e.g., incentive salience). Briefly, it is proposed that negative affect (e.g., anxiety, irritability, sadness, etc.) is a key feature of withdrawal, which comes about rapidly throughout the day, but is also relieved rapidly by drug use. Through classical conditioning, this state is associated with a variety of cues (both internal and external) and contexts, which allow an individual to predict the onset of the aversive emotional state and take steps (i.e., take their drug of choice) in order to prevent it (Baker et al., 2004).

Aside from self-report accounts from individuals with substance use disorders, evidence in support of this idea comes from animal studies involving drug discrimination. For example, rats can be trained to emit one operant response while on one drug and another when under a different drug (or a placebo). In both cases, the only cues available to perform this discrimination task are interoceptive (i.e., internal to the organism). Thus, both the direct effects of drugs and their withdrawal symptoms can serve as discrimination cues (Gauvin, Harland, Criado, Michaelis, & Holloway, 1989). Using this paradigm, it is possible to examine whether anything can substitute for a given drug or for the state of withdrawal—for example, an animal trained to press a lever in response to withdrawal will also press the same lever when given a drug known to produce anxiety in humans, or when exposed to a stressor (e.g., electric shock or a predator) (Baker et al., 2004). This suggests that the internal cues produced by either withdrawal or by stress, are similar and can all serve as stimuli for drug use. Negative affect due to withdrawal, as well as that associated with stress, might be additive and result in sufficient motivation for avoidance behavior (i.e., drug taking), which will then be strengthened via negative reinforcement (Baker et al., 2004).

Experiential avoidance

Consistent with this view, much of the human research on avoidance has been within an emotion regulation framework, where avoidance serves as one of several coping strategies used to reduce aversive emotional experiences. In this context, avoidance is often referred to as "experiential avoidance", which occurs when an individual takes steps toward ensuring that particular aversive private events (e.g., thoughts, emotions, memories, and bodily sensations) with which they are unwilling to remain in contact with, are not elicited (Hayes et al., 1996). This can involve altering their form and frequency, or the conditions under which they are experienced. It also entails what is at times labeled "emotional avoidance" or "cognitive avoidance" when specifically referring to the avoidance of particular emotions or thoughts, respectively (Hayes et al., 1996).

Experiential avoidance may have several components, including behavioral avoidance (avoidance of situations that lead to distress), distress aversion (unwillingness to accept distress), procrastination (attempts to delay predictable or impending distress), distraction and suppression (attempts to ignore or inhibit distress), repression and denial (to distance and dissociate from distress) and distress endurance (thoughts or behaviors that help persevere while in distress) (Buckner, Zvolensky, Farris, & Hogan, 2014; Gámez, Chmielewski, Kotov, Ruggero, & Watson, 2011). A distinction has also been made between deliberate and controlled forms of experiential avoidance (e.g., suppression) and involuntary and automatic forms (e.g., repression) (Chawla & Ostafin, 2007). Despite these attempts to delineate the components of experiential avoidance, this construct has been criticized for being too broad (Chawla & Ostafin, 2007), and it remains unknown whether some of its components play a more important role than others for the development, maintenance and treatment of substance use disorders. It is also subject to debate whether experiential avoidance, and the emotion regulation framework it is a part of, are simply new names for old constructs or models (Zinbarg & Mineka, 2007).

Regardless, one of the reasons the use of avoidance might be a common coping strategy is that it is immediately reinforcing, and the short-term consequences are valued more than the long-term consequences. That might be especially true in addiction, which is associated with more impulsive behavior, guided by immediate rather than long-term reinforcement. However, in the long-term, avoidance can be counterproductive. For example, emotions such as anxiety, anger, or depression, can trigger craving and withdrawal among detoxified opiate addicts (Childress, McLellan, Natale, & O'Brien, 1987). Such private experiences may be the product of classical conditioning, and would be elicited automatically upon encountering appropriate cues or situations that have been previously associated with them (Hayes et al., 1996). This should increase the risk of relapse. In addition, while substance use might provide temporary relief, in doing so, drug-related behavior will be strengthened. However, the drug of choice can also make the problem worse (Stockwell & Bolderston, 1987), which argues against a pure negative reinforcement account of addiction, including that based on dampening aversive emotions. It is also possible for the same drug to have both types of effects, with tension-reducing effects in the short-term but detrimental effects in the long-term. This is well-documented for alcohol. Thus, it is more important to establish the conditions under which a given drug will be effective as a negative reinforcer (Stockwell & Bolderston, 1987).

Coping can also involve approach behavior (Moos, Brennan, Fondacaro, & Moos, 1990) and so it is useful to consider whether this type of coping is more adaptive than avoidance coping. For example, behavioral coping strategies that involve approach include seeking guidance and support, as well as treating the situation as a problem to be solved (i.e., coping based on problem solving). The severity, as well as the type of stressor can influence the type of coping strategy that is selected. McCrae (1984) suggests that people facing a challenge use more problem-focused coping, while those facing threat or loss use more avoidance coping (McCrae, 1984). In general, strategies based on approach are more adaptive than those focused on avoidance. In one study, Moos et al. (1990) found that older problem drinkers (age 55–65) were more likely to use cognitive and behavioral avoidance to cope with stress compared to non-problem drinkers. This was not because the types of stressors self-reported by the two groups of participants differed, nor because they were appraised differently—both groups tended to view the stressors as a challenge. Problem drinkers who scored greater on avoidance coping also had more depression and physical symptoms. In this study, those who had encountered more negative stressful events during the past year were more likely to use both approach and avoidance coping (Moos et al., 1990). However, other research has shown that chronic severe stress can lead to greater reliance on avoidance rather than approach coping (Fondacaro & Moos, 1989).

In another study, experiential avoidance and anxiety sensitivity were examined as predictors of motives for alcohol consumption in a sample of young adults (Stewart, Zvolensky, & Eifert, 2001). The results indicated that experiential avoidance significantly predicted increased alcohol consumption for both negative reinforcement (e.g., coping) and positive reinforcement. However, a similar study using a sample of substance-abusing veterans, did not find experiential avoidance to predict addiction severity even when considered in combination with other psychological variables (Forsyth, Parker, & Finlay, 2003). These discrepant findings may have been due to the high overlap between experiential avoidance and other psychological variables, as well as the inclusion of substances belonging to different drug categories. This is an important consideration given that little is known about the relationship between experiential avoidance and substances belonging to particular drug classes.

Moreover, neither of these studies looked at the impact of experiential avoidance on alcohol dependence as accompanied by a comorbid mental health issue. Experiential avoidance may also help explain the high comorbidity observed between substance use and other disorders, including anxiety and mood disorders (Chawla & Ostafin, 2007; Forsyth et al., 2003; Levin et al., 2012; Stewart et al., 2001; Westrup, 1999). For example, there is at least some degree of avoidance in individuals with alcohol dependence who also suffer from comorbid anxiety and/or depression (Carpenter & Hasin, 1999; Crum et al., 2013; DeMartini & Carey, 2011; Lechner et al., 2014). The same may also be the case for other drugs. For example, Buckner et al. (2014) studied individuals who reported using cannabis to cope with social anxiety. To assess coping motives (i.e., whether drug use is motivated by coping), participants completed a self-report questionnaire about the reasons they use cannabis (e.g., to forget worries, to get high, to enjoy a party, to fit in with a group, and to expand awareness). Behavioral avoidance was found to mediate the relationship between social anxiety and coping motives (Buckner et al., 2014). Finally, consistent with the idea that avoidance contributes to disorders from multiple categories, a meta-analysis of self-report studies (Aldao, Nolen-Hoeksema, & Schweizer, 2010) found a medium to large effect size of avoidance coping across all disorders considered, including anxiety, depression, eating disorders and substance use disorders. However, only a single study including data relevant for substance use (Cooper, Wood, Orcutt, & Albino, 2003) met inclusion criteria in the analysis making it difficult to draw general conclusions about the role of avoidance.

Khantzian (1997) has proposed that subclinical levels of distress can also motivate drug use as a form of self-medication (i.e., avoidance of an aversive emotional state), but that much of the research on the role of distress has focused on clinical samples. In these samples, the issue of what came first—substance abuse, or some other comorbid disorder—is unclear. It is possible that these patients had subclinical levels of distress that motivated drug use to begin with, but as a result of drug use, these problems became worse and met clinical diagnostic criteria (e.g., for depression or mood disorder) (Khantzian, 1997). Thus, subjective states of distress (not clinical levels of distress) may contribute to the development of substance use disorders. These states of distress are often ignored by longitudinal studies that have found drug use to precede, rather than follow, the development of other mental disorders. Such findings have been interpreted, perhaps unjustifiably, as evidence against the self-medication hypothesis (Khantzian, 1997).

In contrast, Levy (2018) argues that self-medication is overused as an explanation for substance abuse—it is relevant for some, but not for others. While patients will often self-report using drugs to medicate some problem, they also report other reasons for use, such as to get high or for pleasure. The reasons given for relapse after a period of abstinence are similar to those reported for why drug use began in the first place—some state it was to deal with emotional problems while others would like to get high again. Regardless of the true reason, citing self-medication can simply represent the individual trying to justify or excuse the substance use behavior (Levy, 2018). In addition, accepting self-medication as an explanation may also be detrimental for treatment. Specifically, if the patient is led to believe, or left to believe, that their substance use is due to dealing with emotional problems, they may assume that these problems must be resolved first before the substance-related behavior can be treated. Thus, they may fail to recognize that their drug use has become independent of these problems, or might even be their source (Levy, 2018).

Avoidance coping and addiction vulnerability

As discussed above, there are other reasons for drug use besides coping, including enhancement of positive affect, social pressure, and celebration (Carpenter & Hasin, 1999). Research has also examined whether drug use to cope with stress is a risk factor for developing substance use disorders, and whether it carries greater importance than other motives. Three models that aim to explain the relationship between alcohol use and coping, but that also apply to other drug use have been described (Cunningham, Sobell, Sobell, Gavin, & Annis, 1995). They include the risk-factor, the generalizing, and the epiphenomena model. The risk-factor model predicts that individuals vulnerable to problem alcohol use can be identified prior to the development of an alcohol use disorder. The generalizing model proposes that individuals who start drinking are more likely to also use alcohol for coping (i.e., generalize alcohol use so it now also serves to manage stress) (Cunningham et al., 1995). These individuals are therefore more likely to endorse coping as a reason for drinking, although this did not cause them to initiate alcohol use in the first place. Thus, unlike the risk-factor model, the generalizing model predicts coping motives for drinking will not exist early on (i.e., prior to the development of problem use) and will not help identify vulnerable individuals (Cunningham et al., 1995). Last, the epiphenomena model proposes that coping motives depend on the current level of alcohol use or on negative affect at the time of testing (i.e., coping motives are merely a byproduct). Unlike the risk-factor model but similar to the generalizing model, it predicts that there should be no differences in coping motives early on that can help identify vulnerable individuals. In contrast to the generalizing model, it predicts that controlling for the severity of alcohol use or for negative affect should eliminate differences in coping motives between problem and non-problem drinkers (Cunningham et al., 1995).

Cross-sectional studies have consistently found that problem drinkers report more coping motives compared to non-problem drinkers (Carey & Carey, 1995; Carpenter & Hasin, 1998; Connors, O'Farrell, & Cutter, 1990). On the other hand, the results have been less consistent for other motives (Carpenter & Hasin, 1999). Prospective studies have also been conducted to establish whether coping motives contribute to vulnerability for problem alcohol use. For example, consistent with the risk-factor model, a study by Carpenter and Hasin (1998) found that those who endorsed greater drinking to cope were more likely to be diagnosed with alcohol dependence at a 1-year follow-up, lending support to the idea that avoidance coping preceded the alcohol problem. In contrast, no relationship was found for drinking to increase positive affect.

However, a major limitation of both cross-sectional and prospective studies is that they typically do not account for the influence of negative affect on coping motives (Carpenter & Hasin, 1999). As predicted by the epiphenomena model, it is possible that any differences in coping motives observed between problem and non-problem drinkers are due to differences in negative affect at the time of testing, as opposed to a pre-existing condition that motivates them to drink prior to the development of problem use. Similarly, it is possible that even if a pre-existing condition did exist, the relationship between coping and alcohol use can be overestimated due to the severity of use and the level of negative affect at the time of testing (Carpenter & Hasin, 1999).

The nature of the relationship between drug use and coping has important implications for treatment. A risk-factor relationship suggests that prevention programs should be developed that target individuals who have started to use alcohol, or other drugs, in order to cope with stress prior to the development of addiction (Carpenter & Hasin, 1999). In contrast, if coping motives only serve as a marker of the current level of substance-related problems, they can help identify individuals who should be recommended for an intervention program (Carpenter & Hasin, 1999).

Treatment implications

In general, treatment approaches have focused on changing the nature of the avoided experience (e.g., via stress reduction or pharmacological treatment) to reduce the need for drug use as the preferred method of coping with these experiences (Hayes et al., 1996; Stockwell & Bolderston, 1987). For example, if alcohol is used to self-medicate an anxiety disorder, alternative treatment methods can be used to address the anxiety symptoms. Another strategy has tried to condition aversion to drugs in order to promote avoidance of the drug itself. Yet another, which has recently gained popularity, is to promote psychological acceptance of aversive experiences rather than changing those experiences. Acceptance can be thought of as the opposite of avoidance and might be particularly beneficial. Many therapies (e.g., mindfulness, acceptance and commitment therapy, dialectical behavioral therapy) include at least some components that focus on acceptance, or on replacing avoidance with acceptance (Chiesa & Serretti, 2014; Gratz & Tull, 2010; Hayes et al., 1996). For example, these approaches may promote the acceptance of cravings or urges, which has been used as part of relapse prevention for substance abuse (Hayes et al., 1996). Another approach that includes acceptance training is mindfulness-based cognitive therapy (Teasdale, Segal, & Williams, 1995). Although there is disagreement about the meaning of the term mindfulness, in general, mindfulness-based therapies try to counter avoidance by training individuals to engage with ongoing experiences, including aversive emotions, in a deliberate and non-judgmental manner (Hayes et al., 1996; Leigh, Bowen, & Marlatt, 2005; Witkiewitz, Marlatt, & Walker, 2005).

Experiential avoidance may be a core psychological process that contributes to both substance abuse and other psychological problems, and so research on this construct could ultimately enhance the efficacy of prevention and treatment programs (Levin et al., 2012). For example, as discussed earlier, experiential avoidance is common to both alcohol-dependence, and comorbid mood and anxiety disorders, and has been implicated in relapse (Brady & Sinha, 2005; Chawla & Ostafin, 2007; Forsyth et al., 2003; Kushner, Abrams, & Borchardt, 2000; Lechner et al., 2014; Levin et al., 2012; Merikangas et al., 1998; Pacek et al., 2013; Robinson, Sareen, Cox, & Bolton, 2009; Westrup, 1999). Augmenting standard treatment for alcohol use disorder (AUD) with elements of cognitive behavioral therapy for anxiety and mood disorders (e.g., emotion regulation skills) is beneficial for individuals with AUD and comorbid anxiety disorders, or depression, when a functional link between these conditions is evident (Anker, Kushner, Thuras, Menk, & Unruh, 2016). Furthermore, although not directly studied, there is evidence to suggest that experiential avoidance may bring about slower extinction learning (Graham & Milad, 2011; Pittig, van den Berg, & Vervliet, 2016). This is of particular significance as extinction response patterns were found to predict treatment outcome in specific anxiety disorders (Pittig et al., 2016).

While, in general, avoidance coping has been associated with worse outcomes, its impact also depends on other factors, such as self-efficacy. Self-efficacy can be defined as the confidence one has in their ability to achieve specific goals and has been previously associated with a number of positive substance abuse treatment outcomes, including decisions to seek treatment, reduction of use during treatment, and successful abstinence (Levin, Ilgen, & Moos, 2007). For example, Levin et al. (2007) focused on two cognitive coping strategies—positive reappraisal and cognitive avoidance—in an all-male alcohol abuse patient group. The participants were assessed at treatment entry, discharge and at a 5-year follow-up (Levin et al., 2007). Positive reappraisal is classified as a type of approach coping that involves reinterpreting a stressor as harmless or even valuable (Garland, Gaylord, & Park, 2009). In the context of substance use, it could involve thinking about how craving is uncomfortable but only temporary (Levin et al., 2007). For those with lower self-efficacy at discharge, reliance on avoidance was associated with worse outcomes at the 5-year follow-up. However, as self-efficacy increased, the influence of avoidance decreased. Thus, regardless of reappraisal or avoidance, higher self-efficacy at discharge was predictive of positive 5-year outcomes, including less alcohol use and fewer alcohol dependence symptoms (Levin et al., 2007). A potential limitation of this study, however, was that self-efficacy was measured through a single self-report item (i.e., a question on how confident participants felt that they would be abstinent a year later).

Experiential avoidance is also associated with relapse (Vuchinich & Tucker, 1996; Westrup, 1999), which is a major hurdle for long-term treatment success. That is, avoidant coping, negative affect, and stressful life events have consistently been found to predict relapse in substance abusing populations (Vuchinich & Tucker, 1996). Westrup (1999) investigated the relationship between experiential avoidance and alcohol relapse by assessing whether experiential avoidance discriminated alcohol-dependent "relapsers" from non-relapsers in a sample of 71 inpatients receiving treatment for alcohol dependence. It was found that only 4% of the variance between relapsers and non-relapsers was accounted for by experiential avoidance alone. However, when experiential avoidance was entered in a model including negative life events, both variables were able to significantly distinguish relapsers from non-relapsers (Westrup, 1999). These findings suggest that alcohol-dependent individuals who are both experientially avoidant and experiencing negative life events are more likely to relapse than those who are less avoidant, but also experience negative life events. Nonetheless, the findings of this study are limited in that they do not allow directional interpretations of the relationship; nor were comorbid mental health issues accounted for. Further, in a recent study of alcohol-dependent individuals with comorbid anxiety disorder, it was found that individuals who endorsed drinking as a coping mechanism consumed larger quantities and were more likely to relapse at 3-year follow-up (Menary, Kushner, Maurer, & Thuras, 2011). These findings also lend support to literature implicating drinking to cope as a potential moderator of the relationship between alcohol-dependence and comorbid psychiatric disorders (Anker et al., 2016).

Overgeneral memory as a type of cognitive avoidance

Individuals with a variety of emotional disorders, such as depression and posttraumatic stress disorder, have a tendency to recall more general memories (i.e., lose details) when asked to retrieve episodic memory content (Moore & Zoellner, 2007; Williams et al., 2007). According to the affect regulation hypothesis, the retrieval of more general memories when prompted to recall specific events may represent a type of cognitive avoidance strategy. Specifically, the memory retrieval process might be interrupted to prevent retrieval of details that may elicit negative emotions as a way to avoid this experience, resulting in a more general memory (Gandolphe, Nandrino, Hancart, & Vosgien, 2013a, 2013b).

A similar tendency has also been observed in individuals with addiction. For example, reduced autobiographical memory specificity has been reported in patients with opioid use disorder (Gandolphe et al., 2013b). This was the case regardless of whether or not the patients had comorbid depression. Similarly, it has also been reported in alcohol dependence (D'Argembeau, Van der Linden, Verbanck, & Noël, 2006) and cannabis use disorder (Gandolphe & Nandrino, 2011). As with other types of avoidance, this may have short-term benefits (e.g., stress reduction), but could increase psychological distress in the long-term (Kashdan, Barrios, Forsyth, & Steger, 2006). Interestingly, Gandolphe et al. (2013b) also found that higher cognitive avoidance was associated with reduced episodic memory specificity in both patients and controls, consistent with the idea that reduced specificity might be a vulnerability factor for the development of substance abuse, as well as a variety of other disorders. Future research should determine whether reduced specificity is limited to emotional content, or represents a more general pre-existing tendency that also influences retrieval of non-emotional memories.

Avoidance and reward sensitivity

Both aversive and appetitive learning are important for guiding behavior. The brain's reward system is thought to play a crucial role in the development and maintenance of drug addiction (Robbins & Everitt, 1999; Volkow, Fowler,

Wang, Ding, & Gatley, 2002) with several lines of evidence converging toward the hypothesis that drug addicts have a deficient reward system, and that drug intake is an attempt to compensate for this deficit (Robbins & Everitt, 1999). Many studies of addiction have investigated reward processes (Hyman, 2005), and showed that addicts overvalue rewarding stimuli (e.g., drugs) compared to controls (i.e., addicts show increased reward sensitivity).

Becker and Murphy's (1988) theory claims that people with addiction are more likely to show a high preference for immediate reward but neglect any associated long-term adverse consequences (Becker & Murphy, 1988). For example, Dong, Huang, and Du (2011) suggest that addicts are insensitive to loss of jobs or relationships, and are more likely to lose jobs due to their drug-seeking behavior (Dong et al., 2011). Consistent with this, reduced sensitivity to monetary loss in cocaine-dependent subjects has been reported, which further demonstrates that changes in how rewards and losses are processed is associated with an inability to control drug use in the face of negative consequences, a core feature of addiction (Hester, Bell, Foxe, & Garavan, 2013).

Reward sensitivity may also moderate the relationship between avoidance coping and substance use. For example, individuals who show deficits in emotion regulation, and are more sensitive to reward, might be more likely to use drugs to regulate emotion (e.g., to reduce the impact of negative emotions). This could then put them at greater risk for developing substance use disorders (Aldao et al., 2010). Indeed, higher reward sensitivity is correlated with earlier onset of drinking (Pardo, Aguilar, Molinuevo, & Torrubia, 2007) and alcohol use and abuse in a non-clinical sample (Loxton & Dawe, 2001). There is also evidence that a deficit in emotion regulation can prime reward systems in the brain (Brady & Sinha, 2005), especially in those with higher reward sensitivity to begin with (Aldao et al., 2010). Thus, a combination of poor emotion regulation and increased sensitivity to reward might constitute another risk factor for the development and maintenance of substance use disorders.

Avoidance in the laboratory

Animal research on avoidance

Most laboratory research on avoidance has been in non-human animals. Avoidance paradigms regularly consist of an aversive event (e.g., electric shock), and some salient cue (e.g., a light or tone) that precedes the event and serves as a warning signal. Unlike the human literature on experiential avoidance, a distinction is also made between escape and avoidance responses. Animals, typically rodents, learn to terminate the aversive event while it is already ongoing (which is scored as an escape response), and later learn to avoid it completely. Typically, there is a transition from escape to avoidance over the course of training, such that escape responses decrease while avoidance responses increase. Studies have shown that various drugs of abuse in humans (e.g., opiates), or the withdrawal of such substances, increase avoidance behavior in animals (Mucha, 1991; Shannon, 1983).

For example, Shannon (1983) found an increase in avoidance behavior when rats were given opiates (morphine or buprenorphine) in a continuous lever-press shock avoidance paradigm. In this paradigm, the rats learn to press a lever in order to delay the delivery of electric shock. The avoidance behavior was also found to be dose-dependent—when dose was increased or reached a certain level, the lever-press behavior decreased or plateaued. Regardless of which drug was used, at the higher doses, rats tended to respond more frequently in bursts, and pause until onset of the next shock. Thus, there was an increase in both the number of shock presentations, as well as in escape responses. This could have been due to the opiates disrupting the timing of responses, relieving pain, or due to a greater preference for the rewarding effects of the drugs (as a result of increased dose) compared to the reward associated with avoiding the aversive consequences.

Other types of paradigms have also been used to assess avoidance in animals. For example, Mucha (1991) employed a test for burying of an object paired with an aversive event. Rats used bedding and other material (e.g., wood shavings/newspaper) to bury an electrode paired with aversive shock. They would also bury food paired with a noxious injection. Similarly, Mucha (1991) found that rats displayed persistent avoidance behavior (i.e., burying) if the test object was paired with precipitated withdrawal. It is important to note that in order to burry an object, rats must first approach it, increasing their exposure to it. If the object itself served as a cue that elicited withdrawal then approaching it should increase, rather than reduce, withdrawal symptoms. Thus, negative reinforcement alone appears unable to account for the approach component of this behavior. Instead, Mucha (1991) proposed avoidance is encoded as a goal that is elicited by the cue object itself. The exact avoidance response selected will then depend on the nature of the cue, as well as the situation. For example, if there are no wood shavings or other material available to bury the object with, then another possible response (such as distancing from the object) would be selected. This idea can also help explain the low correlation between withdrawal symptoms and motivation in patients, as well as in animals (Drummond, Cooper, & Glautier, 1990; Edwards, 1990).

Mowrer's two-factor theory has provided one of the most influential explanations of avoidance (Maia, 2010). According to this theory, animals first learn to escape from an aversive event (e.g., shock) that elicits fear. The escape behavior reduces fear, thus it is negatively reinforced. As described earlier, in the typical avoidance learning paradigm used in the laboratory, some initially neutral but salient cue (e.g., tone or light) immediately precedes the aversive event, and becomes associated with it. Thus, due to classical conditioning, this cue also begins to elicit fear. Animals then learn to escape from this cue through operant conditioning (e.g., by learning to press a lever that delays shock), therefore avoiding the aversive event completely. Although animals behave as though they have learned to anticipate or predict the aversive event, according to the theory, they are still simply escaping from a fear-evoking cue.

While this theory continues to be influential, it does suffer from several issues that revolve around its emphasis on fear as central to avoidance (Lovibond, 2006). For example, once animals learn to avoid reliably, they show little signs of fear (Mineka & Gino, 1980). This suggests that, at least later on, avoidance is no longer dependent on fear and may have become automatic. In addition, avoidance learning can occur in the absence of any explicit cue (i.e., warning signal) preceding the aversive event (Sidman, 1962). At first glance, the two-factor theory does not appear to account for such learning, because the avoidance response is no longer reinforced by the removal of what is conceived to be a fear-evoking cue. However, it could always be argued that the cues in this case are internal to the animal (e.g., involving changes in physiological state over time). Due to these limitations, more recent accounts of avoidance have shifted toward a cognitive view based on expectation or prediction.

As mentioned earlier, studies have shown that drugs of abuse in humans can also modulate avoidance learning in animals. For example, one study employed a paradigm in which one cue was paired with reward while another with punishment (sucrose and electric shock, respectively) (Nguyen, Schumacher, Erb, & Ito, 2015). Subsequently, both cues were presented simultaneously. The authors found cocaine pre-exposure weakened learning of the aversive association. That is, it enhanced preference for locations paired with both cues, regardless of the type of paradigm used (maze or runway). Cocaine pre-exposure may have enhanced approach behavior, reduced avoidance behavior, or some combination of both. In the maze paradigm, one arm was paired with shock, another with sucrose, and another was not paired with either of these outcomes (the neutral arm). The pre-exposed group spent similar amounts of time exploring the shock and neutral arms, but spent significantly longer exploring the sucrose arm. Thus, the pre-exposed rats showed no evidence of learning to avoid the shock-paired arm indicating the drug impaired learning the aversive association (Nguyen et al., 2015). These results suggest that persistent drug-related behavior in humans, despite the associated negative consequences, may at least in part be due to impaired aversive learning as a result of drug exposure. This is also consistent with evidence that sensitivity to rewards or losses is altered in addiction.

Evidence from both animal and human studies has indicated that long-term changes occur in the mesolimbic dopamine system after chronic drug use, which plays a role in both appetitive and aversive learning (Koob & Volkow, 2010; Wheeler & Carelli, 2009). Several brain differences have been described between high- and low-avoidant strains of rats, including differences in this dopamine system. For example, Roman rats have been selectively bred to be either high or low on avoidance learning in a two-way shuttle box (Fattore, Piras, Corda, & Giorgi, 2009). In this paradigm, animals learn to move back and forth two sides of a compartment in order to avoid electric shock administered on either side, typically following a cue signaling shock. Fattore et al. (2009) examined the acquisition, extinction, reinstatement and reacquisition of cocaine self-administration in these strains. The highly avoidant rats displayed more self-administration (i.e., more lever-press responses that result in drug infusion), slower extinction of these responses when they no longer led to drug, and more robust reinstatement of the lever-press behavior. That is, the rats display more lever-press responses following a priming dose of cocaine, even though these responses do not result in drug infusion (i.e., take place under extinction conditions). In addition, only the highly avoidant rats were able to reacquire self-administration at lower doses of cocaine. The highly avoidant rats have also been shown to have greater preference for ethanol (compared to water) and similarly for saccharin over water in a two-bottle test. This suggests the results extend beyond just cocaine, and apply for other drugs, as well as for natural reinforcers (Fattore et al., 2009).

Computer-based tasks

As discussed earlier, much of human avoidance behavior has been assessed by subjective self-report questionnaires (Taylor & Sullman, 2009). These questionnaires are therefore based on the participant's judgment of their behavior rather than measuring the behavior itself. Such approaches are subject to all the limitations of self-report, including people falsifying or withholding information or that people may not be able to accurately or objectively assess their own behavior. Questionnaires that probe past behavior patterns also, of course, are unable to directly assess how these behaviors are acquired.

With the aim of creating situations more akin to the conditioning preparations that have guided the study of nonhuman animal learning, researchers have tried to develop tasks where learning is inferred from participants' behavior, rather than from their judgments (Sheynin, Beck, Servatius, & Myers, 2014). One way of ethically achieving this is by the use of computer-based tasks. Various tasks have been successfully used to test passive avoidance (Arcediano, 1996), active avoidance (Molet, Leconte, & Rosas, 2006), differential effects of reinforcement contingencies and contextual variables (Raia, Shillingford, Miller, & Baier, 2000), discriminative learning and contextual change effects on learning (Byron Nelson & Del Carmen Sanjuan, 2006), conditioned place preference for drugs in healthy individuals (Childs & de Wit, 2009, 2013; Mayo et al., 2013), as well as for non-drug rewards in opioid addicts relative to healthy controls (Radell et al., 2018). These computer tasks are purported to be comparable to the traditional conditioning paradigms used in animal research and are generally recognized as valid measures of avoidance (Molet et al., 2006; Sheynin et al., 2014).

For example, Sheynin et al. (2016) were one of the first in the field of addiction to generalize animal models to humans by examining abnormal acquisition and extinction of avoidance behavior in opioid-dependent human subjects. The study comprised a sample of 26 healthy controls and 27 individuals meeting diagnostic criteria for heroin-dependence and on medication. Participants were required to complete a computer-based task designed to measure avoidance, which was done by calculating hiding duration during different periods of the task (Sheynin et al., 2016). The authors found that although there was no difference between the experimental and control groups in escape responding (i.e., hiding) during an aversive event, heroin-dependent men were observed to make more avoidant responses during the extinction trial which predicted the aversive event. Heroin-dependent men were also slower to extinguish the avoidance response when the aversive event was no longer preceded by the warning signal (Sheynin et al., 2016).

Computer-based tasks have also been used to examine how individuals with substance use disorders learn from reward compared to punishment, and how this learning transfers to new situations. For example, Myers et al. (2016) compared learning and generalization in individuals with opioid use disorder, recruited from a methadone maintenance treatment program, to healthy controls. The subjects completed a probabilistic classification task in which they had to guess the category membership of different stimuli. For some stimuli, correct guesses were rewarded, while for others, incorrect guesses were punished. In this study, there was no difference between how controls and patients learned (Myers et al., 2016). However, a reinforcement learning model was used to understand the trial-by-trial behavior of participants. The model suggested that despite the lack of an overall group difference, the patients were more likely to change their response strategy based on recent feedback, with significantly greater lose-shift behavior compared to controls. This is not an ideal strategy for probabilistic tasks since they require a sense of the overall probability of a particular outcome (Myers et al., 2016). Thus, the patients may have placed too much stock into the result of a single trial.

In another computer-based task, Mahlberg et al. (2017) presented compound stimuli (combinations of shapes and colors) and participants had to learn to pick stimuli paired with reward, and skip (i.e., avoid) those that led to punishment. Unlike Myers et al. (2016), the outcomes in this study were deterministic, and the patients showed acquisition deficits for both the reward and punishment trials compared to control participants (Mahlberg et al., 2017). In addition, in a subsequent test which included new combinations of shapes and colors, controls appeared to use prior knowledge of both shape and color to respond, but patients appeared to rely more on shape. Thus, patients also showed a generalization deficit compared to controls (Mahlberg et al., 2017). Deficits in generalization have also been reported in individuals with alcohol use disorder (Rustemeier et al., 2012).

Deficits in generalization may help reduce avoidance of negative outcomes. This is consistent with one of the main features of addiction—the persistence of drug use despite its detrimental consequences. In another study by Myers et al. (2017), this time using an acquired equivalence task, patients with opioid use disorder and non-addicted controls were found to perform similarly on reward trials. However, the control group was better at learning to avoid punishment than the patient group (Myers et al., 2017). Thus, on the one hand, drug use might represent a type of avoidance used to cope with negative emotional states. However, reduced sensitivity to punishment, coupled with reduced avoidance of such outcomes, may lead to further drug use to cope with their emotional consequences, forming a vicious cycle.

Alternatively, as Myers et al. (2017) point out, successful performance on the punishment trials of their task also depends on learning that the absence of feedback indicates successful avoidance. That is, only incorrect responses on these trials were punished. No feedback was given for correct responses, making this outcome ambiguous. Thus, patients may not have necessarily been deficient at learning from punishment, but may have instead differed from controls in how they process ambiguous outcomes. Indeed, other research has shown that patients with opioid use disorder have greater intolerance of uncertainty (Garami et al., 2017), which may alter how ambiguous (i.e., more uncertain) feedback is processed. Intolerance of uncertainty has also been related to decision making in a computer-based conditioned place preference task in another study comparing patients with opioid use disorder to non-addicted controls (Radell et al., 2018).

Limitations and future directions

The computer-based tasks discussed above are also subject to several limitations that may have influenced the results. For example, although similar to tasks used in previous studies, Sheynin et al. (2016) only provided brief instructions to participants, which stated that the aim of the game was to maximize points. When this and the lack of opportunity for reward during the warning phase (discussed above) are accounted for, participants may have been compelled to make more hiding responses during the avoidance acquisition and extinction trials to minimize point deduction and maximize points gained. Participants were also intentionally not informed about the availability of safe areas, so it is often not clear whether an absence of responding is due to the lack of knowledge about the availability of that response or whether it is the result of a voluntary choice not to execute that response. In future work, a post-hoc questionnaire might be used to ask about the information learned during the task and the purpose or reason for a participant's behavior.

Another explanation for the results may be that the 'aversive' event of the task did not necessarily elicit an emotional response. Research has identified stressful life events and negative affect as precipitants of avoidant behavior (Anker et al., 2016; Keyes, Hatzenbuehler, & Hasin, 2011; Menary et al., 2011; Pacek et al., 2013; Robinson et al., 2009; Vieten, Astin, Buscemi, & Galloway, 2010). Yet, no study to date has measured whether the aversive on-screen event elicits an emotional response. This is largely inferred in this task and similar tasks. Of note, many participants did not realize that they were experiencing a loss of points when hit by an enemy in the task (the aversive event)—often asking what the "- 5" on the screen signified. Therefore, at least in the first few acquisition trials, some participants did not discern the presence of the enemy as an aversive event. Therefore, their responses may have reflected a strategic response to maximizing points rather than experiencing the event as "aversive" at the emotional level and responding in an avoidant manner.

In addition, a question about experience with video games might also be meaningful since such experience might bias performance. The average age and years of education for the patients with opiate addiction was 40.79 and 10.82, respectively, while those for the control group were 39.03 and 12.91. Thus, while in this study both groups were of similar age and education, older participants may have less experience with video games, which could affect the outcome of results. This needs to be considered in future studies.

Future studies would also benefit from several possible manipulations. First, the effect of different response-outcome contingencies could be tested. For instance, in the task, execution of avoidance behavior protects the participant from the aversive event but has no effect on the warning signal. However, it would be possible to have the avoidance behavior also result in the termination of this signal, as in many animal studies. This appears to enhance avoidance learning in rats (Bolles & Grossen, 1969), which also prefer a contingency where responses result in both avoidance of the aversive event and immediate termination of the warning signal (Culbertson & Badia, 1973). One interpretation of these findings is that the termination of the warning provided additional feedback (indicating that the aversive event had been successfully averted) facilitating the acquisition of avoidance behavior. Thus, it would be interesting to see if similar facilitation would occur in participants with opiate addiction.

A potential confound in the studies discussed above (Mahlberg et al., 2017; Myers et al., 2017, 2016; Radell et al., 2018; Sheynin et al., 2016) was that patients were tested immediately after receiving their daily methadone maintenance dose. Thus, it remains possible that the differences in learning and generalization reported across these studies resulted from the acute effects of the drug (Mahlberg et al., 2017). Previous studies have indicated that cognitive performance, particularly working memory, verbal memory, and reaction speed, are altered by the use of methadone and buprenorphine, with various differences observed between these two substances (Rapeli et al., 2007). The opiate group had consumed these substances at various time delays between dosing and completing the task. Thus, future research may consider whether performing avoidance learning tasks prior to dosing, after dosing, and at various times before and after would change the results.

Summary and conclusions

Overall, there is evidence that initial drug use may be motivated by coping, and that avoidant coping in particular may increase the risk for problem use. The tendency to avoid aversive cognitive and emotional events (i.e., experiential avoidance) may also contribute to the maintenance of drug use, as well as the propensity to relapse. As such, targeting avoidance, for example, through interventions that promote opposing tendencies such as acceptance, can provide a fruitful avenue for augmenting long-term treatment success. While greater avoidance appears to increase the risk for addiction, there is evidence that drug use can also impair subsequent aversive learning, perhaps by reducing the sensitivity to the negative consequences of substance use. Furthermore, chronic use may also reduce the sensitivity to non-drug reward, further placing the drug in a position to not only serve as a means of regulating negative cognitions and emotions, but also as a powerful reinforcer on its own.

Nonetheless, there are also several limitations of this work that future research should address. There is a wealth of research on alcohol use and its relationship to avoidance but there is a lack of similar research on other types of drugs. Furthermore, most of this research is based on self-report from addicts or from non-clinical samples. Computer-based tasks can provide a powerful complement to self-report measures, as a more objective means to assess avoidance learning and behavior in humans. These tasks can be designed to mimic those that have been used in non-human animal studies for decades, and provide a way to bridge the two literatures. Furthermore, these tasks can be deployed while imaging the human brain or collecting other physiological measures, in order to gain further insight into the mechanisms of avoidance, as well as how these mechanisms may be altered in addiction. It has become evident that increased avoidance contributes to a diverse set of mental disorders, thus studies in this area have broader relevance beyond substance use disorders.

References

Aldao, A., Nolen-Hoeksema, S., & Schweizer, S. (2010). Emotion-regulation strategies across psychopathology: A meta-analytic review. *Clinical Psychology Review*, *30*(2), 217–237.

American Psychiatric Association. (2013). *Diagnostic and statistical manual of mental disorders* (5th ed.). Washington, DC: American Psychiatric Association.

Anker, J. J., Kushner, M. G., Thuras, P., Menk, J., & Unruh, A. S. (2016). Drinking to cope with negative emotions moderates alcohol use disorder treatment response in patients with co-occurring anxiety disorder. *Drug and Alcohol Dependence*, *159*, 93–100.

Arcediano, F. (1996). A behavioural preparation for the study of human Pavlovian conditioning. *The Quarterly Journal of Experimental Psychology: Section B*, *49*(3), 270–283.

Baker, T. B., Piper, M. E., McCarthy, D. E., Majeskie, M. R., & Fiore, M. C. (2004). Addiction motivation reformulated: An affective processing model of negative reinforcement. *Psychological Review*, *111*(1), 33–51.

Becker, G. S., & Murphy, K. M. (1988). A theory of rational addiction. *Journal of Political Economy*, *96*(4), 675–700.

Bolles, R. C., & Grossen, N. E. (1969). Effects of an informational stimulus on the acquisition of avoidance behavior in rats. *Journal of Comparative and Physiological Psychology*, *68*(1), 90–99.

Brady, K. T., & Sinha, R. (2005). Co-occurring mental and substance use disorders: The neurobiological effects of chronic stress. *American Journal of Psychiatry*, *162*(8), 1483–1493.

Buckner, J. D., Zvolensky, M. J., Farris, S. G., & Hogan, J. (2014). Social anxiety and coping motives for cannabis use: The impact of experiential avoidance. *Psychology of Addictive Behaviors*, *28*(2), 568–574.

Byron Nelson, J., & Del Carmen Sanjuan, M. (2006). A context-specific latent inhibition effect in a human conditioned suppression task. *The Quarterly Journal of Experimental Psychology*, *59*(6), 1003–1020.

Carey, K. B., & Carey, M. P. (1995). Reasons for drinking among psychiatric outpatients: Relationship to drinking patterns. *Psychology of Addictive Behaviors*, *9*(4), 251–257.

Carpenter, K. M., & Hasin, D. S. (1998). Reasons for drinking alcohol: Relationships with DSM-IV alcohol diagnoses and alcohol consumption in a community sample. *Psychology of Addictive Behaviors*, *12*(3), 168–184.

Carpenter, K. M., & Hasin, D. S. (1999). Drinking to cope with negative affect and DSM-IV alcohol use disorders: A test of three alternative explanations. *Journal of Studies on Alcohol*, *60*(5), 694–704.

Chawla, N., & Ostafin, B. (2007). Experiential avoidance as a functional dimensional approach to psychopathology: An empirical review. *Journal of Clinical Psychology*, *63*(9), 871–890.

Chiesa, A., & Serretti, A. (2014). Are mindfulness-based interventions effective for substance use disorders? A systematic review of the evidence. *Substance Use & Misuse*, *49*(5), 492–512.

Childress, A. R., McLellan, A. T., Natale, M., & O'Brien, C. P. (1987). Mood states can elicit conditioned withdrawal and craving in opiate abuse patients. *NIDA Research Monograph*, *76*, 137–144.

Childs, E., & de Wit, H. (2009). Amphetamine-induced place preference in humans. *Biological Psychiatry*, *65*(10), 900–904.

Childs, E., & de Wit, H. (2013). Contextual conditioning enhances the psychostimulant and incentive properties of D-amphetamine in humans. *Addiction Biology*, *18*(6), 985–992.

Connors, G. J., O'Farrell, T. J., & Cutter, H. S. (1990). Using a drinking motivation scale to predict degrees of problematic drinking. *Drug & Alcohol Dependence*, *26*(2), 175–181.

Cooper, M. L., Wood, P. K., Orcutt, H. K., & Albino, A. (2003). Personality and the predisposition to engage in risky or problem behaviors during adolescence. *Journal of Personality and Social Psychology*, *84*(2), 390–410.

Crum, R. M., Mojtabai, R., Lazareck, S., Bolton, J. M., Robinson, J., Sareen, J., … Alvanzo, A. A. (2013). A prospective assessment of reports of drinking to self-medicate mood symptoms with the incidence and persistence of alcohol dependence. *JAMA Psychiatry*, *70*(7), 718–726.

Culbertson, S., & Badia, P. (1973). Choice of a terminating over a non-terminating signal in free-operant avoidance. *Journal of the Experimental Analysis of Behavior*, *20*(2), 235–243.

Cunningham, J. A., Sobell, M. B., Sobell, L. C., Gavin, D. R., & Annis, H. R. (1995). Heavy drinking and negative affective situations in a general population and a treatment sample: Alternative explanations. *Psychology of Addictive Behaviors*, *9*(2), 123–127.

D'Argembeau, A., Van der Linden, M., Verbanck, P., & Noël, X. (2006). Autobiographical memory in non-amnesic alcohol-dependent patients. *Psychological Medicine*, *36*(12), 1707–1715.

DeMartini, K. S., & Carey, K. B. (2011). The role of anxiety sensitivity and drinking motives in predicting alcohol use: A critical review. *Clinical Psychology Review*, *31*(1), 169–177.

Dong, G., Huang, J., & Du, X. (2011). Enhanced reward sensitivity and decreased loss sensitivity in Internet addicts: An fMRI study during a guessing task. *Journal of Psychiatric Research*, *45*(11), 1525–1529.

Drummond, D. C., Cooper, T., & Glautier, S. P. (1990). Conditioned learning in alcohol dependence: Implications for cue exposure treatment. *British Journal of Addiction*, *85*(6), 725–743.

Edwards, G. (1990). Withdrawal symptoms and alcohol dependence: Fruitful mysteries. *British Journal of Addiction, 85*(4), 447–461.

Fattore, L., Piras, G., Corda, M. G., & Giorgi, O. (2009). The Roman high- and low-avoidance rat lines differ in the acquisition, maintenance, extinction, and reinstatement of intravenous cocaine self-administration. *Neuropsychopharmacology, 34*(5), 1091–1101.

Fondacaro, M. R., & Moos, R. H. (1989). Life stressors and coping: A longitudinal analysis among depressed and nondepressed adults. *Journal of Community Psychology, 17*(4), 330–340.

Forsyth, J. P., Parker, J. D., & Finlay, C. G. (2003). Anxiety sensitivity, controllability, and experiential avoidance and their relation to drug of choice and addiction severity in a residential sample of substance-abusing veterans. *Addictive Behaviors, 28*(5), 851–870.

Gámez, W., Chmielewski, M., Kotov, R., Ruggero, C., & Watson, D. (2011). Development of a measure of experiential avoidance: The multidimensional experiential avoidance questionnaire. *Psychological Assessment, 23*(3), 692–713.

Gandolphe, M.-C., & Nandrino, J.-L. (2011). Stratégies de surgénéralisation des souvenirs autobiographiques chez les consommateurs de cannabis et les polyconsommateurs de substances psychoactives. *L'Encéphale, 37*(2), 144–152.

Gandolphe, M.-C., Nandrino, J.-L., Hancart, S., & Vosgien, V. (2013a). Autobiographical memory and differentiation of schematic models in substance-dependent patients. *Journal of Behavior Therapy and Experimental Psychiatry, 44*(1), 114–121.

Gandolphe, M.-C., Nandrino, J.-L., Hancart, S., & Vosgien, V. (2013b). Reduced autobiographical memory specificity as an emotional avoidance strategy in opioid-dependent patients. *Canadian Journal of Behavioural Science/Revue Canadienne Des Sciences Du Comportement, 45*(4), 305–312.

Garami, J., Haber, P., Myers, C. E., Allen, M. T., Misiak, B., Frydecka, D., & Moustafa, A. A. (2017). Intolerance of uncertainty in opioid dependency—Relationship with trait anxiety and impulsivity. *PLoS ONE, 12*(7), 1–12.

Garland, E., Gaylord, S., & Park, J. (2009). The role of mindfulness in positive reappraisal. *Explore: The Journal of Science and Healing, 5*(1), 37–44.

Gauvin, D. V., Harland, R. D., Criado, J. R., Michaelis, R. C., & Holloway, F. A. (1989). The discriminative stimulus properties of ethanol and acute ethanol withdrawal states in rats. *Drug & Alcohol Dependence, 24*(2), 103–113.

Graham, B. M., & Milad, M. R. (2011). The study of fear extinction: Implications for anxiety disorders. *The American Journal of Psychiatry; Washington, 168*(12), 1255–1265.

Gratz, K. L., & Tull, M. T. (2010). Emotion regulation as a mechanism of change in acceptance-and mindfulness-based treatments. In R. A. Baer (Ed.), *Assessing mindfulness and acceptance: Illuminating the processes of change* (pp. 107–133). Oakland, CA: New Harbinger Publication.

Hayes, S. C., Wilson, K. G., Gifford, E. V., Follette, V. M., & Strosahl, K. (1996). Experiential avoidance and behavioral disorders: A functional dimensional approach to diagnosis and treatment. *Journal of Consulting and Clinical Psychology, 64*(6), 1152–1168.

Hester, R., Bell, R. P., Foxe, J. J., & Garavan, H. (2013). The influence of monetary punishment on cognitive control in abstinent cocaine-users. *Drug and Alcohol Dependence, 133*(1), 86–93.

Hutcheson, D. M., Everitt, B. J., Robbins, T. W., & Dickinson, A. (2001). The role of withdrawal in heroin addiction: Enhances reward or promotes avoidance? *Nature Neuroscience, 4*(9), 943–947.

Hyman, S. E. (2005). Addiction: A disease of learning and memory. *American Journal of Psychiatry, 162*(8), 1414–1422.

Jaffe, J. H. (1965). Drug addiction and drug abuse. In L. S. Goodman & A. Gilman (Eds.), *The pharmacological basis of therapeutics* (pp. 284–324). New York: Macmillan.

Jaffe, J. H. (1992). Current concepts of addiction. In C. P. O'Brien & J. Jaffe (Eds.), *Addictive states* (pp. 1–21). New York: Raven Press.

Kashdan, T. B., Barrios, V., Forsyth, J. P., & Steger, M. F. (2006). Experiential avoidance as a generalized psychological vulnerability: Comparisons with coping and emotion regulation strategies. *Behaviour Research and Therapy*, *44*(9), 1301–1320.

Keyes, K. M., Hatzenbuehler, M. L., & Hasin, D. S. (2011). Stressful life experiences, alcohol consumption, and alcohol use disorders: The epidemiologic evidence for four main types of stressors. *Psychopharmacology*, *218*(1), 1–17.

Khantzian, E. J. (1997). The self-medication hypothesis of substance use disorders: A reconsideration and recent applications. *Harvard Review of Psychiatry*, *4*(5), 231–244.

Koob, G. F., & Volkow, N. D. (2010). Neurocircuitry of addiction. *Neuropsychopharmacology*, *35*(1), 217–238.

Kushner, M. G., Abrams, K., & Borchardt, C. (2000). The relationship between anxiety disorders and alcohol use disorders: A review of major perspectives and findings. *Clinical Psychology Review*, *20*(2), 149–171.

Lechner, W. V., Shadur, J. M., Banducci, A. N., Grant, D. M., Morse, M., & Lejuez, C. W. (2014). The mediating role of depression in the relationship between anxiety sensitivity and alcohol dependence. *Addictive Behaviors*, *39*(8), 1243–1248.

Leigh, J., Bowen, S., & Marlatt, G. A. (2005). Spirituality, mindfulness and substance abuse. *Addictive Behaviors*, *30*(7), 1335–1341.

Levin, C., Ilgen, M., & Moos, R. (2007). Avoidance coping strategies moderate the relationship between self-efficacy and 5-year alcohol treatment outcomes. *Psychology of Addictive Behaviors*, *21*(1), 108–113.

Levin, M. E., Lillis, J., Seeley, J., Hayes, S. C., Pistorello, J., & Biglan, A. (2012). Exploring the relationship between experiential avoidance, alcohol use disorders, and alcohol-related problems among first-year college students. *Journal of American College Health*, *60*(6), 443–448.

Levy, M. (2018). The many faces (and potential dangers) of self-medication as an explanatory concept for substance use. *International Journal for the Advancement of Counselling*, 1–10.

Lovibond, P. (2006). *Fear and avoidance: An integrated expectancy model.* In *Fear and learning: From basic processes to clinical implications*. Washington, DC: American Psychological Association.(pp. 117–132).

Loxton, N. J., & Dawe, S. (2001). Alcohol abuse and dysfunctional eating in adolescent girls: The influence of individual differences in sensitivity to reward and punishment. *International Journal of Eating Disorders*, *29*(4), 455–462.

Mahlberg, J., Haber, P., Morley, K., Weidemann, G., Hogarth, L., Beck, K. D., … Moustafa, A. A. (2017). Reward and punishment-based compound cue learning and generalization in opiate dependency. *Experimental Brain Research*, *235*(10), 3153–3162.

Maia, T. V. (2010). Two-factor theory, the actor-critic model, and conditioned avoidance. *Learning & Behavior*, *38*(1), 50–67.

Mayo, L. M., Fraser, D., Childs, E., Momenan, R., Hommer, D. W., de Wit, H., & Heilig, M. (2013). Conditioned preference to a methamphetamine-associated contextual cue in humans. *Neuropsychopharmacology*, *38*(6), 921–929.

McCrae, R. R. (1984). Situational determinants of coping responses: Loss, threat, and challenge. *Journal of Personality and Social Psychology*, *46*(4), 919–928.

Menary, K. R., Kushner, M. G., Maurer, E., & Thuras, P. (2011). The prevalence and clinical implications of self-medication among individuals with anxiety disorders. *Journal of Anxiety Disorders*, *25*(3), 335–339.

Merikangas, K. R., Mehta, R. L., Molnar, B. E., Walters, E. E., Swendsen, J. D., Aguilar-Gaziola, S., … Dewit, D. J. (1998). Comorbidity of substance use disorders with mood and anxiety disorders: Results of the International Consortium in Psychiatric Epidemiology. *Addictive Behaviors*, *23*(6), 893–907.

Mineka, S., & Gino, A. (1980). Dissociation between conditioned emotional response and extended avoidance performance. *Learning and Motivation*, *11*(4), 476–502.

Molet, M., Leconte, C., & Rosas, J. M. (2006). Acquisition, extinction and temporal discrimination in human conditioned avoidance. *Behavioural Processes*, *73*(2), 199–208.

Moore, S. A., & Zoellner, L. A. (2007). Overgeneral autobiographical memory and traumatic events: An evaluative review. *Psychological Bulletin*, *133*(3), 419–437.

Moos, R. H., Brennan, P. L., Fondacaro, M. R., & Moos, B. S. (1990). Approach and avoidance coping responses among older problem and nonproblem drinkers. *Psychology and Aging*, *5*(1), 31–40.

Mucha, R. F. (1991). What is learned during opiate withdrawal conditioning? Evidence for a cue avoidance model. *Psychopharmacology*, *104*(3), 391–396.

Myers, C. E., Rego, J., Haber, P., Morley, K., Beck, K. D., Hogarth, L., & Moustafa, A. A. (2017). Learning and generalization from reward and punishment in opioid addiction. *Behavioural Brain Research*, *317*, 122–131.

Myers, C. E., Sheynin, J., Balsdon, T., Luzardo, A., Beck, K. D., Hogarth, L., … Moustafa, A. A. (2016). Probabilistic reward- and punishment-based learning in opioid addiction: Experimental and computational data. *Behavioural Brain Research*, *296*, 240–248.

Nguyen, D., Schumacher, A., Erb, S., & Ito, R. (2015). Aberrant approach-avoidance conflict resolution following repeated cocaine pre-exposure. *Psychopharmacology*, *232*(19), 3573–3583.

Pacek, L. R., Storr, C. L., Mojtabai, R., Green, K. M., La Flair, L. N., Alvanzo, A. A., … Crum, R. M. (2013). Comorbid alcohol dependence and anxiety disorders: A national survey. *Journal of Dual Diagnosis*, *9*(4), 271–280.

Pardo, Y., Aguilar, R., Molinuevo, B., & Torrubia, R. (2007). Alcohol use as a behavioural sign of disinhibition: Evidence from JA Gray's model of personality. *Addictive Behaviors*, *32*(10), 2398–2403.

Pittig, A., van den Berg, L., & Vervliet, B. (2016). The key role of extinction learning in anxiety disorders: Behavioral strategies to enhance exposure-based treatments. *Current Opinion in Psychiatry*, *29*(1), 39–47.

Radell, M. L., Allen, M. T., Favaloro, B., Myers, C. E., Haber, P., Morley, K., & Moustafa, A. A. (2018). Intolerance of uncertainty and conditioned place preference in opioid addiction. *PeerJ*, *6*, 1–19.

Raia, C. P., Shillingford, S. W., Miller, H. L., Jr., & Baier, P. S. (2000). Interaction of procedural factors in human performance on yoked schedules. *Journal of the Experimental Analysis of Behavior*, *74*(3), 265–281.

Rapeli, P., Fabritius, C., Alho, H., Salaspuro, M., Wahlbeck, K., & Kalska, H. (2007). Methadone vs. buprenorphine/naloxone during early opioid substitution treatment: A naturalistic comparison of cognitive performance relative to healthy controls. *BMC Clinical Pharmacology*, *7*(1), 1–10.

Robbins, T. W., & Everitt, B. J. (1999). Drug addiction: Bad habits add up. *Nature (London)*, *398*(6728), 567–570.

Robinson, J., Sareen, J., Cox, B. J., & Bolton, J. (2009). Self-medication of anxiety disorders with alcohol and drugs: Results from a nationally representative sample. *Journal of Anxiety Disorders*, *23*(1), 38–45.

Rustemeier, M., Römling, J., Czybulka, C., Reymann, G., Daum, I., & Bellebaum, C. (2012). Learning from positive and negative monetary feedback in patients with alcohol dependence. *Alcoholism: Clinical and Experimental Research*, *36*(6), 1067–1074.

Shannon, H. E. (1983). Stimulation of avoidance behavior by buprenorphine in rats. *Psychopharmacology*, *80*(1), 19–23.

Sheynin, J., Beck, K. D., Servatius, R. J., & Myers, C. E. (2014). Acquisition and extinction of human avoidance behavior: Attenuating effect of safety signals and associations with anxiety vulnerabilities. *Frontiers in Behavioral Neuroscience*, *8*, 1–11.

Sheynin, J., Moustafa, A. A., Beck, K. D., Servatius, R. J., Casbolt, P. A., Haber, P., … Myers, C. E. (2016). Exaggerated acquisition and resistance to extinction of avoidance behavior in treated heroin-dependent males. *Journal of Clinical Psychiatry*, *77*(3), 386–394.

Sidman, M. (1962). Classical avoidance without a warning stimulus. *Journal of the Experimental Analysis of Behavior*, *5*(1), 97–104.

Solomon, R. L., & Corbit, J. D. (1974). An opponent-process theory of motivation: I. Temporal dynamics of affect. *Psychological Review*, *81*(2), 119–145.

Stewart, S. H., Zvolensky, M. J., & Eifert, G. H. (2001). Negative-reinforcement drinking motives mediate the relation between anxiety sensitivity and increased drinking behavior. *Personality and Individual Differences*, *31*(2), 157–171.

Stockwell, T., & Bolderston, H. (1987). Alcohol and phobias. *British Journal of Addiction*, *82*(9), 971–979.

Taylor, J. E., & Sullman, M. J. (2009). What does the driving and riding avoidance scale (DRAS) measure? *Journal of Anxiety Disorders*, *23*(4), 504–510.

Teasdale, J. D., Segal, Z., & Williams, J. M. G. (1995). How does cognitive therapy prevent depressive relapse and why should attentional control (mindfulness) training help? *Behaviour Research and Therapy*, *33*(1), 25–39.

Vieten, C., Astin, J. A., Buscemi, R., & Galloway, G. P. (2010). Development of an acceptance-based coping intervention for alcohol dependence relapse prevention. *Substance Abuse*, *31*(2), 108–116.

Volkow, N. D., Fowler, J. S., Wang, G.-J., Ding, Y.-S., & Gatley, S. J. (2002). Role of dopamine in the therapeutic and reinforcing effects of methylphenidate in humans: Results from imaging studies. *European Neuropsychopharmacology*, *12*(6), 557–566.

Vuchinich, R. E., & Tucker, J. A. (1996). Alcoholic relapse, life events, and behavioral theories of choice: A prospective analysis. *Experimental and Clinical Psychopharmacology*, *4*(1), 19–28.

Westrup, D. (1999). Experiential avoidance and alcohol dependence relapse (Doctoral dissertation). *Dissertation Abstracts International*, *62*(568), 871–890.

Wheeler, R. A., & Carelli, R. M. (2009). Dissecting motivational circuitry to understand substance abuse. *Neuropharmacology*, *56*, 149–159.

Williams, J. M. G., Barnhofer, T., Crane, C., Herman, D., Raes, F., Watkins, E., & Dalgleish, T. (2007). Autobiographical memory specificity and emotional disorder. *Psychological Bulletin*, *133*(1), 122–148.

Witkiewitz, K., Marlatt, G. A., & Walker, D. (2005). Mindfulness-based relapse prevention for alcohol and substance use disorders. *Journal of Cognitive Psychotherapy*, *19*(3), 211–228.

Zinbarg, R. E., & Mineka, S. (2007). Is emotion regulation a useful construct that adds to the explanatory power of learning models of anxiety disorders or a new label for old constructs? *American Psychologist*, *62*(3), 259–261.

Chapter 7

Theories of compulsive drug use: A brief overview of learning and motivation processes

Lauren M. Foreman[a], Irina Baetu[a], Janice Rego[b], Lyndsey E. Collins-Praino[a], Ahmed A. Moustafa[c]
[a]*University of Adelaide, Adelaide, SA, Australia,* [b]*School of Social Sciences and Psychology, Western Sydney University, Sydney, NSW, Australia,* [c]*Marcs Institute for Brain, Behaviour, and Development and School of Psychology, Western Sydney University, Penrith, NSW, Australia*

Introduction

The most vulnerable members of society are also the most likely to become trapped in the debilitating cycle of drug and alcohol abuse. A history of childhood maltreatment, physical and sexual abuse, parental neglect, peer pressure, low academic achievement, poor mental health and early exposure to drugs and alcohol are commonly reported risk factors (Mayes & Suchman, 2015; Whitesell, Bachand, Peel, & Brown, 2013). In 2016, an estimated 164 million people worldwide had a substance use disorder with a prevalence of 3–4% across Australia and 5–6% across the United States. Globally, alcohol is the most used substance, followed by opioids, cannabis, cocaine and amphetamine. The disease burden for alcohol abuse is largest in countries with lower socioeconomic development (namely, Eastern Europe and Russia); while the disease burden for drug abuse is largest in countries with high socioeconomic development (namely, high-income North America; Degenhardt et al., 2018). The effects of substance use disorder are far-reaching. They include physical and mental illness, economic burden, relationship distress or dissatisfaction and harm to developing foetuses and children (Daley, 2013). Given its diverse risk factors, high prevalence, and manifold negative consequences, it is clear that substance use disorder requires multi-level analysis and targeted interventions for the individual, their families, communities and wider society.

One level of analysis that continues to yield exciting advancements and possibilities for the treatment of individuals with substance use disorder focuses on underlying behavioral and neurobiological mechanisms through which drug use becomes an addiction (henceforth, the term "drug" is inclusive of both drugs

Cognitive, Clinical, and Neural Aspects of Drug Addiction. https://doi.org/10.1016/B978-0-12-816979-7.00007-8

and alcohol). Substance use disorder has two main characteristics: a persistent compulsion to seek the drug and a loss of control in limiting intake, despite negative consequences. This begs the question, why do compulsions to take drugs continue unabated (and indeed, even intensify), in the face of escalating aversive outcomes? Initially, neuroscientists looked to the pursuit of pleasure to explain the compulsive nature of drug abuse. However, it soon become clear that while pleasure may drive initial drug use, it is not the mechanism through which persistent drug use is maintained (e.g., Berridge & Kringelbach, 2015). This is supported by the simple observations that pleasure markedly decreases as tolerance to the drug develops and as serious drug related harms emerge. Perhaps then, it could be argued that drug use is maintained by the avoidance of withdrawal; that is, continued drug use serves as a form of negative reinforcement. However, this fails to explain why compulsions to use drugs persist long after withdrawal abates or why late relapse, months or even years after an individual has stopped using, is more common than not (Hser, Hoffman, Grella, & Anglin, 2001; Hunt, Barnett, & Branch, 1971). Additionally, it does not account for cases where compulsive drug use and repeated relapse occur, even in the absence of significant withdrawal, as is the case with cocaine or methamphetamine users.

Thus, while the pursuit of pleasure and avoidance of punishment are certainly involved in the cycle of addiction, it appears they are not the driving forces. Consequently, neuroscientists have turned to the neural mechanisms underlying learning and motivation to explain the compulsive nature of drug addiction. This chapter will briefly introduce these mechanisms and survey how they have been studied and interpreted in the context of several influential theories of compulsive drug use.

An introduction to associative learning and underlying neural circuitry

To address the conundrum of compulsions, theorists have drawn on the principles of associative learning theory (Everitt, Dickinson, & Robbins, 2001; Robinson & Berridge, 1993; Schulteis, Ahmed, Morse, Koob, & Everitt, 2000; Stewart, de Wit, & Eikelboom, 1984; Wikler, 1965). It has long been accepted that animals and humans alike do not learn about stimuli in their environments as discrete units; rather, they build complex representations of the world, made up of interconnected units. Specific associations between cues, behaviors and outcomes are thought to form following experience with the co-occurrence of these events (Dickinson, 2012; Warren, 1916). When one unit is activated, other, associated units are also activated. This enables an organism to predict future events on the basis of previous experience and adapt accordingly. Importantly, associative learning processes rely on neurophysiological responses to stimuli that are not necessarily accompanied by conscious reasoning. In the healthy brain, this promotes the efficient uptake of previously adaptive behavior, while allowing higher order cognitive resources to be deployed elsewhere.

Over a century ago, Pavlov famously stumbled upon a procedure for researching associative learning in animals (Pavlov, 1927). He noticed that dogs would salivate in response to stimuli which reliably preceded food, such as the footsteps of his assistant or the sound of a metronome. Pavlov believed that this occurred because the dogs had learnt to associate an unconditioned stimulus (US; food), which naturally elicits an unconditioned response (salivation), with a neutral stimulus (metronome). Over time, the neutral stimulus became a conditioned stimulus (CS) which elicited an automatic, conditioned response (salivation) on its own. Pavlov further discovered the surprising durability of these learnt associations (Pavlov, 1927). Namely, he showed that by repeatedly exposing his dogs to the sound of a metronome without subsequent food delivery (i.e., extinction training), their anticipatory salivation would disappear, only to partially re-emerge after a delay, a phenomenon he termed spontaneous recovery. Moreover, this spontaneous recovery was shown to increase with the passage of time. Thus, it seems it is difficult to completely "undo" or reverse original learning.

Pavlovian conditioning is closely related to response-outcome learning, as stimuli predictive of reward are likely to facilitate behaviors to procure the reward. According to Skinner, who termed response-outcome learning "operant" conditioning (also known as "instrumental"), actions which are followed by desirable outcomes (i.e., they are reinforced) are more likely to be repeated, while actions which are not reinforced (or indeed, punished) are less likely to be performed in the future (Ferster & Skinner, 1957; Skinner, 1938). Evidently, one of the main functions of associative learning is to develop a repertoire of predictive cues (via classical conditioning) and adaptive behaviors (via instrumental conditioning) that maximizes positive outcomes and minimizes negative outcomes. Biologically salient, aversive (e.g., pain, loss) or appetitive (e.g., food, money) events are easily associated with the predictive cues and behaviors that co-occur with them, and this learning is evidenced in both altered behavior and underlying neural circuitry.

While an in-depth discussion of the neural mechanisms of mammalian associative learning is outside the scope of the current manuscript, recent decades have seen enormous strides forward in the understanding of this phenomenon. Associative learning is known to be organized into a series of anatomically distinct neural circuits, involving different hubs depending on the type of learning engaged (Fanselow & Poulos, 2005). For example, while the amygdala serves as a major hub for Pavlovian fear conditioning (i.e., the pairing of neutral stimuli with aversive USs; Maren, 2001), classical eyeblink conditioning (in which an air-puff to the eye serves as the US) is under control of the cerebellum and associated pathways (Christian & Thompson, 2003). During Pavlovian conditioning, these neural circuits undergo a number of complex cellular and molecular alterations, driving the acquisition and consolidation of learning into memory (see Schafe, Nader, Blair, & LeDoux, 2001 for a review). For example, pairing of the CS and US during fear conditioning is known to lead to long-lasting changes in both synaptic transmission and neuronal activity in the amygdala, with long-term potentiation (LTP; persistent increases in synaptic strength) induced at all of the fear conditioning-related sensory input pathways to the amygdala (Blair, Schafe, Bauer, Rodrigues, & LeDoux, 2001; McKernan & Shinnick-Gallagher, 1997; Ota, Monsey, Wu, & Schafe, 2010; Quirk, Armony, & LeDoux, 1997; Quirk, Repa, & LeDoux, 1995; Rogan, Staubli, & LeDoux, 1997). In the same way, other forms of associative learning lead to long-lasting cellular and molecular changes within their own neural circuits. Indeed, as will be discussed in more detail below, such alterations have been demonstrated in several key brain regions associated with addiction (Grueter, Rothwell, & Malenka, 2012; Kauer, 2004; Kauer & Malenka, 2007; Luscher & Malenka, 2011; Mameli & Luscher, 2011).

Drugs of abuse constitute one of the most potent appetitive events available and as a result, a wide variety of environmental and internal stimuli (i.e., CSs) readily become associated with them. Locations, people and paraphernalia present before or during drug use, as well as internal states, such as a particular mood, are prime candidates for developing conditioned responses in anticipation of the drug itself, and thus taking on the properties of CSs. Conditioned responses might include approaching, engaging with, displaying a behavioral preference for, or an attentional bias towards drug-associated cues or contexts (Cavallo, Ruiz, & de Wit, 2016; Mayo et al., 2013; Mayo & de Wit, 2015). Alternatively, an individual might experience subjective liking, craving, arousal or activation of the sympathetic nervous system (Childress et al., 1999; Garavan et al., 2000; Grant et al., 1996).

In animal models, rats have been shown to demonstrate Pavlovian conditioned approach in response to drug-cues (e.g., Peters & De Vries, 2014), whereby presentation of a cue previously associated with a drug reward elicits approach behaviors and engagement with the cue (termed "sign-tracking") or engagement with the location where the reward is expected to appear (termed "goal-tracking"). Rats have also been shown to prefer a location where they have previously received drugs, relative to a location where they received a saline injection (e.g., Badanich, Adler, & Kirstein, 2006). Healthy non-dependent humans have likewise demonstrated drug-induced place preference, rating a location paired with D-amphetamine administration more favorably than a placebo location (Childs & de Wit, 2009). A similar study showed enhanced behavioral preference and attentional bias (measured via eye gaze) towards drug-associated cues relative to placebo-associated cues (Mayo & de Wit, 2015).

A study conducted by Boileau et al. (2007), demonstrated both behavioral and neurophysiological conditioned responding to drug-associated cues. Healthy participants received three amphetamine capsules across the course of several days. Two weeks later, they returned to the drug paired environment and were administered a placebo capsule of identical appearance to the amphetamine capsules taken previously. Placebo elicited the same self-reported ratings of high, euphoria and energy as the drug itself; suggesting that the context and/or capsule had developed conditioned responses similar to the unconditioned responses elicited by the drug (see Table 1). Interestingly, while both the drug and placebo capsule elicited equivalent alterations in behavior, they also elicited equivalent increases in dopamine transmission in the ventral striatum, as estimated from [11C]raclopride binding potential, using the positron emission tomography (PET) method (Boileau et al., 2007).

Indeed, associative learning in the context of drug use is associated with alterations in several neural circuits that receive input from dopaminergic neurons in the ventral tegmental area (VTA) of the midbrain, together termed the mesolimbic dopamine system (Berke & Hyman, 2000; Everitt & Robbins, 2005; Hyman, 2005). The VTA is the source of dopaminergic input to the amygdala, hippocampus, prefrontal cortex (PFC) and, perhaps most significantly, the nucleus accumbens (NAc; located in the ventral striatum; Swanson, 1982). Dopamine within the NAc acts to modulate multiple aspects of motivated behavior and effort-related decision making (Salamone et al., 2016; Salamone, Correa, Mingote, & Weber, 2005; Salamone, Correa, Yang, Rotolo, & Presby, 2018). Dopamine has also been shown to play a key role in reward learning (Schultz, 1998, 2016). Dopamine acting on the amygdala and PFC is known to be important for the valuation of rewards (Everitt, Cardinal, Parkinson, & Robbins, 2003) and the formation of reward-associated memories (Everitt et al., 2007; Goldstein & Volkow, 2011). Significantly, whether directly or indirectly, all drugs of abuse act to increase levels of synaptic dopamine within the mesolimbic circuitry (e.g., Di Chiara & Imperato, 1988; Drevets et al., 2001; Laruelle et al., 1995; Lecca et al., 2006; Volkow et al., 1999; Weiss, Lorang, Bloom, & Koob, 1993; Wise, Leone, Rivest, & Leeb, 1995; but see Nutt, Lingford-Hughes, Erritzoe, and Stokes, 2015 for an alternative perspective). This increase in dopamine not only acts to mediate the immediate reinforcing effects of drugs, but also leads to long-lasting alterations in the mesolimbic dopamine circuit associated with alterations in synaptic plasticity, the putative neural substrate of learning.

Initially, exposure to a drug of addiction leads to synaptic potentiation in the VTA, which, with continued drug exposure, triggers changes in downstream targets of the VTA, including the NAc and PFC (van Huijstee & Mansvelder, 2014).

TABLE 1 Example of a classical conditioning procedure (Boileau et al., 2007)

Before conditioning		During conditioning		After conditioning		Result
US	UR	NS US	UR	CS	CR	The context becomes a CS which elicits a CR (euphoria) due to its association with the US (drug)
→		→		→		
Drug	Euphoria	Context + Drug	Euphoria	Context	Euphoria	

Note: NS, neutral stimulus; *US*, unconditioned stimulus; *UR*, unconditioned response; *CS*, conditioned stimulus; *CR*, conditioned response.

Incredibly, even a single in vivo exposure to an addictive drug, such as cocaine, can lead to LTP at glutamatergic synapses onto VTA dopaminergic neurons (Ungless, Whistler, Malenka, & Bonci, 2001). Similar effects have now been demonstrated within 24 hours of a single exposure to opiates, nicotine, ethanol and benzodiazepines, with all leading to a potentiation of AMPA receptor transmission in VTA dopaminergic neurons (Saal, Dong, Bonci, & Malenka, 2003). These changes, coupled with a concomitant loss of LTP at inhibitory GABAergic synapses onto VTA dopaminergic neurons (Niehaus, Murali, & Kauer, 2010) and a potentiation of GABAergic input to inhibitory interneurons of the VTA (thereby disinhibiting VTA dopaminergic neurons; Bocklisch et al., 2013), act to enhance the excitability of dopaminergic neurons of the VTA. This, in turn, increases dopaminergic neurotransmission onto downstream targets and, after repeated exposure, triggers synaptic plasticity in these areas, particularly the NAc and PFC. It is these later stages of synaptic remodeling in downstream targets (elegantly reviewed in van Huijstee & Mansvelder, 2014) that likely drive the behavioral changes seen in addiction. These include anhedonia (the inability to feel pleasure associated with natural rewards) and incubation of craving (progressive increases in post-abstinence craving, mediated by changes in the NAc) and relapse of drug-seeking (mediated by changes in the PFC).

In animals, the presentation of drug-associated cues has been shown to preserve cocaine-seeking for up to a year after a single self-administration experience. This is in comparison with a highly palatable food reward, for which cue-induced responding disappeared within 3 months (Ciccocioppo, Martin-Fardon, & Weiss, 2004). As might be expected, studies have also shown that drug users' subjective experience of craving increases when exposed to drug-related images. This occurs even after a 12-month period of abstinence and intensive inpatient treatment (Franken, de Haan, van der Meer, Haffmans, & Hendriks, 1999; Sinha, Fuse, Aubin, & O'Malley, 2000). Extinction therapy—repeated, controlled exposure to CSs without subsequent exposure to the US—is the go-to method for breaking down unhelpful CS-US associations, such as those between drug-related stimuli and drug receipt. However, attempts to attenuate cue-induced drug craving with extinction therapy have proven largely unsuccessful, further emphasizing the resistant nature of drug-related associations (see Kantak & Nic Dhonnchadha, 2011 for a review; Konova et al., 2017). This is exacerbated by the spontaneous recovery effect (Pavlov, 1927), which predicts longer delays or periods of abstinence will result in heightened reactivity to drug-cues, and thus a greater risk of relapse.

The enduring, learning-related changes in behavior described here, are paralleled by changes in the mesolimbic dopamine circuit (Namba, Tomek, Olive, Beckmann, & Gipson, 2018). In support of this, even after 3 months of abstinence from self-administration of cocaine, drug-induced synaptic potentiation could still be detected in VTA dopaminergic neurons in rats (Chen et al., 2008). Within the NAc, potentiation of AMPA receptor transmission due to an increased AMPA/NMDA ratio after a period of prolonged drug withdrawal

(Jedynak et al., 2016; Kourrich, Rothwell, Klug, & Thomas, 2007) may enhance the responsiveness of NAc medium spiny neurons (the target of VTA dopaminergic projections) to drug-associated cues, leading to enhanced craving and drug-seeking behavior (van Huijstee & Mansvelder, 2014). In fact, Dong and Nestler (2014) have proposed that such synaptic changes represent a rejuvenation of the critical period of synaptic development within the mesolimbic circuitry, leading to unusually strong and long-lasting synaptic changes and, consequently, the durability of drug-related associations. As evidence of this, in rats with a history of cocaine self-administration, the presentation of drug-associated cues elicited increases in both the AMPA/NMDA ratio and dendritic spine size in the NAc, even in the absence of the drug (Gipson et al., 2013). The potentiation of drug-seeking behaviors elicited by drug-related cues may be further enhanced by increased glutamate release in response to drug-related cues, due to basal decreases within the NAc during withdrawal and subsequent disinhibition of presynaptic glutamate release (Gipson, Kupchik, & Kalivas, 2014).

Cue-induced relapse is demonstrated in rats and monkeys using the reinstatement model, in which repeated self-administration of a drug is extinguished (i.e., the response that used to result in a reliable drug dose is no longer followed by drug administration), before being reinstated via exposure to cues previously predictive of the drug (see Bossert, Marchant, Calu, & Shaham, 2013; Epstein, Preston, Stewart, & Shaham, 2006; Shaham, Shalev, Lu, de Wit, & Stewart, 2003 for reviews). Using this reinstatement procedure, Tunstall and Kearns (2016) tested the hypothesis that drug-predicting cues would be more powerful than cues predictive of natural rewards (see Table 2). Food restricted rats were

TABLE 2 Example of a cue-induced reinstatement procedure (Tunstall & Kearns, 2016)

Instrumental training			Free choice	Extinction	Reinstatement	Result
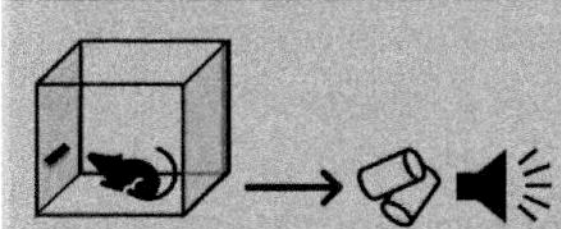 Left lever	Grain pellets	White noise	 Rats could choose a reward by pressing either the left or right lever	 Lever presses had no consequence. They were not followed by grain, cocaine or associated cues.	 White noise?	During free choice, rats chose grain on 70–80% of trials. After extinction, presentation of the cues elicited more lever pressing for cocaine than for pellets, suggesting stronger cue-drug associations
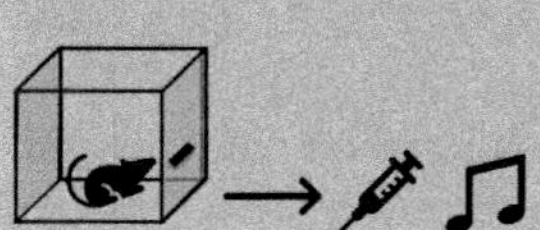 Right lever	Cocaine	Tone			 Tone?	

Note: During reinstatement, lever pressing elicited the associated cue but not the reward.

able to lever press for grain pellets or cocaine infusions. Upon each response, the designated reward was delivered along with a distinct audio-visual cue. The responses were then extinguished, such that lever pressing led to neither reward nor cue presentation. The following day presentation of both cues was used to assess the extent to which cue-induced reinstatement occurred. Interestingly, before extinction rats sought food rewards more than drug rewards; whereas, after extinction, exposure to drug cues generated more drug-seeking than food cues generated food-seeking. This was likewise demonstrated at 3 and 8-week follow-ups. Thus, it appears, even though food rewards are initially more reinforcing that drug rewards, CSs associate with drug-rewards more readily than nondrug-rewards. Moreover, in line with the spontaneous recovery effect, other studies have shown that cue-induced reinstatement of drug-seeking progressively increases with time; with maximal drug-seeking recorded months after the acute withdrawal phase (e.g., Grimm, Hope, Wise, & Shaham, 2001).

These behavioral effects are in line with the long-lasting changes in synaptic plasticity in the mesolimbic dopaminergic circuitry seen in addiction and may be especially dependent upon alterations of the PFC, particularly the medial PFC (mPFC). In recent years, the mPFC has received particular attention for its role in memory extinction, leading to hope that manipulation of the mPFC may have utility for the extinction of drug memories and the subsequent prevention of relapse (Zhang et al., 2018). During cue-induced reinstatement of drug-seeking behavior in rats with a history of cocaine self-administration, rapid synaptic potentiation in the NAc was dependent on activity levels in the prelimbic cortex, a dorsal subregion of the mPFC that sends glutamatergic projections to the NAc. Inhibition of the prelimbic cortex via infusion of a GABA agonist was able to prevent both cue-induced synaptic changes and drug-seeking behavior (Gipson et al., 2013). Conversely, activity in the infralimbic cortex, a ventral subregion of the mPFC, can suppress drug-seeking responses (Jamie Peters, LaLumiere, & Kalivas, 2008). This may be due to cue-induced synaptic depression in the ventral mPFC. In support of this, while re-exposure to drug-associated cues after 3 weeks of abstinence from heroin self-administration normally induces rapid synaptic depression in the mPFC, thereby impairing response inhibition, blocking endocytosis of GluA2-containing AMPA receptors attenuated cue-induced relapse of heroin-seeking behavior in a rodent model (Van den Oever et al., 2008).

So far, we have introduced the principles of stimulus-outcome and stimulus-response associative learning and discussed their applicability to drug addiction. We have learnt that different types of associative learning employ different neural circuits, with drug-related reward learning centering on the mesolimbic dopamine circuit (specifically the VTA, amygdala, hippocampus, PFC and NAc). Changes in behavior, driven by drug-related learning, are paralleled by changes in the function of this circuitry. Behavioral studies demonstrate that drug associations are more easily acquired, longer lasting and more difficult to extinguish than associations with natural rewards. Unusually strong and long-lasting synaptic changes in the mesolimbic dopamine system may be able to account for this. Moreover, manipulation of these neural processes may provide avenues for the development of more effective treatment protocols. We now turn to several theories of compulsive drug use which attempt to explain the precise mechanisms by which drugs of abuse become addictive.

Prediction error theory of compulsive drug use

Primate studies conducted by Schultz, Dayan, and Montague (1997) provided seminal evidence for neurophysiological responsivity to classically conditioned stimuli. It was demonstrated that following Pavlovian conditioning, dopaminergic neurons in the VTA, which once activated in response to a juice reward, instead responded to a sensory cue that predicted reward availability (Schultz et al., 1997). The activity of these neurons in response to predictive cues therefore seemed to reflect the expectation of the reward. Specifically, Schultz and colleagues (1997) demonstrated that larger than expected rewards elicit phasic dopamine firing (i.e., a rapid burst in action potentials), while anticipated rewards lead to little or no change. Smaller than expected rewards elicit a decrease in dopamine firing below baseline tonic activity. Thus, the sign and magnitude of phasic dopamine firing appears to reflect how surprising reward receipt or omission is. As learning proceeds, this neural activation is gradually transferred from reward onset to cue onset, such that cue-elicited spikes in dopamine reflect the size of the expected reward.[1] The discrepancy between the amount of dopamine release at cue onset and at reward onset, represents the error between the size of an expected and actual outcome. This error appears to act like a "teaching signal", which, over time, corrects the animal's expectations such that the outcome becomes fully anticipated and the error disappears. In light of this, theorists have proposed that dopamine transmission encodes reward "prediction errors" (PEs), which are a longstanding, central tenet in formal learning theory (Barto, Sutton, & Brouwer, 1981; Eshel et al., 2015; Rescorla & Wagner, 1972; Waelti, Dickinson, & Schultz, 2001).

PEs—the error between an expected and actual outcome—operate as the key driver in models of learning. Mechanisms for predicting an upcoming event or outcome, and altering expectations on the basis of surprising outcomes, are vital to successful learning. Indeed, algorithmic representations of PE processing have been very successful in simulating a wide variety of learning phenomena (Miller, Barnet, & Grahame, 1995), as well as the dopaminergic transmission

[1]According to the most widely applied perspective, put forward by Sutton and Barto (1987, 1998), the value inherent in a reward backpropagates to the predictive cue, which in turn promotes selection of the response associated with that cue. However, recent findings indicate that phasic dopamine encodes more than a simple cache-value signal. Predictions about upcoming events can be made on the basis of direct relationships (e.g., A→X) or indirect relationships mediated by a wider associative network (e.g., if A→X and X→Y then A→Y), and this complexity appears to be reflected in dopaminergic activity. Specific findings are discussed in Sharpe and Schoenbaum (2018).

that occurs under these same learning conditions (e.g., Tobler, Dickinson, & Schultz, 2003; Waelti et al., 2001). Electroencephalographic and neuroimaging studies have also shown PE-like striatal activity in humans using various appetitive outcomes (Chase, Kumar, Eickhoff, & Dombrovski, 2015; Holroyd & Coles, 2002), and there is preliminary evidence that this activity reflects changes in dopaminergic transmission (Ferenczi et al., 2016; Knutson & Gibbs, 2007; but see Lohrenz, Kishida, & Montague, 2016).

According to the PE theory of compulsive drug use, if the increase in mesolimbic dopamine seen in response to drugs of abuse (see previous section) involves rapid firing in close proximity to reward receipt, this would create an artificial signal that the drug reward is always "larger than expected" (positive PE), regardless of whether the experienced effects of the drug were fully expected (no PE), or even "smaller than expected" (negative PE). That is, drugs of addiction may mimic positive PEs and thus usurp normal learning mechanisms by fuelling ever-stronger cue-outcome associations. By contrast, the strength of the associations involved in natural reward learning begin to plateau or wane when the error in one's prediction diminishes, or the outcome becomes less rewarding. In the context of drug use, a PE signal that fails to respond to such changes in outcome is hypothesized to result in a cycle of pathological overlearning, which in turn biases behavior towards further drug-seeking (Hyman, Malenka, & Nestler, 2006; Keiflin & Janak, 2015; Redish, 2004; Schultz, 2016).

There are a number of findings which lend support to this hypothesis. Firstly, in the context of natural reward, Steinberg et al. (2013) demonstrated that artificial optogenetic activation of dopamine neurons during the delivery of an already expected reward can mimic a PE, leading to a sustained increase in cue-induced reward-seeking behaviors. This supports the basic assumption that artificial PEs can alter learning processes and subsequent behavior. Specifically, a phasic increase in dopamine, carefully timed to coincide with the delivery of a reward that no longer elicited a dopamine response, appears to have caused the cue-outcome relationship to strengthen beyond its usual scope. The question is, do drugs of abuse produce similar artificial PEs, and would the timing of these PEs cause alterations in learning?

Employing fast-scan cyclic voltammetry, Phillips, Stuber, Heien, Wightman, and Carelli (2003) demonstrated that rapid, real-time increases in NAc dopamine transmission coincide with the presentation of cues previously paired with cocaine delivery. Thus, transient dopamine release events (putative phasic firing) indexed drug reward expectations, in line with PE processing. However, the effects of drug receipt on dopamine transmission, as measured by Stuber, Wightman, and Carelli (2005), did not reflect normal PE processing. Across the course of instrumental conditioning and extinction, cocaine infusion was shown to initiate long-lasting, unsynchronized spontaneous dopamine transients; the frequency of which correlated with changes in cocaine concentration, rather than changes in expectation or surprise (Stuber et al., 2005). Such a surge in dopamine signaling may be sufficient to imitate positive PEs and fuel excessive learning about predictive cues. Theoretically, such activity would generate ongoing feedback that everything in the environment is surprising and more rewarding than expected. As such, the drug would be conferred a higher and higher reward value upon each administration, and this would transfer to the cues predictive of drug receipt. The stronger the associative relationship between a cue and reward (or the more value the cue is endowed; Sutton & Barto, 1987, 1998), the more likely it will initiate reward seeking behaviors in the future. Studies have demonstrated similar increases in dopamine transients with other drugs of abuse (see Covey, Roitman, & Garris, 2014; Keiflin & Janak, 2015 for reviews).

Despite the elegance of this theory, the PE model of compulsive drug use is limited in a few regards. Evidence for the acute effects of drugs on phasic dopamine activity are still mixed, with some studies demonstrating increased and others decreased activity in response to various drug classes and conditions (see Wanat, Willuhn, Clark, & Phillips, 2009 for a review). Surprisingly, one study found a progressive decrement in phasic dopamine with an escalation in amphetamine self-administration (Willuhn, Burgeno, Groblewski, & Phillips, 2014). There are also at least a couple of behavioral studies that demonstrate normal PE processing during drug receipt. For example, one study showed that rats were able to learn when a cocaine reward was smaller than expected (Marks, Kearns, Christensen, Silberberg, & Weiss, 2010) and, unlike with optogenetic activation of dopamine signaling (see above; Steinberg et al., 2013), drug-receipt was unable to increase cue-induced responding for an already expected reward (Panlilio, Thorndike, & Schindler, 2007).

Interestingly, while Stuber et al. (2005) demonstrated an increase in spontaneous dopamine transients that correlated with drug concentrations (see above), dopamine transients time-locked to the reinforced response were also evident. These time-locked transients diminished in size throughout extinction, as if to reflect the weakening relationship between operant responses and drug-receipt. Thus, it seems the effect of cocaine on phasic dopamine was superimposed over normal error processing. The ramifications of this for learning would depend on how the timing of drug-induced transients coincide with various stimulus, response and outcome events. Indeed, Daberkow et al. (2013) demonstrated amphetamine-induced phasic dopamine hyperactivity led to failed operant behavior. It was argued that this resulted from a decoupling of dopamine transients and learning events. Going forward, proponents of the PE theory of compulsive drug use should attempt to account for these complexities.

It should also be noted that opponents of the PE theory of compulsive drug use have argued that the close relationship between PE processing and dopamine signaling does not provide evidence of causation. That is, just because dopamine signaling codes for PEs, does not mean it *causes/drives* learning, it could equally be a downstream reflection of learning mechanisms operating elsewhere. This is an important distinction because many of the dopamine hypotheses set forth to explain clinical phenomena rely on the assumption of causation (Berridge, 2007). This includes the proposal that pharmacologically heightened dopamine release causes "run-away" reinforcement of drug-seeking cues and behaviors by strengthening associative links beyond their usual scope.

Specifically, the PE hypothesis of compulsive drug use has been drawn into question due to the observation that mutant mice with a genetically engineered inability to produce dopamine, as well as rats with neurochemical lesions that deplete mesolimbic dopamine, have both been shown to engage in some forms of normal reward learning (despite severe Parkinsonian symptoms). For example, rats with neurochemical lesions could learn the new values of sweet tastes after they had been paired with illness (Berridge & Robinson, 1998) and mutant mice were able to show a preference for and choose a spout which delivers sucrose over a spout that delivers water (Cannon & Palmiter, 2003), as well as choose a place previously paired with a drug reward over a control location (i.e., conditioned place preference; Hnasko, Sotak, & Palmiter, 2005). Mutant mice were also able to find food rewards and learn about them in a maze task (Robinson, Sandstrom, Denenberg, & Palmiter, 2005); though, this learning was not evidenced until they were later tested under dopamine replacement medication. Together, this evidence suggests that implementation of learning (perhaps via diminished "wanting" or motivation to perform, as per the incentive sensitization theory discussed in the next section), but not learning itself was compromised by a lack of dopamine. These studies cast doubt over the causal role of dopamine in learning and, as a result, some theorists have looked elsewhere to explain the compulsive nature of drug use (e.g., Berridge, 2007, 2012; FitzGerald, Dolan, & Friston, 2015; Holton & Berridge, 2017).

Researchers continue to grapple with these seemingly paradoxical findings, that dopamine signals obey the rules of associative learning and yet are not necessary for some types of learning to occur. The recent emergence of optogenetic techniques has widened the scope for testing the causal effects of phasic dopamine on learning. Optogenetics allow for temporally precise and selective control of putative dopamine neurons, and thus provides the means by which artificial bursts in dopamine can be introduced or natural bursts can be suppressed in animals engaging in learning procedures. If the animal's behavior changes in line with the assertions of PE theory, this provides further evidence that phasic dopamine acts as a "teaching signal" from which the animal learns or updates cue-outcome associations and alters their behavior accordingly. However, it could equally be asserted that learning these associations occurred via alternate circuitry and the artificially induced dopamine signal simply played a permissive role in allowing that learning to be implemented.

More interesting to this debate are optogenetic manipulations imposed on classical conditioning procedures, which do not involve a performance element. By divorcing learning from performance, it is more readily argued that a manipulation of phasic dopamine specifically altered learning; indeed, this was the approach taken by Sharpe et al. (2017). Sharpe and colleagues combined a sensory preconditioning procedure with a blocking procedure (see Table 3). In the

TABLE 3 A simplified depiction of Sharpe et al.'s (2017) study design

Preconditioning		Conditioning	Test	Result
A → X Tone → Light	A C → X Tone + Siren → Light (boxed)	X → US Light → Sucrose	C? Siren	Food cup entries in response to cue C were increased because optogenetic activation unblocked learning about C
	A D → X Tone + White noise → Light		D? White noise	Food cup entries in response to the control cue D were minimal because learning about D was blocked during preconditioning

Note: In the full design several other controls were implemented. *A, C, D and X,* counterbalanced neutral conditioned stimuli; *US,* unconditioned stimulus; *Box,* optogenetic activation during stimulus presentation; D and C were presented at test under extinction and the number of food cup entries were assessed.

preconditioning phase, rats were trained to associate neutral cues A → X, before learning to associate neutral cues AC → X. Normally, learning about C would be blocked, because A fully predicts X and there is consequently no error between the expected and actual outcome on AC → X trials. However, in order to mimic a PE, and thus create the means by which to build an association between cues C and X, dopamine neurons were optogenetically activated directly before presentation of X on AC → X trials. That is, optogenetic activation was expected to cause artificial PE signals that would have allowed the otherwise blocked C → X association to form. Once these neutral cues had received sufficient training, the conditioning phase commenced, in which X was repeatedly paired with a food reward. Finally, rats were tested on Pavlovian approach behaviors (food cup entries) elicited by cue C (presented alone and without reward). The number of food cup entries for C were compared to the number of entries for a control cue which had likewise been blocked during preconditioning but was not manipulated optogenetically. C was found to elicit significantly more entries than the control cue, suggesting that optogenetic activation had successfully mimicked a PE and a C → X association had formed; C then acquired an association with the food reward via its relationship with X.

Importantly, rats were shown to have learnt the association between cue C and the food reward, without the need to perform an instrumental response. It is therefore difficult to argue that optogenetic stimulation affected motivation or instrumental performance, as the critical manipulation occurred while the animals were exposed to neutral cues. Moreover, given optogenetic activation indirectly produced a relationship between C and the food outcome, via their shared association with X, it seems increasingly plausible that dopamine transmission plays a role in the learning of associative relationships that become integrated in a broader representational model of the world. This contrasts with an incentive-sensitization perspective (outlined in the next section), which would argue that optogenetic activation caused cue X to gain salience, making it more attractive and able to incentivize or motivate behavior. However, increasing the salience of X would have also facilitated conditioning in the subsequent phase (X → food); yet, learning proceeded no faster in the optogenetic group than in a control group.

Evidently, the use of optogenetics provides an exciting avenue for testing some of the long-standing assumptions of the PE theory of dopamine transmission and, by extension, the PE theory of compulsive drug use. Applied carefully, within more complex behavioral paradigms, it is also possible to contrast the assertions of PE theory with those of other theories. In doing so, some of the ongoing debates faced by researchers applying learning theory to addiction may be resolved.

Incentive-sensitization theory of compulsive drug use

While the PE theory of compulsive drug use focuses solely on the associative learning component of reward, and the manner in which this is compromised, the incentive sensitization theory considers motivation as equally important (Robinson & Berridge, 1993). Indeed, it is argued that compulsive drug use results from heightened motivation rather than learning. Instead of encoding PEs which drive new learning, proponents of the incentive sensitization theory argue that dopaminergic transmission encodes "incentive salience" which acts as a gateway for reward-eliciting behaviors. Incentive salience represents the intensity of a triggered desire or the degree to which a reward is "wanted". The reward itself or stimuli predictive of the reward (i.e., CSs) trigger these desires; becoming "motivational magnets" which command attention, induce approach and guide behavior towards the reward. The more salient/attractive a cue or reward, the more incentivized the animal is to maintain interest and perform the actions necessary to obtain the reward (Berridge, 2012).

Incentive salience has two inputs: Pavlovian learning on the one hand and neurobiological state factors on the other. That is, both the strength of learnt associations, as well as the current state of the mesolimbic-related brain system, are integrated to determine how large the incentive is to perform a reward-eliciting behavior. For example, an animal may have learnt that performing an instrumental response leads to a food reward. However, if they are already satiated, they are likely to be less incentivized to demonstrate this learning. Conversely, if an animal is in a state of hunger, they will be more incentivized, and this will potentiate performance. Importantly, if the animal is in the same state (e.g., of moderate hunger) throughout a learning procedure, the incentive salience computation reflects only the learning input. This is argued to disguise incentive salience and to give the impression that dopamine mediates pure learning (as argued in the previous section). This confound is said to have given rise to the many reported findings that dopamine signaling encodes reward PEs. That is, it is said that dopamine activity only appears to encode reward surprise and expectation because the physiological state is held constant and thus motivation tracks learning. Indeed, the incentive sensitization theory asserts that dopamine activity reflects motivation (i.e., incentive salience) and learning is merely an input computed elsewhere (Berridge & O'Doherty, 2014). These conclusions are supported by evidence presented in the previous section that animals with almost no brain dopamine can still exhibit normal learning under certain conditions.

Further support is offered by a broad body of work on the role of dopamine in motivation and effort related decision making (most recently reviewed in Salamone et al., 2018). Using behavioral tasks designed to offer a choice between a high-effort option leading to a more preferred reinforcer and a low-effort option leading to a less preferred reinforcer, Salamone and colleagues have demonstrated that interference with mesolimbic dopamine transmission produces a bias towards the lower effort option (e.g., Cousins, Atherton, Turner, & Salamone, 1996; Cousins, Wei, & Salamone, 1994; Salamone et al., 1991), without altering primary food preference or reinforcement (Aberman & Salamone, 1999; Salamone, Arizzi, Sandoval, Cervone, & Aberman, 2002). Conversely, pharmacologically reversing the effects of dopamine depletion within the NAc is associated with a shift back to the high-effort/high-reward option (e.g., Farrar et al., 2008; Farrar et al., 2010). More recently, these findings have been corroborated by a wealth of evidence from studies utilizing optogenetic, chemogenetic and physiological techniques (see Salamone, Correa, Yang, et al., 2018 for a review).

Berridge and colleagues developed a proof of principal paradigm to test the assertion that dopamine transmission is involved in motivation rather than learning (see Berridge, 2012 for a review). Rats are trained to associate a CS with an intensely salty US squirted into their mouths. Upon receipt of the US, they tend to respond with disgust—gaping and flailing their forelimbs. It does not take long for the rats to learn the CS-US pairing. Mesolimbic-related brain circuitry is subsequently stimulated in such a way as to induce a robust appetite for salt, which causes the animal to act as if it is in a state of severe sodium deficiency (indeed, if they are given the US now, they respond with signs of pleasure; Tindell, Smith, Pecina, Berridge, & Aldridge, 2006). The animal is then presented with the CS again (alone and in extinction). Rather than demonstrating the previously learnt value of the cue and turning away from the CS, the rat instead engages in approach behaviors, and tries to "consume" the stimulus (i.e., sign-tracking behavior). It is argued that the CS was transformed into a motivational magnet, reversing a learned revulsion and inducing a sudden attraction (Robinson & Berridge, 2013). Moreover, in the salt deficient state, ventral pallidum neurons, which did not previously fire to the salt-associated CS, will now fire as robustly as if the animal were presented with a sucrose-associated

CS. This occurs with no further learning and can thus only be attributed to the change in neurobiological state factors. According to this logic, the incentive salience of the salt-associated CS was adaptively recalculated by integrating the novel physiological state with the pre-existing cue-outcome association. This in turn elicited behavioral and neural reactions very different from those expected from a CS-US input alone (Tindell, Smith, Berridge, & Aldridge, 2009).

The same logic behind the salt deficiency paradigm can be applied to drug addiction. As discussed previously, drugs of abuse enhance dopaminergic signaling in the mesolimbic system. The incentive-sensitization theory asserts that these signals cause a change in neurobiological state factors that amplify the incentive salience of drug-rewards and drug-cues. That is, the degree to which the drug is "wanted" is intensified (similar to how a desire for the salty US was induced). This is distinct from the degree to which the drug is "liked". Namely, the actual hedonic impact of the drug (i.e., the extent to which it is "liked") remains unchanged (or even diminishes), but the motivation to acquire the drug is heightened (e.g., Berridge & Robinson, 2016). There is an over-attribution of incentive salience such than an animal or human "wants" the reward to a greater degree than their "liking" of the reward warrants. Indeed, increasing dopaminergic transmission via a variety of manipulations (including drug administration) has been found to potentiate "wanting," but not "liking," in animals. For example, rats with heightened dopamine activity will work harder and perform more proficiently for a sucrose reward (indicative of heightened "wanting"), despite the reward itself inducing the same hedonic orofacial expressions (indicating that "liking" had not changed; Cagniard, Balsam, Brunner, & Zhuang, 2006; Pecina, Cagniard, Berridge, Aldridge, & Zhuang, 2003; Wyvell & Berridge, 2000).

Separable "liking" and "wanting" circuitry has also been demonstrated. For example, animal studies provide evidence for a network of small "hedonic hotspots" distributed throughout the limbic system, which can be manipulated to enhance "liking". Hotspots have been located in the NAc, ventral pallidum, PFC and brainstem. In each instance, neurochemical stimulation of these hotspots with an appropriate drug microinjection will enhance orofacial "liking" expressions (note that none of these manipulations induce heightened dopaminergic activity). Beyond these specific sites, microinjections fail to enhance, and may even suppress, "liking" reactions. Moreover, stimulating these hotspots fails to enhance "wanting" (see Berridge & Kringelbach, 2015 for a review).

Evidence for the "liking" and "wanting" dissociation in humans is sparse and findings more inconsistent than in animals (Arulkadacham et al., 2017; Grigutsch, Lewe, Rothermund, & Koranyi, 2019; Kalapatapu et al., 2012; Ostafin, Marlatt, & Troop-Gordon, 2010; Willner, James, & Morgan, 2005). This may be due to the methodological limitations implicated in measuring self-report "liking" and "wanting" which, according to the incentive sensitization theory, are preconscious states. Efforts to develop and refine implicit measures are underway (see Tibboel, De Houwer, & Van Bockstaele, 2015 for a review). Perhaps the most compelling evidence can be drawn from studies employing

Parkinson's disease (PD) patients. PD patients (who have severely depleted brain dopamine) are administered medications, such as levodopa, in order to increase dopamine levels. A small subset of patients become addicted to their medication, despite the drug lacking euphoric effects or dysphoric withdrawal. Using the [11C]raclopride PET method, Evans et al. (2006) demonstrated that an equivalent drug-dose will induce greater ventral striatal dopamine release in PD patients who have developed compulsions, compared to patients who have not. Moreover, the extent to which dopaminergic neurotransmission was enhanced by medication positively correlated with ratings for "wanting", but not "liking", their medication. Liking was equivalent between groups.

The question remains, if heightened dopamine transmission, induced by drug administration, potentiates incentive salience and thus causes an animal or human to "want" and pursue a drug-reward more fervently, what mechanism causes compulsions to persist after drug use has stopped? Why are individuals with substance use disorder so prone to relapse, even after months or years of abstinence? This takes us to the second part of the incentive-sensitization theory.

A parallel body of research emerged alongside the "liking" and "wanting" literature of the last three decades. In the late 1980s, Terry Robinson and colleagues began to uncover evidence that the dopamine system can be "sensitized" by a variety of addictive drugs. Repeated use of addictive drugs has been shown to result in enduring neuroadaptations, whereby the mesolimbic system becomes hyper-reactive (i.e., it will produce an amplified dopaminergic response) to drugs and drug-cues, which in turn magnifies their incentive salience (Paulson, Camp, & Robinson, 1991). This could be likened to how an individual with a genetic inability to generate leptin, the satiety hormone, has an enduring brain state that amplifies motivations to elicit food-rewards (Farooqi et al., 2007). Compared to an individual who is satiated, an individual who is constantly hungry will experience much stronger "wanting" reactions to food advertisements or to the smell of food when walking past a café. Likewise, an individual with a drug-sensitized mesolimbic dopamine system, will experience much stronger temptations to take drugs when in the presence of drugs or drug-cues (Berridge, 2017; Robinson, Robinson, & Berridge, 2014).

Sensitization of drugs has been demonstrated both in behavior and relevant brain activity. For example, previous noncontingent exposure to drugs has been shown to amplify subsequent conditioned place preference, as well as potentiate the acquisition of drug self-administration responses. These behavioral effects are mirrored by increased dopamine release (in the VTA, NAc and medial PFC) in response to an acute drug dose in animals with a history of drug exposure, compared to drug-naïve animals (see Steketee & Kalivas, 2011 for a review). Even prenatal exposure to drugs can result in sensitization evidenced in adulthood (Macuchova & Slamberova, 2017). Thus, previous drug exposure appears to amplify the impact of subsequent drug exposure on dopamine-related brain systems and this is coupled with an increase in motivated behavior, presumably due to heightened "wanting".

Sensitization of drug-cues (as opposed to the drug itself) was evidenced in a study conducted by Wyvell and Berridge (2001); see Table 4). Rats were trained to lever-press for a sucrose reward. On a separate occasion they were trained to associate a CS (e.g., tone) with the same reward. After training, a subset of rats were exposed to repeated noncontingent amphetamine administration (i.e., sensitization). Ten days after this period of drug exposure, and in a drug free state, rats were tested on the instrumental response under extinction conditions. Cue-triggered lever-pressing for the reward was potentiated to the same degree as it was in rats responding in an intoxicated state (with no history of drug exposure). Moreover, both groups responded for the reward significantly more than drug-naive rats, even though baseline responding was equivalent. A subsequent study demonstrated that ventral pallidal neurons mirror these behavioral effects, equivalently increasing their cue-triggered firing rate in both the sensitized and acute conditions (Tindell, Berridge, Zhang, Pecina, & Aldridge, 2005). These findings demonstrate that exposure to drugs produces long-lasting effects that continue to influence the attribution of incentive salience to drug-cues in the drug-free state. Indeed, the sensitized animals responded to the drug-cues as if they were intoxicated. As discussed in detail, in the upcoming Pavlovian to instrumental transfer section, these results cannot be attributed to a learning effect as learning was complete before drug administration.

Both behavioral and dopaminergic drug and drug-cue sensitization have been evidenced in humans (see Leyton, 2007; Robinson et al., 2014 for reviews). Using PET and [11C]raclopride, Boileau et al. (2006) demonstrated that three single doses of amphetamine across five days is sufficient to induce enduring

TABLE 4 Example of a sensitization of drug-cues paradigm (Wyvell & Berridge, 2001)

Pavlovian training	Instrumental training	Drug exposure	Test	Result
Lever press → Sucrose	Tone → Sucrose	Amphetamine microinjection directly before test	Tone?	Acute drug exposure (i.e., intoxication) significantly increased cue-induced lever pressing when compared to a drug-naïve control group
		Amphetamine microinjections. 1/day for 6 days 10 day withdrawal period		Repeated exposure to the drug prior to test (i.e., sensitization), increased cue-induced lever pressing to the same degree as intoxication

Note: Test was conducted under extinction conditions.

sensitization. Two weeks and one-year post drug exposure, amphetamine administration elicited a larger dopamine response in ventral striatum, dorsal caudate and putamen when compared to the initial dose (indeed, it was largest at one year). This was accompanied by a proportional increase in psychomotor responses, energy and alertness. The sustained or even increasing effects of drug sensitization post-abstinence may be able to explain why late relapse is so common among individuals with substance use disorder. A further study with a group of PD patients whose medication had caused them to develop compulsions for alternate rewards (food, sex, gambling, etc.) showed larger phasic increases in NAc and ventral striatal dopamine release in response to visual reward cues, compared to those who had not developed compulsions. This indicates that dopamine replacement medication had induced sensitization of the neural response to reward cues and thus also increased the incentive salience of those cues in PD patients who had developed compulsions compared to those who had not (O'Sullivan et al., 2011).

One of the underlying assumptions of the incentive sensitization theory is that dopamine transmission mediates motivation and not learning. However, this remains a point of contention. While there are several studies which indicate animals with severely depleted dopamine can still demonstrate learning, other studies indicate some forms of learning are not evidenced in a dopamine depleted state (Darvas, Fadok, & Palmiter, 2011; Robinson, Rainwater, Hnasko, & Palmiter, 2007). Moreover, as discussed in the previous section, optogenetic manipulations of dopamine activity can directly alter learning in the absence of any motivation or performance confounds (Sharpe et al., 2017). This may indicate that dopamine is not exclusively involved in either motivation or learning but is involved to varying degrees in both. It is also worth noting that a particular brain system may be able to induce an aspect of cognition or behavior, while not being necessary for it to occur. For example, selective stimulation of the pleasure "hotspot" in orbitofrontal cortex can significantly heighten pleasure and, yet, even after widespread damage of this area, pleasure is still evidenced (Berridge & Kringelbach, 2015). In the same way, it is possible that dopamine-related systems are not necessary for certain types of learning and yet can still be manipulated to facilitate learning. Otherwise, it may be that only some types of learning and not others are reliant on dopaminergic processing. By distinguishing between different types of learning, as has been proposed by Daw and O'Doherty (2014), we may be able to discriminate between these possibilities.

Goal directed and habitual control

Another explanation for the development of pathological patterns of behavior, including addiction, focuses on the transition from goal-directed behavior to habitual behavior (see Gillan, Robbins, Sahakian, van den Heuvel, & van Wingen, 2016 for a review). The two theories described above have focused on quantitative changes in performance in order to explain compulsive drug use (i.e., a quantitative increase in conditioned responding which derives from either artificial positive PEs or heightened incentive salience). This theory, on the other hand, places emphasis on a qualitative transition in the source of behavior. For example, while responding for a drug-reward may remain quantitatively the same (e.g., an animal keeps pressing a lever to acquire a reward), responding may at first reflect a desire to obtain the reward and later reflect a mere learnt habit.

Indeed, Everitt (2014) notes that the conventional view of positive reinforcement conflates the two dissociable processes of goal-directed behavior and habitual behavior. That is, the simple act of performing an instrumental response can be derived from either a response-outcome (R-O) association or a stimulus-response (S-R) association. The former, a declarative process, is deemed goal-directed because R-O associations are sensitive to the current value of the outcome. Thus, if an instrumental behavior no longer elicits the reward it once did, the R-O association is weakened, and behavior diminishes. The latter, a preconscious process, is deemed habitual because S-R associations are elicited automatically by antecedent stimuli, bypassing stored representations of the goal. Therefore, under habitual control, changes in the outcome do not initiate immediate changes in the response.

The outcome devaluation test (see Table 5) is used to assess whether behavior derives from goal-directed or habitual control. Either animals or humans are trained to perform an instrumental response (e.g., lever press or button press) for a reward (e.g., food or money). Once performance has plateaued, indicating the response has been acquired, the reward is devalued. This may involve allowing the reward to be consumed until satiety or pairing the reward with illness. The instrumental response is then assessed under extinction conditions (i.e., without rewards). If there is a reduction in responding during the extinction phase to reflect the diminished value of the reward, behavior is deemed goal-directed.

TABLE 5 The outcome devaluation task (elegantly discussed in Corbit, 2018)

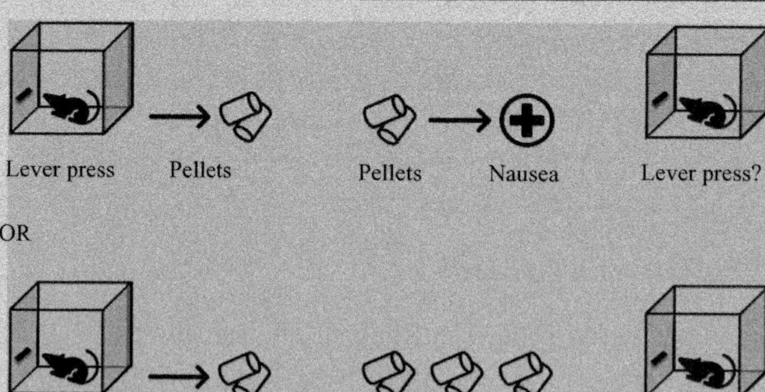

Instrumental training	Devaluation	Test	Result
Lever press → Pellets	Pellets → Nausea	Lever press?	After devaluation (when the reward is no longer desirable) the animal demonstrates goal-directed behaviour if they reduce lever pressing. If the animal continues to respond for the devalued reward, their behaviour is deemed habitual
OR			
Lever press → Pellets	Satiety	Lever press?	

Note: Devaluation involves either conditioned taste aversion (the food reward leads to sickness) or sensory specific satiety (the food reward is consumed freely).

Conversely, responding that continues regardless of the new value of the outcome, is deemed habitual (i.e., triggered by the environment).

Goal-directed and habitual learning processes appear to operate in a dynamic balance, with one or the other predominating depending on the specific conditions of training or testing (Hogarth, Balleine, Corbit, & Killcross, 2013). The former, goal-directed process shows a relative dominance during early performance and the latter, habitual process predominates during later performance. Indeed, rat studies using outcome devaluation tests demonstrate that instrumental responding to acquire food is flexibly modified in response to changes in outcome value after minimal training but becomes insensitive to devaluation after more extensive training (e.g., Adams, 1982; Adams & Dickinson, 1981; Sage & Knowlton, 2000). This has also been replicated in a human study using button presses to acquire food rewards. Once satiated, participants who had been overtrained demonstrated habitual control, while participants who had received moderate training reduced responding as per the new outcome value (Tricomi, Balleine, & O'Doherty, 2009). In principle, allowing tried-and-true behaviors to be governed by S-R associations is adaptive in a stable environment, as it frees more resources for goal-directed activities. However, if the outcome is no longer desirable, continued habitual responding can become detrimental.

A body of evidence from basic neuroscience studies employing lesion, inactivation and electrophysiological recording techniques in animals, as well as structural and functional imaging in humans, supports the hypothesis for two dissociable R-O and S-R mechanisms, controlled by distinct corticostriatal pathways (see Balleine & O'Doherty, 2009; Malvaez & Wassum, 2018; O'Hare, Calakos, & Yin, 2018 for reviews). The dorsomedial (DMS) region (with inputs from the medial PFC) is central to goal-directed actions, while the dorsolateral (DLS) region (with inputs from sensorimotor cortices) is central to habitual actions.[2] For example, DMS lesion or inactivation during devaluation procedures suppresses goal-directed behavior but promotes habitual control (Gremel et al., 2016; Gremel & Costa, 2013; Yin, Knowlton, & Balleine, 2005; Yin, Ostlund, Knowlton, & Balleine, 2005). Conversely, lesions to DLS abnormally preserves goal-directed behavior, preventing the transition to habitual control (Yin, Knowlton, & Balleine, 2004).

Studies have demonstrated that drug-seeking also relies on DMS and DLS (Clemens, Castino, Cornish, Goodchild, & Holmes, 2014; Zapata, Minney, & Shippenberg, 2010). For example, using a traditional devaluation test, Corbit, Nie, and Janak (2012) demonstrated that inactivation of DMS attenuated alcohol seeking responses after minimal training (when goal-directed control usually predominates) but had no effect on performance after extensive training. Inactivation of DLS, on the other hand, had no effect after limited training, but

[2]The localization is somewhat different in humans, with goal-directed control centering on the caudate nucleus and ventromedial prefrontal cortex, and habitual control on the posterior putamen (e.g., Wunderlich, Dayan, & Dolan, 2012).

restored goal directed responding after extensive training (when habitual control usually predominates). Thus, DMS and DLS are implicated in goal-directed and habitual drug-seeking, respectively (see Corbit, 2018 for a review).

Like natural rewards, drug-seeking undergoes a transition from goal-directed to habitual control after prolonged exposure. However, unlike natural rewards, drug-seeking undergoes an abnormally rapid shift. For example, despite equal training, food-seeking responses diminish following outcome devaluation while alcohol-seeking and cocaine-seeking responses are maintained, suggesting that alcohol- and cocaine-seeking responses become habitual faster than food-seeking responses (Dickinson, Wood, & Smith, 2002; Miles, Everitt, & Dickinson, 2003). Repeated noncontingent exposure to drugs such as cocaine, amphetamine and methamphetamine has also been shown to promote an accelerated transition from R-O to S-R control during subsequent responding for natural rewards (Corbit, Chieng, & Balleine, 2014; Furlong, Corbit, Brown, & Balleine, 2018; Nelson & Killcross, 2006; Nordquist et al., 2007). This occurs despite evidence that drug-exposed and control animals alike develop an aversion to the reinforcer during devaluation (Nelson & Killcross, 2006).

The acute effects of alcohol have been shown to impair goal-directed responding and promote habitual control for devalued food rewards in humans, despite equivalent appraisal of reward value between experimental and control participants (Hogarth, Attwood, Bate, & Munafo, 2012). Vollstadt-Klein et al. (2010) further demonstrated greater cue-induced dorsal striatal activation (implicated in habitual control) in heavy drinkers compared to light drinkers, who instead showed greater activation of the ventral striatum. Similarly, Voon et al. (2014) demonstrated that individuals with a history of methamphetamine exposure employ a habitual strategy when responding for monetary rewards. This shift towards habit formation occurred during the early stages of learning and was associated with lower grey matter volume in caudate and medial orbitofrontal cortex, which are thought to be involved in goal-directed control. Together, this evidence suggests that compulsive drug use is characterized by an aberrant emergence of automatic behavior at the expense of intentional regulation, driven by an impaired ability to utilize representations of outcome identities (Hogarth et al., 2013).

Interestingly, Voon et al. (2014) also demonstrated that the duration of abstinence among individuals with a history of alcohol abuse correlated with a shift towards habitual control, such that more recent alcohol use was associated with greater habit learning about monetary rewards. This may indicate that heightened habitual control is a product of drug exposure rather than a vulnerability factor leading to drug abuse. A number of other studies have shown that individuals with substance use disorder who are tested in a non-intoxicated state are no faster than controls at transitioning to habitual control in paradigms that used secondary reinforcers (e.g., pictures of cigarettes) or food rewards (e.g., Hogarth et al., 2018; Hogarth & Chase, 2012), further casting doubt on the hypothesis that a lack of goal-directed control or enhanced habitual responding is a marker of vulnerability to addiction. Nonetheless, whether a dispositional precursor to drug abuse or a result of drug exposure, heightened habitual control remains a viable candidate mechanism for the emergence of drug-related compulsions.

While it appears drugs of abuse cause habits to emerge prematurely, it is still not clear whether this would result from impaired goal-directed control or strengthened habitual control, or both. It may be that diminished DMS activity unmasks the expression of habits, or alternatively, heightened DLS activity suppresses goal-directed processing. However, it is difficult to assess this neurobiologically, as drugs of abuse produce changes in both DMS and DLS. For example, methamphetamine, amphetamine, cocaine and alcohol have all been shown to upregulate glutamatergic activity (excitatory neurotransmission) in DLS (Cuzon Carlson et al., 2011; Furlong et al., 2018; Ghasemzadeh, Mueller, & Vasudevan, 2009; Li, Kolb, & Robinson, 2003); which presumably leads to an increase in DLS output and thus enhanced S-R habit formation. However, reductions in glutamate receptor and vesicular proteins in DMS have also been evidenced, indicative of diminished R-O, goal-directed, learning (Furlong et al., 2018). In a methamphetamine-associated context, reduced goal-directed behavior was associated with changes in DMS but not DLS, and a DMS-specific pharmacological manipulation was sufficient to recover goal-directed control (Furlong, Supit, Corbit, Killcross, & Balleine, 2017). Yet, goal-directed behavior has also been rescued in alcohol seeking rats via a DLS-specific manipulation (Corbit, Nie, & Janak, 2014). Thus, it seems the bias towards habitual responding in drug addiction reflects changes in both circuits. Because DMS and DLS appear to operate in competition for behavioral control, and drugs of addiction tend to favor the latter over the former, it may be that interventions to upregulate DMS and/or downregulate DLS (or their respective counterpart circuits in humans) will be a valuable contribution to the treatment of addiction. However, the exact mechanisms by which DMS and DLS determine behavioral control are still largely unknown and a clearer understanding of this will be important moving forward.

While a transfer from goal-directed to habitual control offers a parsimonious explanation for why drug use becomes compulsive, there are also other factors to consider. The devaluation test assesses responding for the devalued outcome under extinction (no reward is given for the operant response), which does not reflect naturalistic settings. It is very rare that a drug user would perform the operant responses required to obtain a drug and yet not experience the drug reward. If the drug-reward has been devalued (e.g., it is no longer providing the hedonic effect it once did or it is producing other negative effects), and the individual continues to experience the devalued reward, the S-R associations should eventually extinguish and habitual responding cease. The outcome devaluation paradigm alone cannot explain why the behavior of individuals with substance use disorder does not reflect extinction. That is, it cannot explain why addicts keep seeking a drug that has been devalued and repeatedly experienced in its devalued state. Given this limitation, the theory of goal-directed and habitual control may need to be integrated with other theories in order to fully explain compulsive drug-seeking.

It has been shown that drugs of abuse alter the ability of synapses in the NAc to undergo subsequent synaptic plasticity. For example, repeated cocaine use has been shown to impair NMDA receptor-dependent LTP and LTD at glutamatergic synapses within the NAc core, even after a period of abstinence, in animal models (Kasanetz et al., 2010; Martin, Chen, Hopf, Bowers, & Bonci, 2006). This persistent impairment in synaptic plasticity may help to explain why those with substance use disorder continue to inflexibly and compulsively seek the drug, even in the face of negative consequences (van Huijstee & Mansvelder, 2014). Another possibility is that drug addiction is also characterized by a lack of responsivity to the punishment that accompanies compulsive drug use; an idea that is explored in an upcoming section.

It is also worth noting that the outcome devaluation paradigm fails to capture the interactions that occur between Pavlovian and instrumental conditioning. There is evidently more to the development of drug addiction than a faster transition of instrumental responding from goal-directed to habitual control. Classically conditioned stimuli and operant responses interact in diverse ways, an example of which is demonstrated in the Pavlovian to instrumental transfer effect (PIT), discussed in the next section.

Pavlovian-instrumental-transfer

The PIT procedure is a useful demonstration of stimulus-controlled responding whereby a classically conditioned stimulus can potentiate an instrumental response (Everitt & Robbins, 2005; van den Bos et al., 2004). PIT occurs when a CS, already associated with a reward, influences instrumental responses towards the same or a different reward (Cartoni, Puglisi-Allegra, & Baldassarre, 2013). A typical PIT procedure involves training an animal to learn an association between a CS (e.g., a sound) and an outcome (e.g., food). This produces a Pavlovian conditioned response (e.g., approach behavior to the food) whenever the CS is presented. During a separate training phase, the animal learns that the same outcome (food) can be obtained by performing an instrumental response (e.g., lever pressing). The previously trained CS is never presented during this instrumental phase. In the subsequent test phase, the animal is assessed on lever pressing both in the presence and the absence of the CS under extinction conditions, that is, without the food reward. The animal typically engages in more lever-pressing in the presence of the CS than in its absence, even though the CS and the instrumental response were never previously paired (Cartoni et al., 2013; Holmes, Marchand, & Coutureau, 2010). The extinction test procedure ensures that the CS and the instrumental response are not reinforced simultaneously, eliminating the possibility that the animal learns that the CS is a discriminative stimulus for the instrumental response (i.e., a stimulus that signals the response will be reinforced). Thus, the PIT effect demonstrates that the CS's rewarding properties transfer to the instrumental response in the absence of new learning, so a classically conditioned stimulus can invigorate an independently acquired instrumental response.

According to Everitt and Robbins (2005), an understanding of the PIT procedure can shed light on cue-controlled behaviors like drug-seeking. For example, a drug addict may have learnt that the instrumental response of visiting their drug dealer will lead to a drug reward. However, they are also exposed to a variety of environmental stimuli en route to their destination, as well as other stimuli such as the paraphernalia involved in drug administration and the drug itself. These all have the potential to become classically conditioned. As previously noted, subsequent exposure to these stimuli increases the likelihood of drug-seeking responses and relapse. Indeed, cocaine addicts have been known to engage in frantic behaviors to obtain a drug-reward when in the presence of drug cues. For example, upon seeing a white speck on the ground, they might scramble on hands and knees to obtain it. They may even put what they find into a pipe and try to smoke it (Berridge, 2012; Rosse et al., 1993). Such cue-elicited increases in drug-seeking may be mediated by mechanisms like PIT.

Wyvell and Berridge (2001) used the PIT paradigm to validate the incentive sensitization theory of addiction. The rats first learnt that a sucrose pellet could be obtained by lever-pressing. They were then trained to associate a tone with a sucrose pellet outcome. After completion of both instrumental and Pavlovian training and prior to the test phase, the animals in the experimental group were injected with amphetamine once a day for 6 days (i.e., sensitization), while control animals were injected with saline. Ten days later, the animals were tested on instrumental responding in the presence and absence of the tone, under extinction conditions. The experimental group showed increased cue-triggered lever-pressing compared to control animals (i.e., a larger PIT effect). As both instrumental and Pavlovian training were completed prior to drug sensitization, this drug-induced enhancement of PIT cannot be attributed to associative learning mechanisms. That is, drug administration could not have altered learning during the training phases as it occurred after learning was complete. Because of this, Wyvell and Berridge argued that drug-induced sensitization as a result of prior amphetamine administration enhanced the incentive salience of the CS, or the extent to which the reward was "wanted", causing increased reward-seeking even in the drug-free state. Increased "wanting" behaviors in the sensitized group were equivalent to those demonstrated in a control group which received a single amphetamine microinjection directly before test. Thus, previous exposure to the drug (i.e. sensitization) had a similar effect on PIT to an acute dose. Both results demonstrate that exposure to amphetamine can amplify the expression of PIT in the absence of new learning.

Interestingly, another study incorporating the PIT procedure found evidence of a reduced transfer effect (decreased lever-pressing) when amphetamine or methamphetamine was administered immediately after the Pavlovian phase, during putative consolidation of learning (Hall & Gulley, 2011). Thus, it may be the case that administration of a drug immediately after learning a CS-US association disrupts consolidation processes and the ability of the CS to control instrumental responding. Together, the studies conducted by Wyvell and Berridge (2001) and Hall and Gulley (2011) imply that the timing of drug administration in relation to the acquisition of the CS-US association is vital for its influence on instrumental responding. This points towards the complexity of Pavlovian and instrumental interactions in real-world settings. Some classically conditioned stimuli may potentiate instrumental responding while others may not, depending on their temporal relationship with drug administration.

The PIT procedure can also be adapted to test generalization effects. Generalization refers to the tendency of a learning experience to extend to other similar situations. The ability to generalize is typically adaptive, but certain mental states might restrict or enhance generalization, which can have a profound impact on behavior. In the context of conditioning, generalization might involve the transfer of learned responses from one stimulus to another similar stimulus. The PIT paradigm offers a way of investigating generalization (or transfer) effects by allowing the distinction between general and outcome-specific forms of PIT. Outcome-specific PIT occurs when a CS only potentiates an instrumental response to obtain the same reward, but not a response to obtain an alternative reward. General PIT, however, occurs when a CS influences instrumental responding for a different reward, and therefore represents a form of generalization. This latter form of PIT implies that the general motivating properties of a rewarding outcome, instead of its specific properties, are sufficient to elicit instrumental responding in the presence of a CS.

Corbit, Janak, and Balleine (2007) have developed a paradigm that dissociates between general and outcome-specific PIT (see Table 6). It employs three distinct Pavlovian CSs, each paired with a different outcome (S1-O1, S2-O2 and S3-O3). During the instrumental phase, the animals learn to perform two instrumental responses to obtain two of the outcomes from the previous Pavlovian phase (R1-O1 and R2-O2). Subsequently, the test phase assesses presentation of all three CSs. Outcome-specific PIT is demonstrated by enhanced responding upon presentation of stimuli with the same trained outcome; that is, increased R1 responding upon presentation of S1, and increased R2 responding in the presence of S2. However, the same subjects can also exhibit general PIT, wherein both responses (R1 and R2) increase upon presentation of S3.

It appears that the two types of PIT, outcome-specific and general, rely on different neural circuits. PIT is thought to be a manifestation of a stimulus-outcome-response (S-O-R) chain, whereby the CS retrieves the memory of the reward outcome, which in turn retrieves the memory of its associated instrumental response (Balleine & Ostlund, 2007; Corbit & Balleine, 2016; but see Cartoni, Balleine, & Baldassarre, 2016 for a review of alternative perspectives). The classically conditioned S-O and the instrumentally conditioned R-O (or O-R) associations are learnt separately, thus posing the question of how they become integrated such that a PIT effect is observed in the later test phase.

TABLE 6 Dissociating between general and outcome-specific PIT (Corbit et al., 2007)

Pavlovian training		Instrumental training		Test	Result
S1 Tone	→ O1 Sucrose	R1 Right lever	→ O1 Sucrose	S1? Tone	Presentation of S1 invigorated R1. This demonstrates specific transfer
S2 Siren	→ O2 Pellet	R2 Left lever	→ O2 Pellet	S2? Siren	Presentation of S2 invigorated R2. This demonstrates specific transfer
S3 White noise	→ O3 Polycose			S3? White noise	Presentation of S3 invigorated both R1 and R2. This demonstrates general transfer

Note: S1, S2 and S3, stimulus 1, 2 and 3; *R1, R2, R3*, response 1, 2 and 3; *O1, O2 and O3*, outcome 1, 2 and 3.

Previous research suggests that the NAc, basolateral amygdala (BLA) and the central subnuclei of the amygdala (CeA) play a critical role in this integration. Whereas the shell of the NAc and the BLA seem to be critical for outcome-specific PIT (Corbit, Muir, & Balleine, 2001; Shiflett & Balleine, 2010), the core of the NAc and the CeA seem to be critical for general PIT (Corbit & Balleine, 2005, 2011; Hall, Parkinson, Connor, Dickinson, & Everitt, 2001). Imaging studies in humans have similarly found the involvement of the NAc and the amygdala in PIT (Mendelsohn, Pine, & Schiller, 2014; Talmi, Seymour, Dayan, & Dolan, 2008).

Furthermore, the influence of the NAc on PIT is exerted, at least partially, via dopaminergic neurotransmission. Dopamine agonist infusions in the NAc enhance PIT (Peciña & Berridge, 2013) and dopamine activity in the NAc correlates with the magnitude of PIT (Wassum, Ostlund, Loewinger, & Maidment, 2013). Moreover, neurotransmission via dopamine D1 receptors seems to be critical for PIT (especially outcome-specific PIT; Laurent, Bertran-Gonzalez, Chieng, & Balleine, 2014; Lex & Hauber, 2008), suggesting a key role for the direct (striatonigral) pathway within the basal ganglia.

General and outcome-specific PIT not only differ, at least partially, in their neural substrates, but also in their sensitivity to current outcome values, in a way similar to goal-directed and habitual behavior. For instance, Corbit et al. (2007) found that a change in motivational state differentially affects general and outcome-specific PIT. They introduced an outcome devaluation manipulation via a motivational shift from hunger to satiety, and subsequently assessed both forms of PIT. The shift from hunger to satiety reduced general, but not outcome-specific, transfer. Thus, specific transfer effects seem impervious to a devaluation of the outcome. When a reward is devalued, its specific sensory properties remain unchanged (e.g., its appearance or smell), while its general motivating properties do change (i.e., the reward becomes less desirable). This implies that CSs that trigger specific PIT potentiate instrumental responses to obtain an outcome independent of its current value, and most likely rely on the ability of the CS to retrieve the specific sensory properties of the outcome, which in turn retrieve the memory of and invigorate the specific instrumental response associated with it. Conversely, general transfer effects are subject to outcome devaluation, as responding is dependent on the outcome's current value, and most likely rely on the ability of the CS to retrieve the general motivating properties of the outcome, which invigorate responses for other rewards as well (Cartoni et al., 2013; Corbit & Balleine, 2016).

A parallel can be drawn between general versus specific PIT and goal-directed versus habitual actions, as both general PIT and goal-directed actions are sensitive to outcome devaluation, whereas specific PIT and habitual actions are insensitive to outcome devaluation (Corbit & Balleine, 2016). Drug-seeking may initially be a goal-directed action, as the drug reward may be sought for its value, thereby providing a general source of motivation for the instrumental drug-seeking response (Cartoni et al., 2013). However, when drug-seeking responses become habitual, PIT effects are likely to be specific to the drug reward and no longer dependent on its current value. Consistent with this, a few studies have found enhanced PIT effects and habit-driven performance at the expense of goal-directed behavior in individuals with substance use disorder (Garbusow et al., 2016; Sjoerds, van den Brink, Beekman, Penninx, & Veltman, 2014), although a number of failures to find such differences have also been reported (Hogarth et al., 2018; Hogarth & Chase, 2012; Hogarth, Maynard, & Munafo, 2015).

Sensitivity to reward and punishment and its association with cognitive control and mood

As discussed earlier, the goal-directed and habitual control theory fails to explain why habits do not gradually extinguish when the outcomes experienced are no longer rewarding. That is, it does not explain why drug addiction is characterized by not only compulsive use, but also by an inability to extinguish drug-seeking behaviors despite repeated experience with the immediate negative consequences of the drug (e.g., nausea).

The research reviewed so far has focused on appetitive learning, in which an animal or human learns about the rewarding events that follow neutral stimuli or actions. Yet, learning about the occurrence of aversive events is also fundamental to successful adaptation, as this type of learning may help inhibit actions that lead to unwanted consequences. Addiction seems to be characterized not only by aberrant pursuit of drug rewards, but also by deficits in cognitive and inhibitory control mechanisms (Garavan & Hester, 2007; Garavan & Stout, 2005). Transition from voluntary drug use to compulsive habit appears to involve the disruption of executive functions that leads to the loss of behavioral inhibition (Berke, 2003; Everitt & Robbins, 2005). This loss of behavioral inhibition may lead to excessive pursuit of drug-rewards and an inability to control this pursuit. Thus, drug addiction may be described in terms of the relative strength between the motivation to seek the drug, and the capability to resist it, which may rely on learning about its negative consequences (Jentsch & Pennington, 2014).

Research assessing cognitive deficits within addiction has been focused on behavioral monitoring, error detection and behavioral control (Garavan & Hester, 2007; Jentsch & Pennington, 2014). Such studies suggest that individuals with addiction have difficulty detecting behavioral errors, leading to an impaired ability to correct their responses based on positive and negative feedback (Bechara & Damasio, 2002). One particular paradigm employed to study this is the Iowa Gambling Task. On each trial, participants are required to select a card from a choice of several decks. Crucially, participants receive feedback on their performance (the amount they have gained or lost following their choice), but they are not made aware of the value associated with the different decks of cards. Hence, they must learn from positive and negative feedback which decks are more advantageous in order to maximize their winnings (Garavan & Stout, 2005). Studies incorporating this task have found that drug users make more detrimental choices compared to controls (Bechara & Damasio, 2002). Furthermore, drug users seem to be hypersensitive to rewards, even at the cost of future consequences. That is, they make maladaptive choices, such as skipping smaller, reliable rewards associated with smaller penalties, and instead opting for choices that are associated with larger rewards but also larger penalties, and which are disadvantageous in the long term (Bechara et al., 2001; Bechara & Damasio, 2002; Bechara, Dolan, & Hindes, 2002; Bickel, Yi, Landes, Hill, & Baxter, 2011). To explain this pattern of results, Anderson, Faulkner, Rilee, Yantis, and Marvel (2013) suggested that rewards (both drug, and non-drug related) become more salient in addiction and are more likely to attract attention.

This abnormal reward sensitivity in drug users could be accompanied by a hyposensitivity to punishment (Bechara et al., 2002). For example, Hester, Bell, Foxe, and Garavan (2013) conducted a study to explore monetary punishment on inhibitory control in drug-dependent individuals. Both drug-dependent and control participants completed a Go/NoGo response inhibition task where monetary fines were imposed on failed response inhibition. Compared to controls, drug-dependent individuals not only showed significant impairment of inhibitory control, but also a diminished ability to improve their performance following negative feedback. This is consistent with the idea that substance use disorder is associated with a hyposensitivity to punishment.

Furthermore, Myers et al. (2017) recently reported intact reward learning, but impaired punishment learning, in heroin-dependent individuals on a methadone maintenance program. They used an experimental design that allowed them to independently assess learning from positive feedback versus learning from negative feedback. Participants learned by trial-and-error which of two options to choose in the presence of different stimuli. On reward-based trials, correct choices resulted in positive feedback whereas incorrect choices resulted in no feedback. In contrast, on punishment-based trials, correct choices resulted in no feedback whereas incorrect choices resulted in negative feedback. Performance on reward-based trials was similar in the heroin-dependent group and a control group, but the heroin-dependent group made more errors on punishment-based trials, suggesting a specific impairment in learning from negative feedback.

Conversely, a study conducted by Mintzer and Stitzer (2002) demonstrated that, despite significant impairment in decision-making, participants on a methadone maintenance treatment showed increased salience to punishment in the Iowa Gambling Task. That is, the performance impairment observed in the methadone group was a function of the frequency of penalties, such that there was poorer performance on the decks with fewer penalties compared to the decks with a higher frequency of penalties. Consistent with these findings, and despite seemingly impaired learning from punishment (e.g., Banca, Harrison, & Voon, 2016; Hester et al., 2013; Myers et al., 2017), substance use disorder seems to be paradoxically associated with a heightened sensitivity to stressors and dysphoria (Baskin-Sommers & Foti, 2015; Hellberg, Russell, & Robinson, 2018).

This overall pattern of enhanced sensitivity to rewards, especially drug rewards and their associated cues, and reduced learning from punishment yet potentially increased sensitivity to stressors, may be due to the differential effects of dopamine on D1 and D2 receptor types. The phasic dopamine release in the NAc caused by drugs of abuse differentially affects medium spiny neurons (MSNs) in the direct (striatonigral) and indirect (striatopallidal) basal ganglia pathways (reviewed in Volkow & Morales, 2015). Direct pathway MSNs preferentially express low-affinity dopamine D1 receptors and are involved in learning to repeat actions previously associated with reward, whereas indirect pathway MSNs preferentially express high-affinity dopamine D2 receptors and are involved in learning to inhibit actions that were previously punished (Kravitz & Kreitzer, 2012). Surges of dopamine release activate D1 receptors but inhibit D2 receptors, resulting in activation of the direct pathway and inhibition of the indirect pathway.

Thus, abnormally high dopamine release in response to drug consumption may have different effects on learning from rewards than on learning from punishments. D1-expressing direct pathway MSNs are thought to be responsive to appetitive PEs signaled by phasic dopamine release. Such appetitive PE signaling, which is normally caused by a surprising reward, would induce neuroplasticity in the direct pathway and increase the likelihood of performing the action that resulted in the rewarding state. In contrast, D2-expressing indirect pathway MSNs are thought to be responsive to aversive PEs signaled by phasic dopamine dips. Such dips in neuronal firing are usually generated by an aversive event or the unexpected omission of an appetitive event, and are thought to strengthen the indirect pathway resulting in a reduced likelihood of performing the action which resulted in the undesirable state. Thus dopamine-induced neuroplasticity in the direct and indirect pathways would shape the behavioral repertoire of an organism such that actions that were followed by positive consequences would be more likely to be repeated whereas actions that were followed by negative consequences would gradually become less likely to be repeated (Cohen & Frank, 2009; Frank, Seeberger, & O'Reilly, 2004). High bursts of dopamine caused by drugs of abuse could, however, shift the balance in favor of neuroplasticity in the direct pathway and against neuroplasticity in the indirect pathway. This is because higher dopamine levels might artificially create or enhance appetitive PE signals, but prevent dopamine dips in response to aversive events from occurring. This would result in enhanced learning to perform actions that result in reward, and a reduced ability to learn to avoid performing actions that result in punishment, a prediction that has received support from pharmacological manipulations in healthy individuals (Frank & O'Reilly, 2006).

This may explain why environmental cues and behaviors occurring in close proximity to drug intake rapidly become associated with the drug reward, whereas the experienced negative consequences have a smaller relative influence. This theory could, in principle, explain why habitual drug-seeking does not extinguish despite repeated experience with the negative consequences of drug consumption. That is, reduced learning from aversive PEs would prevent an organism from extinguishing a habitual response for a devalued drug outcome, even if the devalued outcome is repeatedly experienced.

Consistent with this theory, a higher ratio of D1-to-D2 signaling during cocaine intoxication has been reported (Park, Volkow, Pan, & Du, 2013), and repeated exposure to most drugs results in down-regulation of striatal dopamine D2 receptors (Volkow & Baler, 2014), which might contribute to compulsive drug taking by shifting the balance between reward and punishment learning in favor of reward seeking despite large costs. This would explain why a higher preference for risky choices that result in long-term loss has been observed in individuals with substance use disorder in laboratory tasks (e.g., Bechara et al., 2001).

Furthermore, this shift towards higher D1 signaling could also enhance cue-induced drug-seeking responses, as measured via the PIT paradigm (reviewed in the previous section). As both outcome-specific and general PIT have been shown to depend on neurotransmission via dopamine D1 receptors, a shift toward enhanced D1-mediated transmission could result in potentiated PIT effects, further contributing to the compulsive nature of addiction.

Finally, the large dopamine surges caused by drugs of abuse also activate dopamine D2 autoreceptors, which inhibit further dopamine release (Bello et al., 2011). In the longer term, this could contribute to the down-regulation of experienced reward triggered by natural reinforcers, which might reduce motivation for non-drug rewards and "re-calibrate" the reward system so that only rewards that produce abnormally high surges of dopamine, such as drugs of abuse, activate it (Zijlstra, Veltman, Booij, van den Brink, & Franken, 2009). This could explain why addiction is associated with an enhanced susceptibility to stressors and dysphoric states (Baskin-Sommers & Foti, 2015; Koob, 2013), despite reduced cognitive, goal-directed control and impaired learning from punishment.

Conclusion

Here we have reviewed a number of theories that propose different potential mechanisms underlying the development and maintenance of addiction. Although neither theory fully explains all aspects of this complex disorder, each nevertheless provides useful insights into different facets of addiction. Moreover, rather than being contradictory, many aspects of these theories are complementary, and together they may provide a more complete understanding of addiction and its potential treatment avenues.

A particular strength of this field is the long history of computational modeling whereby mathematical implementations have been developed to capture the specific computations assumed to drive learning and behavior. So far, there have been excellent efforts to formalize the theories we have presented in mathematical terms, allowing researchers to generate and test precise predictions, and to refine these mathematical implementations in light of new data. For instance, the incentive salience theory has been formalized by Zhang, Berridge, Tindell, Smith, and Aldridge (2009) and Anselme (2015). Both mathematical models attribute incentive salience to CSs as a function of reinforcement history and current motivation state. The addition of the latter component allows these models to explain cue-triggered "wanting" behaviors, one of the potential mechanisms altered in drug addiction.

Furthermore, goal-directed and habitual behavior have been simulated with model-based and model-free reinforcement learning algorithms, respectively (Daw, Niv, & Dayan, 2005). Model-based algorithms build a complex representation of the environment that includes the likely transitions from a given state to possible actions, and from possible actions to the likely ensuing outcomes. Decisions are therefore based on a search through all possible futures for the trajectory that yields the most desired outcome. Because choices are ultimately driven by a representation of the outcome, these algorithms are sensitive to sudden changes in outcome value and hence mimic goal-directed control. In contrast, model-free algorithms learn the accumulated value (the predicted future reward) of an action in a given state. Because the algorithm only remembers the value of each experienced state-action pair, decisions to perform an action are driven by these learnt values rather than by a representation of the outcome. Model-free algorithms are therefore more computationally efficient, yet also more inflexible, as sudden changes in outcome value will not affect the stored state-action values. Consequently, these algorithms predict that outcome devaluation procedures will not influence instrumental choices, as is the case when behavior is under habitual control. Daw et al. (2005) further propose that both algorithms operate simultaneously, but the organism arbitrates between them depending on the uncertainty of their predictions. This allows the model to predict a shift from goal-directed behavior to habitual behavior, or a lack thereof, for different experimental manipulations (Daw et al., 2005).

Reinforcement learning models have also been implemented more precisely as mathematical models that simulate learning in the direct and indirect pathways of the basal ganglia. These models can make precise predictions regarding learning from positive and negative feedback under different tonic dopamine levels, as well as their reliance on neuroplasticity via dopamine D1 or D2 receptors (see Schroll & Hamker, 2013 for a review). Consequently, this type of model can explain why drugs that affect the dopamine system impair learning from negative feedback, and as a result prevent the extinction of acquired habitual drug-seeking behaviors.

Future endeavors could integrate complementary aspects of these models, for example by allowing both reinforcement learning and motivation states to influence behavior (e.g., Dayan & Berridge, 2014). Furthermore, the model proposed by Daw et al. (2005) is, in principle, compatible with some implementations of learning in the basal ganglia. Developing more complex models would allow one to test whether various assumptions assumed by different theories are necessary or sufficient to explain behavior, for example, by comparing model versions that incorporate or deduct various theoretical assumptions via model selection. A similar approach has been used in other contexts, for example, to investigate the role of reinforcement learning and working memory in performance (Collins & Frank, 2012).

Finally, most research so far has focused on documenting and understanding performance differences observed in individuals with substance use disorder, but future research could also investigate individual differences in learning that might predict vulnerability to substance use disorder, such as a pre-morbid propensity to learn more from rewarding experiences than from punishment, or abnormal salience attribution to reward-predicting stimuli. Longitudinal studies would be particularly useful in this regard, coupled with computational modeling that could shed light on the underlying mechanisms generating different trajectories in different individuals. Such knowledge would not only provide a better understanding of the factors that could trigger compulsive drug use but may also help identify individuals at risk of developing an addiction and enable early intervention.

References

Aberman, J. E., & Salamone, J. D. (1999). Nucleus accumbens dopamine depletions make rats more sensitive to high ratio requirements but do not impair primary food reinforcement. *Neuroscience*, *92*(2), 545–552. https://doi.org/10.1016/S0306-4522(99)00004-4.

Adams, C. D. (1982). Variations in the sensitivity of instrumental responding to reinforcer devaluation. *The Quarterly Journal of Experimental Psychology Section B*, *34*(2), 77–98. https://doi.org/10.1080/14640748208400878.

Adams, C. D., & Dickinson, A. (1981). Instrumental responding following reinforcer devaluation. *The Quarterly Journal of Experimental Psychology Section B*, *33*(2b), 109–121. https://doi.org/10.1080/14640748108400816.

Anderson, B. A., Faulkner, M. L., Rilee, J. J., Yantis, S., & Marvel, C. L. (2013). Attentional bias for nondrug reward is magnified in addiction. *Experimental and Clinical Psychopharmacology*, *21*(6), 499–506. https://doi.org/10.1037/a0034575.

Anselme, P. (2015). Incentive salience attribution under reward uncertainty: A pavlovian model. *Behavioural Processes*, *111*, 6–18. https://doi.org/10.1016/j.beproc.2014.10.016.

Arulkadacham, L. J., Richardson, B., Staiger, P. K., Kambouropoulos, N., O'Donnell, R. L., & Ling, M. (2017). Dissociation between wanting and liking for alcohol and caffeine: A test of the incentive sensitisation theory. *Journal of Psychopharmacology*, *31*(7), 927–933. https://doi.org/10.1177/0269881117711711.

Badanich, K. A., Adler, K. J., & Kirstein, C. L. (2006). Adolescents differ from adults in cocaine conditioned place preference and cocaine-induced dopamine in the nucleus accumbens septi. *European Journal of Pharmacology*, *550*(1), 95–106. https://doi.org/10.1016/j.ejphar.2006.08.034.

Balleine, B. W., & O'Doherty, J. P. (2009). Human and rodent homologies in action control: Corticostriatal determinants of goal-directed and habitual action. *Neuropsychopharmacology*, *35*(1), 48–69. https://doi.org/10.1038/npp.2009.131.

Balleine, B. W., & Ostlund, S. B. (2007). Still at the choice-point: Action selection and initiation in instrumental conditioning. *Annals of the New York Academy of Sciences*, *1104*(1), 147–171. https://doi.org/10.1196/annals.1390.006.

Banca, P., Harrison, N. A., & Voon, V. (2016). Compulsivity across the pathological misuse of drug and non-drug rewards. *Frontiers in Behavioral Neuroscience*, *10*, 154. https://doi.org/10.3389/fnbeh.2016.00154.

Barto, A. G., Sutton, R. S., & Brouwer, P. S. (1981). Associative search network: A reinforcement learning associative memory. *Biological Cybernetics*, *40*(3), 201–211. https://doi.org/10.1007/bf00453370.

Baskin-Sommers, A. R., & Foti, D. (2015). Abnormal reward functioning across substance use disorders and major depressive disorder: Considering reward as a transdiagnostic mechanism. *International Journal of Psychophysiology*, *98*(2 Pt 2), 227–239. https://doi.org/10.1016/j.ijpsycho.2015.01.011.

Bechara, A., & Damasio, H. (2002). Decision-making and addiction (part i): Impaired activation of somatic states in substance dependent individuals when pondering decisions with negative future consequences. *Neuropsychologia*, *40*(10), 1675–1689. https://doi.org/10.1016/S0028-3932(02)00015-5.

Bechara, A., Dolan, S., Denburg, N., Hindes, A., Anderson, S. W., & Nathan, P. E. (2001). Decision-making deficits, linked to a dysfunctional ventromedial prefrontal cortex, revealed in alcohol and stimulant abusers. *Neuropsychologia*, *39*(4), 376–389. https://doi.org/10.1016/S0028-3932(00)00136-6.

Bechara, A., Dolan, S., & Hindes, A. (2002). Decision-making and addiction (part ii): Myopia for the future or hypersensitivity to reward? *Neuropsychologia*, *40*(10), 1690–1705. https://doi.org/10.1016/S0028-3932(02)00016-7.

Bello, E. P., Mateo, Y., Gelman, D. M., Noaín, D., Shin, J. H., Low, M. J., … Rubinstein, M. (2011). Cocaine supersensitivity and enhanced motivation for reward in mice lacking dopamine d2 autoreceptors. *Nature Neuroscience*, *14*(8), 1033. https://doi.org/10.1038/nn.2862.

Berke, J. D. (2003). Learning and memory mechanisms involved in compulsive drug use and relapse. In J. Q. Wang (Ed.), *Drugs of abuse: Neurological reviews and protocols* (pp. 75–101). Totowa, NJ: Humana Press.

Berke, J. D., & Hyman, S. E. (2000). Addiction, dopamine, and the molecular mechanisms of memory. *Neuron, 25*(3), 515–532. https://doi.org/10.1016/S0896-6273(00)81056-9.

Berridge, K. C. (2007). The debate over dopamine's role in reward: The case for incentive salience. *Psychopharmacology, 191*(3), 391–431. https://doi.org/10.1007/s00213-006-0578-x.

Berridge, K. C. (2012). From prediction error to incentive salience: Mesolimbic computation of reward motivation. *European Journal of Neuroscience, 35*(7), 1124–1143. https://doi.org/10.1111/j.1460-9568.2012.07990.x.

Berridge, K. C. (2017). Is addiction a brain disease? *Neuroethics, 10*(1), 29–33. https://doi.org/10.1007/s12152-016-9286-3.

Berridge, K. C., & Kringelbach, M. L. (2015). Pleasure systems in the brain. *Neuron, 86*(3), 646–664. https://doi.org/10.1016/j.neuron.2015.02.018.

Berridge, K. C., & O'Doherty, J. P. (2014). Chapter 18: From experienced utility to decision utility. In P. W. Glimcher & E. Fehr (Eds.), *Neuroeconomics* (2nd ed, pp. 335–351). San Diego: Academic Press.

Berridge, K. C., & Robinson, T. E. (1998). What is the role of dopamine in reward: Hedonic impact, reward learning, or incentive salience? *Brain Research Reviews, 28*(3), 309–369. https://doi.org/10.1016/S0165-0173(98)00019-8.

Berridge, K. C., & Robinson, T. E. (2016). Liking, wanting, and the incentive-sensitization theory of addiction. *The American Psychologist, 71*(8), 670–679. https://doi.org/10.1037/amp0000059.

Bickel, W. K., Yi, R., Landes, R. D., Hill, P. F., & Baxter, C. (2011). Remember the future: Working memory training decreases delay discounting among stimulant addicts. *Biological Psychiatry, 69*(3), 260–265. https://doi.org/10.1016/j.biopsych.2010.08.017.

Blair, H. T., Schafe, G. E., Bauer, E. P., Rodrigues, S. M., & LeDoux, J. E. (2001). Synaptic plasticity in the lateral amygdala: A cellular hypothesis of fear conditioning. *Learning and Memory, 8*(5), 229–242. https://doi.org/10.1101/lm.30901.

Bocklisch, C., Pascoli, V., Wong, J. C., House, D. R., Yvon, C., de Roo, M., ... Luscher, C. (2013). Cocaine disinhibits dopamine neurons by potentiation of gaba transmission in the ventral tegmental area. *Science, 341*(6153), 1521–1525. https://doi.org/10.1126/science.1237059.

Boileau, I., Dagher, A., Leyton, M., Gunn, R. N., Baker, G. B., Diksic, M., & Benkelfat, C. (2006). Modeling sensitization to stimulants in humans: An [11c]raclopride/positron emission tomography study in healthy men. *Archives of General Psychiatry, 63*(12), 1386–1395. https://doi.org/10.1001/archpsyc.63.12.1386.

Boileau, I., Dagher, A., Leyton, M., Welfeld, K., Booij, L., Diksic, M., & Benkelfat, C. (2007). Conditioned dopamine release in humans: A positron emission tomography [11c]raclopride study with amphetamine. *Journal of Neuroscience, 27*(15), 3998–4003. https://doi.org/10.1523/jneurosci.4370-06.2007.

Bossert, J. M., Marchant, N. J., Calu, D. J., & Shaham, Y. (2013). The reinstatement model of drug relapse: Recent neurobiological findings, emerging research topics, and translational research. *Psychopharmacology, 229*(3), 453–476. https://doi.org/10.1007/s00213-013-3120-y.

Cagniard, B., Balsam, P. D., Brunner, D., & Zhuang, X. (2006). Mice with chronically elevated dopamine exhibit enhanced motivation, but not learning, for a food reward. *Neuropsychopharmacology, 31*(7), 1362–1370. https://doi.org/10.1038/sj.npp.1300966.

Cannon, C. M., & Palmiter, R. D. (2003). Reward without dopamine. *Journal of Neuroscience, 23*(34), 10827–10831. https://doi.org/10.1523/JNEUROSCI.23-34-10827.2003.

Cartoni, E., Balleine, B. W., & Baldassarre, G. (2016). Appetitive pavlovian-instrumental transfer: A review. *Neuroscience and Biobehavioral Reviews*, *71*, 829–848. https://doi.org/10.1016/j.neubiorev.2016.09.020.

Cartoni, E., Puglisi-Allegra, S., & Baldassarre, G. (2013). The three principles of action: A pavlovian-instrumental transfer hypothesis. *Frontiers in Behavioral Neuroscience*, *7*(153), 1–11. https://doi.org/10.3389/fnbeh.2013.00153.

Cavallo, J. S., Ruiz, N. A., & de Wit, H. (2016). Extinction of conditioned responses to methamphetamine-associated stimuli in healthy humans. *Psychopharmacology*, *233*(13), 2489–2502. https://doi.org/10.1007/s00213-016-4297-7.

Chase, H. W., Kumar, P., Eickhoff, S. B., & Dombrovski, A. Y. (2015). Reinforcement learning models and their neural correlates: An activation likelihood estimation meta-analysis. *Cognitive, Affective, & Behavioral Neuroscience*, *15*(2), 435–459. https://doi.org/10.3758/s13415-015-0338-7.

Chen, B. T., Bowers, M. S., Martin, M., Hopf, F. W., Guillory, A. M., Carelli, R. M., … Bonci, A. (2008). Cocaine but not natural reward self-administration nor passive cocaine infusion produces persistent ltp in the vta. *Neuron*, *59*(2), 288–297. https://doi.org/10.1016/j.neuron.2008.05.024.

Childress, A. R., Mozley, P. D., McElgin, W., Fitzgerald, J., Reivich, M., & O'Brien, C. P. (1999). Limbic activation during cue-induced cocaine craving. *American Journal of Psychiatry*, *156*(1), 11–18. https://doi.org/10.1176/ajp.156.1.11.

Childs, E., & de Wit, H. (2009). Amphetamine-induced place preference in humans. *Biological Psychiatry*, *65*(10), 900–904. https://doi.org/10.1016/j.biopsych.2008.11.016.

Christian, K. M., & Thompson, R. F. (2003). Neural substrates of eyeblink conditioning: Acquisition and retention. *Learning and Memory*, *10*(6), 427–455. https://doi.org/10.1101/lm.59603.

Ciccocioppo, R., Martin-Fardon, R., & Weiss, F. (2004). Stimuli associated with a single cocaine experience elicit long-lasting cocaine-seeking. *Nature Neuroscience*, *7*(5), 495–496. https://doi.org/10.1038/nn1219.

Clemens, K. J., Castino, M. R., Cornish, J. L., Goodchild, A. K., & Holmes, N. M. (2014). Behavioral and neural substrates of habit formation in rats intravenously self-administering nicotine. *Neuropsychopharmacology*, *39*(11), 2584. https://doi.org/10.1038/npp.2014.111.

Cohen, M. X., & Frank, M. J. (2009). Neurocomputational models of basal ganglia function in learning, memory and choice. *Behavioural Brain Research*, *199*(1), 141–156. https://doi.org/10.1016/j.bbr.2008.09.029.

Collins, A. G., & Frank, M. J. (2012). How much of reinforcement learning is working memory, not reinforcement learning? A behavioral, computational, and neurogenetic analysis. *European Journal of Neuroscience*, *35*(7), 1024–1035. https://doi.org/10.1111/j.1460-9568.2011.07980.x.

Corbit, L. H. (2018). Understanding the balance between goal-directed and habitual behavioral control. *Current Opinion in Behavioral Sciences*, *20*, 161–168. https://doi.org/10.1016/j.cobeha.2018.01.010.

Corbit, L. H., & Balleine, B. W. (2005). Double dissociation of basolateral and central amygdala lesions on the general and outcome-specific forms of pavlovian-instrumental transfer. *Journal of Neuroscience*, *25*(4), 962–970. https://doi.org/10.1523/jneurosci.4507-04.2005.

Corbit, L. H., & Balleine, B. W. (2011). The general and outcome-specific forms of pavlovian-instrumental transfer are differentially mediated by the nucleus accumbens core and shell. *Journal of Neuroscience*, *31*(33), 11786–11794. https://doi.org/10.1523/jneurosci.2711-11.2011.

Corbit, L. H., & Balleine, B. W. (2016). Learning and motivational processes contributing to pavlovian-instrumental transfer and their neural bases: Dopamine and beyond. In E. H. Simpson & P. D. Balsam (Eds.), *Behavioral neuroscience of motivation* (pp. 259–289). Cham: Springer International Publishing.

Corbit, L. H., Chieng, B. C., & Balleine, B. W. (2014). Effects of repeated cocaine exposure on habit learning and reversal by n-acetylcysteine. *Neuropsychopharmacology*, *39*(8), 1893–1901. https://doi.org/10.1038/npp.2014.37.

Corbit, L. H., Janak, P. H., & Balleine, B. W. (2007). General and outcome-specific forms of pavlovian-instrumental transfer: The effect of shifts in motivational state and inactivation of the ventral tegmental area. *European Journal of Neuroscience*, *26*(11), 3141–3149. https://doi.org/10.1111/j.1460-9568.2007.05934.x.

Corbit, L. H., Muir, J. L., & Balleine, B. W. (2001). The role of the nucleus accumbens in instrumental conditioning: Evidence of a functional dissociation between accumbens core and shell. *Journal of Neuroscience*, *21*(9), 3251–3260. https://doi.org/10.1523/jneurosci.21-09-03251.2001.

Corbit, L. H., Nie, H., & Janak, P. H. (2012). Habitual alcohol seeking: Time course and the contribution of subregions of the dorsal striatum. *Biological Psychiatry*, *72*(5), 389–395. https://doi.org/10.1016/j.biopsych.2012.02.024.

Corbit, L. H., Nie, H., & Janak, P. H. (2014). Habitual responding for alcohol depends upon both ampa and d2 receptor signaling in the dorsolateral striatum. *Frontiers in Behavioral Neuroscience*, *8*(301), 1–9. https://doi.org/10.3389/fnbeh.2014.00301.

Cousins, M. S., Atherton, A., Turner, L., & Salamone, J. D. (1996). Nucleus accumbens dopamine depletions alter relative response allocation in a t-maze cost/benefit task. *Behavioural Brain Research*, *74*(1), 189–197. https://doi.org/10.1016/0166-4328(95)00151-4.

Cousins, M. S., Wei, W., & Salamone, J. D. (1994). Pharmacological characterization of performance on a concurrent lever pressing/feeding choice procedure: Effects of dopamine antagonist, cholinomimetic, sedative and stimulant drugs. *Psychopharmacology*, *116*(4), 529–537. https://doi.org/10.1007/BF02247489.

Covey, D. P., Roitman, M. F., & Garris, P. A. (2014). Illicit dopamine transients: Reconciling actions of abused drugs. *Trends in Neurosciences*, *37*(4), 200–210. https://doi.org/10.1016/j.tins.2014.02.002.

Cuzon Carlson, V. C., Seabold, G. K., Helms, C. M., Garg, N., Odagiri, M., Rau, A. R., … Grant, K. A. (2011). Synaptic and morphological neuroadaptations in the putamen associated with long-term, relapsing alcohol drinking in primates. *Neuropsychopharmacology*, *36*(12), 2513–2528. https://doi.org/10.1038/npp.2011.140.

Daberkow, D. P., Brown, H. D., Bunner, K. D., Kraniotis, S. A., Doellman, M. A., Ragozzino, M. E., … Roitman, M. F. (2013). Amphetamine paradoxically augments exocytotic dopamine release and phasic dopamine signals. *Journal of Neuroscience*, *33*(2), 452–463. https://doi.org/10.1523/jneurosci.2136-12.2013.

Daley, D. C. (2013). Family and social aspects of substance use disorders and treatment. *Journal of Food and Drug Analysis*, *21*(4), 414–420. https://doi.org/10.1016/j.jfda.2013.09.038.

Darvas, M., Fadok, J. P., & Palmiter, R. D. (2011). Requirement of dopamine signaling in the amygdala and striatum for learning and maintenance of a conditioned avoidance response. *Learning and Memory*, *18*(3), 136–143.

Daw, N. D., Niv, Y., & Dayan, P. (2005). Uncertainty-based competition between prefrontal and dorsolateral striatal systems for behavioral control. *Nature Neuroscience*, *8*(12), 1704–1711. https://doi.org/10.1038/nn1560.

Daw, N. D., & O'Doherty, J. P. (2014). Chapter 21: Multiple systems for value learning. In P. W. Glimcher & E. Fehr (Eds.), *Neuroeconomics* (2nd ed, pp. 393–410). San Diego: Academic Press.

Dayan, P., & Berridge, K. C. (2014). Model-based and model-free pavlovian reward learning: Revaluation, revision, and revelation. *Cognitive, Affective, & Behavioral Neuroscience*, *14*(2), 473–492. https://doi.org/10.3758/s13415-014-0277-8.

Degenhardt, L., Charlson, F., Ferrari, A., Santomauro, D., Erskine, H., Mantilla-Herrara, A., … Griswold, M. (2018). The global burden of disease attributable to alcohol and drug use in 195 countries and territories, 1990–2016: A systematic analysis for the global burden of disease study 2016. *Lancet Psychiatry*, *5*(12), 987–1012. https://doi.org/10.1016/S2215-0366(18)30337-7.

Di Chiara, G., & Imperato, A. (1988). Drugs abused by humans preferentially increase synaptic dopamine concentrations in the mesolimbic system of freely moving rats. *Proceedings of the National Academy of Sciences of the United States of America*, *85*(14), 5274–5278. https://doi.org/10.1073/pnas.85.14.5274.

Dickinson, A. (2012). Associative learning and animal cognition. *Philosophical Transactions of the Royal Society, B: Biological Sciences*, *367*(1603), 2733–2742. https://doi.org/10.1098/rstb.2012.0220.

Dickinson, A., Wood, N., & Smith, J. W. (2002). Alcohol seeking by rats: Action or habit? *The Quarterly Journal of Experimental Psychology: Section B*, *55*(4), 331–348. https://doi.org/10.1080/0272499024400016.

Dong, Y., & Nestler, E. J. (2014). The neural rejuvenation hypothesis of cocaine addiction. *Trends in Pharmacological Sciences*, *35*(8), 374–383. https://doi.org/10.1016/j.tips.2014.05.005.

Drevets, W. C., Gautier, C., Price, J. C., Kupfer, D. J., Kinahan, P. E., Grace, A. A., … Mathis, C. A. (2001). Amphetamine-induced dopamine release in human ventral striatum correlates with euphoria. *Biological Psychiatry*, *49*(2), 81–96. https://doi.org/10.1016/S0006-3223(00)01038-6.

Epstein, D. H., Preston, K. L., Stewart, J., & Shaham, Y. (2006). Toward a model of drug relapse: An assessment of the validity of the reinstatement procedure. *Psychopharmacology*, *189*(1), 1–16. https://doi.org/10.1007/s00213-006-0529-6.

Eshel, N., Bukwich, M., Rao, V., Hemmelder, V., Tian, J., & Uchida, N. (2015). Arithmetic and local circuitry underlying dopamine prediction errors. *Nature*, *525*(7568), 243–246. https://doi.org/10.1038/nature14855.

Evans, A. H., Pavese, N., Lawrence, A. D., Tai, Y. F., Appel, S., Doder, M., … Piccini, P. (2006). Compulsive drug use linked to sensitized ventral striatal dopamine transmission. *Annals of Neurology*, *59*(5), 852–858. https://doi.org/10.1002/ana.20822.

Everitt, B. J. (2014). Neural and psychological mechanisms underlying compulsive drug seeking habits and drug memories—Indications for novel treatments of addiction. *European Journal of Neuroscience*, *40*(1), 2163–2182. https://doi.org/10.1111/ejn.12644.

Everitt, B. J., Cardinal, R. N., Parkinson, J. A., & Robbins, T. W. (2003). Appetitive behavior: Impact of amygdala-dependent mechanisms of emotional learning. *Annals of the New York Academy of Sciences*, *985*, 233–250. https://doi.org/10.1111/j.1749-6632.2003.tb07085.x.

Everitt, B. J., Dickinson, A., & Robbins, T. W. (2001). The neuropsychological basis of addictive behaviour. *Brain Research Reviews*, *36*(2), 129–138. https://doi.org/10.1016/S0165-0173(01)00088-1.

Everitt, B. J., Hutcheson, D. M., Ersche, K. D., Pelloux, Y., Dalley, J. W., & Robbins, T. W. (2007). The orbital prefrontal cortex and drug addiction in laboratory animals and humans. *Annals of the New York Academy of Sciences*, *1121*, 576–597. https://doi.org/10.1196/annals.1401.022.

Everitt, B. J., & Robbins, T. W. (2005). Neural systems of reinforcement for drug addiction: From actions to habits to compulsion. *Nature Neuroscience*, *8*(11), 1481–1489. https://doi.org/10.1038/nn1579.

Fanselow, M. S., & Poulos, A. M. (2005). The neuroscience of mammalian associative learning. *Annual Review of Psychology*, *56*(1), 207–234. https://doi.org/10.1146/annurev.psych.56.091103.070213.

Farooqi, I. S., Bullmore, E., Keogh, J., Gillard, J., O'Rahilly, S., & Fletcher, P. C. (2007). Leptin regulates striatal regions and human eating behavior. *Science*, *317*(5843), 1355. https://doi.org/10.1126/science.1144599.

Farrar, A. M., Font, L., Pereira, M., Mingote, S., Bunce, J. G., Chrobak, J. J., & Salamone, J. D. (2008). Forebrain circuitry involved in effort-related choice: Injections of the gabaa agonist muscimol into ventral pallidum alter response allocation in food-seeking behavior. *Neuroscience*, *152*(2), 321–330. https://doi.org/10.1016/j.neuroscience.2007.12.034.

Farrar, A. M., Segovia, K. N., Randall, P. A., Nunes, E. J., Collins, L. E., Stopper, C. M., ... Salamone, J. D. (2010). Nucleus accumbens and effort-related functions: Behavioral and neural markers of the interactions between adenosine a2a and dopamine d2 receptors. *Neuroscience*, *166*(4), 1056–1067. https://doi.org/10.1016/j.neuroscience.2009.12.056.

Ferenczi, E. A., Zalocusky, K. A., Liston, C., Grosenick, L., Warden, M. R., Amatya, D., ... Ramakrishnan, C. (2016). Prefrontal cortical regulation of brainwide circuit dynamics and reward-related behavior. *Science*, *351*(6268), 41–53.

Ferster, C. B., & Skinner, B. F. (1957). *Schedules of reinforcement*. East Norwalk, CT: Appleton-Century-Crofts.

FitzGerald, T. H. B., Dolan, R. J., & Friston, K. (2015). Dopamine, reward learning, and active inference. *Frontiers in Computational Neuroscience*, *9*(136), 1–16. https://doi.org/10.3389/fncom.2015.00136.

Frank, M. J., & O'Reilly, R. C. (2006). A mechanistic account of striatal dopamine function in human cognition: Psychopharmacological studies with cabergoline and haloperidol. *Behavioral Neuroscience*, *120*(3), 497–517. https://doi.org/10.1037/0735-7044.120.3.497.

Frank, M. J., Seeberger, L. C., & O'Reilly, R. C. (2004). By carrot or by stick: Cognitive reinforcement learning in parkinsonism. *Science*, *306*(5703), 1940–1943. https://doi.org/10.1126/science.1102941.

Franken, I. H., de Haan, H. A., van der Meer, C. W., Haffmans, P. M., & Hendriks, V. M. (1999). Cue reactivity and effects of cue exposure in abstinent posttreatment drug users. *Journal of Substance Abuse Treatment*, *16*(1), 81–85. https://doi.org/10.1016/S0740-5472(98)00004-X.

Furlong, T. M., Corbit, L. H., Brown, R. A., & Balleine, B. W. (2018). Methamphetamine promotes habitual action and alters the density of striatal glutamate receptor and vesicular proteins in dorsal striatum. *Addiction Biology*, *23*(3), 857–867. https://doi.org/10.1111/adb.12534.

Furlong, T. M., Supit, A. S. A., Corbit, L. H., Killcross, S., & Balleine, B. W. (2017). Pulling habits out of rats: Adenosine 2a receptor antagonism in dorsomedial striatum rescues methamphetamine-induced deficits in goal-directed action. *Addiction Biology*, *22*(1), 172–183. https://doi.org/10.1111/adb.12316.

Garavan, H., & Hester, R. (2007). The role of cognitive control in cocaine dependence. *Neuropsychology Review*, *17*(3), 337–345. https://doi.org/10.1007/s11065-007-9034-x.

Garavan, H., Pankiewicz, J., Bloom, A., Cho, J. K., Sperry, L., Ross, T. J., ... Stein, E. A. (2000). Cue-induced cocaine craving: Neuroanatomical specificity for drug users and drug stimuli. *American Journal of Psychiatry*, *157*(11), 1789–1798. https://doi.org/10.1176/appi.ajp.157.11.1789.

Garavan, H., & Stout, J. C. (2005). Neurocognitive insights into substance abuse. *Trends in Cognitive Sciences*, *9*(4), 195–201. https://doi.org/10.1016/j.tics.2005.02.008.

Garbusow, M., Schad, D. J., Sebold, M., Friedel, E., Bernhardt, N., Koch, S. P., ... Heinz, A. (2016). Pavlovian-to-instrumental transfer effects in the nucleus accumbens relate to relapse in alcohol dependence. *Addiction Biology*, *21*(3), 719–731. https://doi.org/10.1111/adb.12243.

Ghasemzadeh, M. B., Mueller, C., & Vasudevan, P. (2009). Behavioral sensitization to cocaine is associated with increased glutamate receptor trafficking to the postsynaptic density after extended withdrawal period. *Neuroscience*, *159*(1), 414–426. https://doi.org/10.1016/j.neuroscience.2008.10.027.

Gillan, C. M., Robbins, T. W., Sahakian, B. J., van den Heuvel, O. A., & van Wingen, G. (2016). The role of habit in compulsivity. *European Neuropsychopharmacology*, *26*(5), 828–840. https://doi.org/10.1016/j.euroneuro.2015.12.033.

Gipson, C. D., Kupchik, Y. M., & Kalivas, P. W. (2014). Rapid, transient synaptic plasticity in addiction. *Neuropharmacology*, *76*(Pt B), 276–286. https://doi.org/10.1016/j.neuropharm.2013.04.032.

Gipson, C. D., Kupchik, Y. M., Shen, H., Reissner, K. J., Thomas, C. A., & Kalivas, P. W. (2013). Relapse induced by cues predicting cocaine depends on rapid, transient synaptic potentiation. *Neuron*, *77*(5), 867–872. https://doi.org/10.1016/j.neuron.2013.01.005.

Goldstein, R. Z., & Volkow, N. D. (2011). Dysfunction of the prefrontal cortex in addiction: Neuroimaging findings and clinical implications. *Nature Reviews Neuroscience*, *12*(11), 652–669. https://doi.org/10.1038/nrn3119.

Grant, S., London, E. D., Newlin, D. B., Villemagne, V. L., Liu, X., Contoreggi, C., … Margolin, A. (1996). Activation of memory circuits during cue-elicited cocaine craving. *Proceedings of the National Academy of Sciences of the United States of America*, *93*(21), 12040–12045. https://doi.org/10.1073/pnas.93.21.12040.

Gremel, C. M., Chancey, J. H., Atwood, B. K., Luo, G., Neve, R., Ramakrishnan, C., … Costa, R. M. (2016). Endocannabinoid modulation of orbitostriatal circuits gates habit formation. *Neuron*, *90*(6), 1312–1324. https://doi.org/10.1016/j.neuron.2016.04.043.

Gremel, C. M., & Costa, R. M. (2013). Orbitofrontal and striatal circuits dynamically encode the shift between goal-directed and habitual actions. *Nature Communications*, *4*, 2264. https://doi.org/10.1038/ncomms3264.

Grigutsch, L. A., Lewe, G., Rothermund, K., & Koranyi, N. (2019). Implicit 'wanting' without implicit 'liking': A test of incentive-sensitization-theory in the context of smoking addiction using the wanting-implicit-association-test (w-iat). *Journal of Behavior Therapy and Experimental Psychiatry*, *64*, 9–14. https://doi.org/10.1016/j.jbtep.2019.01.002.

Grimm, J. W., Hope, B. T., Wise, R. A., & Shaham, Y. (2001). Neuroadaptation. Incubation of cocaine craving after withdrawal. *Nature*, *412*(6843), 141–142. https://doi.org/10.1038/35084134.

Grueter, B. A., Rothwell, P. E., & Malenka, R. C. (2012). Integrating synaptic plasticity and striatal circuit function in addiction. *Current Opinion in Neurobiology*, *22*(3), 545–551. https://doi.org/10.1016/j.conb.2011.09.009.

Hall, D. A., & Gulley, J. M. (2011). Disruptive effect of amphetamines on pavlovian to instrumental transfer. *Behavioural Brain Research*, *216*(1), 440–445. https://doi.org/10.1016/j.bbr.2010.08.040.

Hall, J., Parkinson, J. A., Connor, T. M., Dickinson, A., & Everitt, B. J. (2001). Involvement of the central nucleus of the amygdala and nucleus accumbens core in mediating pavlovian influences on instrumental behaviour. *European Journal of Neuroscience*, *13*(10), 1984–1992. https://doi.org/10.1046/j.0953-816x.2001.01577.x.

Hellberg, S. N., Russell, T. I., & Robinson, M. J. F. (2018). Cued for risk: Evidence for an incentive sensitization framework to explain the interplay between stress and anxiety, substance abuse, and reward uncertainty in disordered gambling behavior. *Cognitive, Affective & Behavioral Neuroscience*, 1–22. https://doi.org/10.3758/s13415-018-00662-3.

Hester, R., Bell, R. P., Foxe, J. J., & Garavan, H. (2013). The influence of monetary punishment on cognitive control in abstinent cocaine-users. *Drug and Alcohol Dependence*, *133*(1), 86–93. https://doi.org/10.1016/j.drugalcdep.2013.05.027.

Hnasko, T. S., Sotak, B. N., & Palmiter, R. D. (2005). Morphine reward in dopamine-deficient mice. *Nature*, *438*(7069), 854–857. https://doi.org/10.1038/nature04172.

Hogarth, L., Attwood, A. S., Bate, H. A., & Munafo, M. R. (2012). Acute alcohol impairs human goal-directed action. *Biological Psychology*, *90*(2), 154–160. https://doi.org/10.1016/j.biopsycho.2012.02.016.

Hogarth, L., Balleine, B. W., Corbit, L. H., & Killcross, S. (2013). Associative learning mechanisms underpinning the transition from recreational drug use to addiction. *Annals of the New York Academy of Sciences*, *1282*, 12–24. https://doi.org/10.1111/j.1749-6632.2012.06768.x.

Hogarth, L., & Chase, H. W. (2012). Evaluating psychological markers for human nicotine dependence: Tobacco choice, extinction, and pavlovian-to-instrumental transfer. *Experimental and Clinical Psychopharmacology*, *20*(3), 213–224. https://doi.org/10.1037/a0027203.

Hogarth, L., Lam-Cassettari, C., Pacitti, H., Currah, T., Mahlberg, J., Hartley, L., & Moustafa, A. (2018). Intact goal-directed control in treatment-seeking drug users indexed by outcome-devaluation and pavlovian to instrumental transfer: Critique of habit theory. *European Journal of Neuroscience*, *9*(1), 1–13. https://doi.org/10.1111/ejn.13961.

Hogarth, L., Maynard, O. M., & Munafo, M. R. (2015). Plain cigarette packs do not exert pavlovian to instrumental transfer of control over tobacco-seeking. *Addiction*, *110*(1), 174–182. https://doi.org/10.1111/add.12756.

Holmes, N. M., Marchand, A. R., & Coutureau, E. (2010). Pavlovian to instrumental transfer: A neurobehavioural perspective. *Neuroscience and Biobehavioral Reviews*, *34*(8), 1277–1295. https://doi.org/10.1016/j.neubiorev.2010.03.007.

Holroyd, C. B., & Coles, M. G. H. (2002). The neural basis of human error processing: Reinforcement learning, dopamine, and the error-related negativity. *Psychological Review*, *109*(4), 679–709. https://doi.org/10.1037/0033-295x.109.4.679.

Holton, R., & Berridge, K. C. (2017). Compulsion and choice in addiction. In N. Heather & G. Segal (Eds.), *Addiction and choice: Rethinking the relationship* (pp. 153–170). Oxford University Press. https://doi.org/10.1093/acprof:oso/9780198727224.001.0001. Retrieved from Oxford Scholarship Online.

Hser, Y. I., Hoffman, V., Grella, C. E., & Anglin, M. D. (2001). A 33-year follow-up of narcotics addicts. *Archives of General Psychiatry*, *58*(5), 503–508. https://doi.org/10.1001/archpsyc.58.5.503.

Hunt, W. A., Barnett, L. W., & Branch, L. G. (1971). Relapse rates in addiction programs. *Journal of Clinical Psychology*, *27*(4), 455–456. https://doi.org/10.1002/1097-4679(197110)27:4<455::AID-JCLP2270270412>3.0.CO;2-R.

Hyman, S. E. (2005). Addiction: A disease of learning and memory. *American Journal of Psychiatry*, *162*(8), 1414–1422. https://doi.org/10.1176/appi.ajp.162.8.1414.

Hyman, S. E., Malenka, R. C., & Nestler, E. J. (2006). Neural mechanisms of addiction: The role of reward-related learning and memory. *Annual Review of Neuroscience*, *29*(1), 565–598. https://doi.org/10.1146/annurev.neuro.29.051605.113009.

Jedynak, J., Hearing, M., Ingebretson, A., Ebner, S. R., Kelly, M., Fischer, R. A., … Thomas, M. J. (2016). Cocaine and amphetamine induce overlapping but distinct patterns of ampar plasticity in nucleus accumbens medium spiny neurons. *Neuropsychopharmacology*, *41*(2), 464–476. https://doi.org/10.1038/npp.2015.168.

Jentsch, J. D., & Pennington, Z. T. (2014). Reward, interrupted: Inhibitory control and its relevance to addictions. *Neuropharmacology*, *76*, 479–486. https://doi.org/10.1016/j.neuropharm.2013.05.022.

Kalapatapu, R. K., Bedi, G., Haney, M., Evans, S. M., Rubin, E., & Foltin, R. W. (2012). The subjective effects of cocaine: Relationship to years of cocaine use and current age. *The American Journal of Drug and Alcohol Abuse*, *38*(6), 530–534. https://doi.org/10.3109/00952990.2012.704461.

Kantak, K. M., & Nic Dhonnchadha, B.Á. (2011). Pharmacological enhancement of drug cue extinction learning: Translational challenges. *Annals of the New York Academy of Sciences*, *1216*, 122–137. https://doi.org/10.1111/j.1749-6632.2010.05899.x.

Kasanetz, F., Deroche-Gamonet, V., Berson, N., Balado, E., Lafourcade, M., Manzoni, O., & Piazza, P. V. (2010). Transition to addiction is associated with a persistent impairment in synaptic plasticity. *Science*, *328*(5986), 1709–1712. https://doi.org/10.1126/science.1187801.

Kauer, J. A. (2004). Learning mechanisms in addiction: Synaptic plasticity in the ventral tegmental area as a result of exposure to drugs of abuse. *Annual Review of Physiology*, *66*, 447–475. https://doi.org/10.1146/annurev.physiol.66.032102.112534.

Kauer, J. A., & Malenka, R. C. (2007). Synaptic plasticity and addiction. *Nature Reviews Neuroscience*, *8*(11), 844–858. https://doi.org/10.1038/nrn2234.

Keiflin, R., & Janak, P. H. (2015). Dopamine prediction errors in reward learning and addiction: From theory to neural circuitry. *Neuron*, *88*(2), 247–263. https://doi.org/10.1016/j.neuron.2015.08.037.

Knutson, B., & Gibbs, S. E. (2007). Linking nucleus accumbens dopamine and blood oxygenation. *Psychopharmacology*, *191*(3), 813–822. https://doi.org/10.1007/s00213-006-0686-7.

Konova, A. B., Parvaz, M. A., Bernstein, V., Zilverstand, A., Moeller, S. J., Delgado, M. R., … Goldstein, R. Z. (2017). Neural mechanisms of extinguishing drug and pleasant cue associations in human addiction: Role of the vmpfc. *Addiction Biology*, *24*(1). https://doi.org/10.1111/adb.12545.

Koob, G. F. (2013). Negative reinforcement in drug addiction: The darkness within. *Current Opinion in Neurobiology*, *23*(4), 559–563. https://doi.org/10.1016/j.conb.2013.03.011.

Kourrich, S., Rothwell, P. E., Klug, J. R., & Thomas, M. J. (2007). Cocaine experience controls bidirectional synaptic plasticity in the nucleus accumbens. *Journal of Neuroscience*, *27*(30), 7921–7928. https://doi.org/10.1523/jneurosci.1859-07.2007.

Kravitz, A. V., & Kreitzer, A. C. (2012). Striatal mechanisms underlying movement, reinforcement, and punishment. *Physiology (Bethesda)*, *27*(3), 167–177. https://doi.org/10.1152/physiol.00004.2012.

Laruelle, M., Abi-Dargham, A., van Dyck, C. H., Rosenblatt, W., Zea-Ponce, Y., Zoghbi, S. S., et al. (1995). Spect imaging of striatal dopamine release after amphetamine challenge. *Journal of Nuclear Medicine*, *36*(7), 1182–1190.

Laurent, V., Bertran-Gonzalez, J., Chieng, B. C., & Balleine, B. W. (2014). Δ-opioid and dopaminergic processes in accumbens shell modulate the cholinergic control of predictive learning and choice. *Journal of Neuroscience*, *34*(4), 1358–1369. https://doi.org/10.1523/jneurosci.4592-13.2014.

Lecca, D., Cacciapaglia, F., Valentini, V., Gronli, J., Spiga, S., & Di Chiara, G. (2006). Preferential increase of extracellular dopamine in the rat nucleus accumbens shell as compared to that in the core during acquisition and maintenance of intravenous nicotine self-administration. *Psychopharmacology*, *184*(3), 435–446. https://doi.org/10.1007/s00213-005-0280-4.

Lex, A., & Hauber, W. (2008). Dopamine d1 and d2 receptors in the nucleus accumbens core and shell mediate pavlovian-instrumental transfer. *Learning and Memory*, *15*(7), 483–491. https://doi.org/10.1101/lm.978708.

Leyton, M. (2007). Conditioned and sensitized responses to stimulant drugs in humans. *Progress in Neuropsychopharmacology and Biological Psychiatry*, *31*(8), 1601–1613. https://doi.org/10.1016/j.pnpbp.2007.08.027.

Li, Y., Kolb, B., & Robinson, T. E. (2003). The location of persistent amphetamine-induced changes in the density of dendritic spines on medium spiny neurons in the nucleus accumbens and caudate-putamen. *Neuropsychopharmacology*, *28*(6), 1082–1085. https://doi.org/10.1038/sj.npp.1300115.

Lohrenz, T., Kishida, K. T., & Montague, P. R. (2016). Bold and its connection to dopamine release in human striatum: A cross-cohort comparison. *Philosophical Transactions of the Royal Society, B: Biological Sciences*, *371*(1705), 20150352. https://doi.org/10.1098/rstb.2015.0352.

Luscher, C., & Malenka, R. C. (2011). Drug-evoked synaptic plasticity in addiction: From molecular changes to circuit remodeling. *Neuron*, *69*(4), 650–663. https://doi.org/10.1016/j.neuron.2011.01.017.

Macuchova, E., & Slamberova, R. (2017). Does prenatal methamphetamine exposure induce sensitization to drugs in adulthood? *Physiological Research*, *66*(Suppl. 4), S457–S467.

Malvaez, M., & Wassum, K. M. (2018). Regulation of habit formation in the dorsal striatum. *Current Opinion in Behavioral Sciences*, *20*, 67–74. https://doi.org/10.1016/j.cobeha.2017.11.005.

Mameli, M., & Luscher, C. (2011). Synaptic plasticity and addiction: Learning mechanisms gone awry. *Neuropharmacology*, *61*(7), 1052–1059. https://doi.org/10.1016/j.neuropharm.2011.01.036.

Maren, S. (2001). Neurobiology of pavlovian fear conditioning. *Annual Review of Neuroscience*, *24*(1), 897–931. https://doi.org/10.1146/annurev.neuro.24.1.897.

Marks, K. R., Kearns, D. N., Christensen, C. J., Silberberg, A., & Weiss, S. J. (2010). Learning that a cocaine reward is smaller than expected: A test of redish's computational model of addiction. *Behavioural Brain Research*, *212*(2), 204–207. https://doi.org/10.1016/j.bbr.2010.03.053.

Martin, M., Chen, B. T., Hopf, F. W., Bowers, M. S., & Bonci, A. (2006). Cocaine self-administration selectively abolishes ltd in the core of the nucleus accumbens. *Nature Neuroscience*, *9*, 868. https://doi.org/10.1038/nn1713.

Mayes, L. C., & Suchman, N. E. (2015). Developmental pathways to substance abuse. In D. Cicchetti & D. J. Cohen (Eds.), *Developmental psychopathology: Vol. 3: Risk, disorder, and adaptation*. (pp. 599–619). https://doi.org/10.1002/9780470939406. Retrieved from Wiley Online Library.

Mayo, L. M., & de Wit, H. (2015). Acquisition of responses to a methamphetamine-associated cue in healthy humans: Self-report, behavioral, and psychophysiological measures. *Neuropsychopharmacology*, *40*(7), 1734–1741. https://doi.org/10.1038/npp.2015.21.

Mayo, L. M., Fraser, D., Childs, E., Momenan, R., Hommer, D. W., de Wit, H., & Heilig, M. (2013). Conditioned preference to a methamphetamine-associated contextual cue in humans. *Neuropsychopharmacology*, *38*(6), 921–929. https://doi.org/10.1038/npp.2013.3.

McKernan, M. G., & Shinnick-Gallagher, P. (1997). Fear conditioning induces a lasting potentiation of synaptic currents in vitro. *Nature*, *390*(6660), 607–611. https://doi.org/10.1038/37605.

Mendelsohn, A., Pine, A., & Schiller, D. (2014). Between thoughts and actions: Motivationally salient cues invigorate mental action in the human brain. *Neuron*, *81*(1), 207–217. https://doi.org/10.1016/j.neuron.2013.10.019.

Miles, F. J., Everitt, B. J., & Dickinson, A. (2003). Oral cocaine seeking by rats: Action or habit? *Behavioral Neuroscience*, *117*(5), 927–938. https://doi.org/10.1037/0735-7044.117.5.927.

Miller, R. R., Barnet, R. C., & Grahame, N. J. (1995). Assessment of the rescorla-wagner model. *Psychological Bulletin*, *117*(3), 363–386.

Mintzer, M. Z., & Stitzer, M. L. (2002). Cognitive impairment in methadone maintenance patients. *Drug and Alcohol Dependence*, *67*(1), 41–51. https://doi.org/10.1016/S0376-8716(02)00013-3.

Myers, C. E., Rego, J., Haber, P., Morley, K., Beck, K. D., Hogarth, L., & Moustafa, A. A. (2017). Learning and generalization from reward and punishment in opioid addiction. *Behavioural Brain Research*, *317*, 122–131. https://doi.org/10.1016/j.bbr.2016.09.033.

Namba, M. D., Tomek, S. E., Olive, M. F., Beckmann, J. S., & Gipson, C. D. (2018). The winding road to relapse: Forging a new understanding of cue-induced reinstatement models and their associated neural mechanisms. *Frontiers in Behavioral Neuroscience*, *12*(17), 1–22. https://doi.org/10.3389/fnbeh.2018.00017.

Nelson, A., & Killcross, S. (2006). Amphetamine exposure enhances habit formation. *Journal of Neuroscience*, *26*(14), 3805–3812. https://doi.org/10.1523/jneurosci.4305-05.2006.

Niehaus, J. L., Murali, M., & Kauer, J. A. (2010). Drugs of abuse and stress impair LTP at inhibitory synapses in the ventral tegmental area. *European Journal of Neuroscience*, *32*(1), 108–117. https://doi.org/10.1111/j.1460-9568.2010.07256.x.

Nordquist, R., Voorn, P., de Mooij, A., Joosten, R., Pennartz, C., & Vanderschuren, L. (2007). Augmented reinforcer value and accelerated habit formation after repeated amphetamine treatment. *European Neuropsychopharmacology*, *17*(8), 532–540. https://doi.org/10.1016/j.euroneuro.2006.12.005.

Nutt, D. J., Lingford-Hughes, A., Erritzoe, D., & Stokes, P. R. A. (2015). The dopamine theory of addiction: 40 years of highs and lows. *Nature Reviews Neuroscience*, *16*(5), 305. https://doi.org/10.1038/nrn3939.

O'Hare, J., Calakos, N., & Yin, H. H. (2018). Recent insights into corticostriatal circuit mechanisms underlying habits. *Current Opinion in Behavioral Sciences*, *20*, 40–46. https://doi.org/10.1016/j.cobeha.2017.10.001.

Ostafin, B. D., Marlatt, G. A., & Troop-Gordon, W. (2010). Testing the incentive-sensitization theory with at-risk drinkers: Wanting, liking, and alcohol consumption. *Psychology of Addictive Behaviors*, *24*(1), 157–162. https://doi.org/10.1037/a0017897.

O'Sullivan, S. S., Wu, K., Politis, M., Lawrence, A. D., Evans, A. H., Bose, S. K., … Piccini, P. (2011). Cue-induced striatal dopamine release in parkinson's disease-associated impulsive-compulsive behaviours. *Brain*, *134*(4), 969–978. https://doi.org/10.1093/brain/awr003.

Ota, K. T., Monsey, M. S., Wu, M. S., & Schafe, G. E. (2010). Synaptic plasticity and no-cgmp-pkg signaling regulate pre- and postsynaptic alterations at rat lateral amygdala synapses following fear conditioning. *PLoS ONE*, *5*(6), e11236. https://doi.org/10.1371/journal.pone.0011236.

Panlilio, L. V., Thorndike, E. B., & Schindler, C. W. (2007). Blocking of conditioning to a cocaine-paired stimulus: Testing the hypothesis that cocaine perpetually produces a signal of larger-than-expected reward. *Pharmacology Biochemistry and Behavior*, *86*(4), 774–777. https://doi.org/10.1016/j.pbb.2007.03.005.

Park, K., Volkow, N. D., Pan, Y., & Du, C. (2013). Chronic cocaine dampens dopamine signaling during cocaine intoxication and unbalances d1 over d2 receptor signaling. *Journal of Neuroscience*, *33*(40), 15827–15836. https://doi.org/10.1523/jneurosci.1935-13.2013.

Paulson, P. E., Camp, D. M., & Robinson, T. E. (1991). Time course of transient behavioral depression and persistent behavioral sensitization in relation to regional brain monoamine concentrations during amphetamine withdrawal in rats. *Psychopharmacology*, *103*(4), 480–492. https://doi.org/10.1007/BF02244248.

Pavlov, I. P. (1927). Lecture v internal inhibition (continued): (b) conditioned inhibition. In G. V. Anrep (Ed.), *Conditioned reflexes: An investigation of the physiological activity of the cerebral cortex* (pp. 68–87). London: Oxford University Press.

Peciña, S., & Berridge, K. C. (2013). Dopamine or opioid stimulation of nucleus accumbens similarly amplify cue-triggered 'wanting' for reward: Entire core and medial shell mapped as substrates for pit enhancement. *European Journal of Neuroscience*, *37*(9), 1529–1540. https://doi.org/10.1111/ejn.12174.

Pecina, S., Cagniard, B., Berridge, K. C., Aldridge, J. W., & Zhuang, X. (2003). Hyperdopaminergic mutant mice have higher "wanting" but not "liking" for sweet rewards. *Journal of Neuroscience*, *23*(28), 9395–9402. https://doi.org/10.1523/JNEUROSCI.23-28-09395.2003.

Peters, J., & De Vries, T. J. (2014). Pavlovian conditioned approach, extinction, and spontaneous recovery to an audiovisual cue paired with an intravenous heroin infusion. *Psychopharmacology, 231*(2), 447–453. https://doi.org/10.1007/s00213-013-3258-7.

Peters, J., LaLumiere, R. T., & Kalivas, P. W. (2008). Infralimbic prefrontal cortex is responsible for inhibiting cocaine seeking in extinguished rats. *Journal of Neuroscience, 28*(23), 6046–6053. https://doi.org/10.1523/jneurosci.1045-08.2008.

Phillips, P. E., Stuber, G. D., Heien, M. L., Wightman, R. M., & Carelli, R. M. (2003). Subsecond dopamine release promotes cocaine seeking. *Nature, 422*(6932), 614–618. https://doi.org/10.1038/nature01476.

Quirk, G. J., Armony, J. L., & LeDoux, J. E. (1997). Fear conditioning enhances different temporal components of tone-evoked spike trains in auditory cortex and lateral amygdala. *Neuron, 19*(3), 613–624. https://doi.org/10.1016/S0896-6273(00)80375-X.

Quirk, G. J., Repa, C., & LeDoux, J. E. (1995). Fear conditioning enhances short-latency auditory responses of lateral amygdala neurons: Parallel recordings in the freely behaving rat. *Neuron, 15*(5), 1029–1039. https://doi.org/10.1016/0896-6273(95)90092-6.

Redish, A. D. (2004). Addiction as a computational process gone awry. *Science, 306*(5703), 1944–1947. https://doi.org/10.1126/science.1102384.

Rescorla, R. A., & Wagner, A. R. (1972). A theory of pavlovian conditioning: Variations in the effectiveness of reinforcement and nonreinforcement. In A. H. Black & W. F. Prokasy (Eds.), *Classical conditioning II: Current research and theory*. New York: Appleton-Century-Crofts.

Robinson, T. E., & Berridge, K. C. (1993). The neural basis of drug craving: An incentive-sensitization theory of addiction. *Brain Research Reviews, 18*(3), 247–291. https://doi.org/10.1016/0165-0173(93)90013-P.

Robinson, M. J., & Berridge, K. C. (2013). Instant transformation of learned repulsion into motivational “wanting”. *Current Biology, 23*(4), 282–289. https://doi.org/10.1016/j.cub.2013.01.016.

Robinson, S., Rainwater, A. J., Hnasko, T. S., & Palmiter, R. D. (2007). Viral restoration of dopamine signaling to the dorsal striatum restores instrumental conditioning to dopamine-deficient mice. *Psychopharmacology, 191*(3), 567–578. https://doi.org/10.1007/s00213-006-0579-9.

Robinson, M. J., Robinson, T. E., & Berridge, K. C. (2014). Incentive salience in addiction and over-consumption. In S. D. Preston, M. L. Kringelbach, & B. Knutson (Eds.), *The interdisciplinary science of consumption* (pp. 185–197). Cambridge, MA: MIT Press.

Robinson, S., Sandstrom, S. M., Denenberg, V. H., & Palmiter, R. D. (2005). Distinguishing whether dopamine regulates liking, wanting, and/or learning about rewards. *Behavioural Neuroscience, 119*(1), 5–15. https://doi.org/10.1037/0735-7044.119.1.5.

Rogan, M. T., Staubli, U. V., & LeDoux, J. E. (1997). Fear conditioning induces associative long-term potentiation in the amygdala. *Nature, 390*(6660), 604–607. https://doi.org/10.1038/37601.

Rosse, R. B., Fay-McCarthy, M., Collins, J. P., Jr., Risher-Flowers, D., Alim, T. N., & Deutsch, S. I. (1993). Transient compulsive foraging behavior associated with crack cocaine use. *American Journal of Psychiatry, 150*(1), 155–156. https://doi.org/10.1176/ajp.150.1.155.

Saal, D., Dong, Y., Bonci, A., & Malenka, R. C. (2003). Drugs of abuse and stress trigger a common synaptic adaptation in dopamine neurons. *Neuron, 37*(4), 577–582. https://doi.org/10.1016/S0896-6273(03)00021-7.

Sage, J. R., & Knowlton, B. J. (2000). Effects of us devaluation on win–stay and win–shift radial maze performance in rats. *Behavioral Neuroscience, 114*(2), 295–306. https://doi.org/10.1037/0735-7044.114.2.295.

Salamone, J. D., Arizzi, M., Sandoval, M., Cervone, K., & Aberman, J. (2002). Dopamine antagonists alter response allocation but do not suppress appetite for food in rats: Contrast between the effects of skf 83566, raclopride, and fenfluramine on a concurrent choice task. *Psychopharmacology*, *160*(4), 371–380. https://doi.org/10.1007/s00213-001-0994-x.

Salamone, J. D., Correa, M., Ferrigno, S., Yang, J.-H., Rotolo, R. A., & Presby, R. E. (2018). The psychopharmacology of effort-related decision making: Dopamine, adenosine, and insights into the neurochemistry of motivation. *Pharmacological Reviews*, *70*(4), 747–762. https://doi.org/10.1124/pr.117.015107.

Salamone, J. D., Correa, M., Mingote, S. M., & Weber, S. M. (2005). Beyond the reward hypothesis: Alternative functions of nucleus accumbens dopamine. *Current Opinion in Pharmacology*, *5*(1), 34–41. https://doi.org/10.1016/j.coph.2004.09.004.

Salamone, J. D., Correa, M., Yang, J. H., Rotolo, R., & Presby, R. (2018). Dopamine, effort-based choice, and behavioral economics: Basic and translational research. *Frontiers in Behavioral Neuroscience*, *12*, 52. https://doi.org/10.3389/fnbeh.2018.00052.

Salamone, J. D., Pardo, M., Yohn, S. E., Lopez-Cruz, L., SanMiguel, N., & Correa, M. (2016). Mesolimbic dopamine and the regulation of motivated behavior. *Current Topics in Behavioral Neurosciences*, *27*, 231–257. https://doi.org/10.1007/7854_2015_383.

Salamone, J. D., Steinpreis, R. E., McCullough, L. D., Smith, P., Grebel, D., & Mahan, K. (1991). Haloperidol and nucleus accumbens dopamine depletion suppress lever pressing for food but increase free food consumption in a novel food choice procedure. *Psychopharmacology*, *104*(4), 515–521. https://doi.org/10.1007/BF02245659.

Schafe, G. E., Nader, K., Blair, H. T., & LeDoux, J. E. (2001). Memory consolidation of pavlovian fear conditioning: A cellular and molecular perspective. *Trends in Neurosciences*, *24*(9), 540–546. https://doi.org/10.1016/S0166-2236(00)01969-X.

Schroll, H., & Hamker, F. (2013). Computational models of basal-ganglia pathway functions: Focus on functional neuroanatomy. *Frontiers in Systems Neuroscience*, *7*(122). https://doi.org/10.3389/fnsys.2013.00122.

Schulteis, G., Ahmed, S. H., Morse, A. C., Koob, G. F., & Everitt, B. J. (2000). Conditioning and opiate withdrawal. *Nature*, *405*(6790), 1013. https://doi.org/10.1038/35016630.

Schultz, W. (1998). Predictive reward signal of dopamine neurons. *Journal of Neurophysiology*, *80*(1), 1–27. https://doi.org/10.1152/jn.1998.80.1.1.

Schultz, W. (2016). Dopamine reward prediction error coding. *Dialogues in Clinical Neuroscience*, *18*(1), 23–32.

Schultz, W., Dayan, P., & Montague, P. R. (1997). A neural substrate of prediction and reward. *Science*, *275*(5306), 1593–1599. https://doi.org/10.1126/science.275.5306.1593.

Shaham, Y., Shalev, U., Lu, L., de Wit, H., & Stewart, J. (2003). The reinstatement model of drug relapse: History, methodology and major findings. *Psychopharmacology*, *168*(1–2), 3–20. https://doi.org/10.1007/s00213-002-1224-x.

Sharpe, M. J., Chang, C. Y., Liu, M. A., Batchelor, H. M., Mueller, L. E., Jones, J. L., … Schoenbaum, G. (2017). Dopamine transients are sufficient and necessary for acquisition of model-based associations. *Nature Neuroscience*, *20*(5), 735. https://doi.org/10.1038/nn.4538.

Sharpe, M. J., & Schoenbaum, G. (2018). Chapter 11: Does the dopaminergic error signal act like a cached-value prediction error? In R. Morris, A. Bornstein, & A. Shenhav (Eds.), *Goal-directed decision making: Computations and neural circuits* (pp. 243–258). Cambridge, MA: Academic Press.

Shiflett, M. W., & Balleine, B. W. (2010). At the limbic–motor interface: Disconnection of basolateral amygdala from nucleus accumbens core and shell reveals dissociable components of incentive motivation. *European Journal of Neuroscience*, *32*(10), 1735–1743. https://doi.org/10.1111/j.1460-9568.2010.07439.x.

Sinha, R., Fuse, T., Aubin, L. R., & O'Malley, S. S. (2000). Psychological stress, drug-related cues and cocaine craving. *Psychopharmacology*, *152*(2), 140–148. https://doi.org/10.1007/s002130000499.

Sjoerds, Z., van den Brink, W., Beekman, A. T. F., Penninx, B.W.J.H., & Veltman, D. J. (2014). Cue reactivity is associated with duration and severity of alcohol dependence: An fmri study. *PLoS ONE*, *9*(1), e84560. https://doi.org/10.1371/journal.pone.0084560.

Skinner, B. F. (1938). *The behavior of organisms: An experimental analysis*. Oxford, England: Appleton-Century.

Steinberg, E. E., Keiflin, R., Boivin, J. R., Witten, I. B., Deisseroth, K., & Janak, P. H. (2013). A causal link between prediction errors, dopamine neurons and learning. *Nature Neuroscience*, *16*(7), 966–973. https://doi.org/10.1038/nn.3413.

Steketee, J. D., & Kalivas, P. W. (2011). Drug wanting: Behavioral sensitization and relapse to drug-seeking behavior. *Pharmacological Reviews*, *63*(2), 348–365. https://doi.org/10.1124/pr.109.001933.

Stewart, J., de Wit, H., & Eikelboom, R. (1984). Role of unconditioned and conditioned drug effects in the self-administration of opiates and stimulants. *Psychological Review*, *91*(2), 251–268. https://doi.org/10.1037/0033-295X.91.2.251.

Stuber, G. D., Wightman, R. M., & Carelli, R. M. (2005). Extinction of cocaine self-administration reveals functionally and temporally distinct dopaminergic signals in the nucleus accumbens. *Neuron*, *46*(4), 661–669. https://doi.org/10.1016/j.neuron.2005.04.036.

Sutton, R. S., & Barto, A. G. (1987). *A temporal-difference model of classical conditioning*. In *Paper presented at the Proceedings of the ninth annual conference of the cognitive science society*.

Sutton, R. S., & Barto, A. G. (1998). *Reinforcement learning: An introduction*. (Vol. 1). Cambridge, MA: The MIT Press.

Swanson, L. W. (1982). The projections of the ventral tegmental area and adjacent regions: A combined fluorescent retrograde tracer and immunofluorescence study in the rat. *Brain Research Bulletin*, *9*(1), 321–353. https://doi.org/10.1016/0361-9230(82)90145-9.

Talmi, D., Seymour, B., Dayan, P., & Dolan, R. J. (2008). Human pavlovian–instrumental transfer. *Journal of Neuroscience*, *28*(2), 360–368. https://doi.org/10.1523/jneurosci.4028-07.2008.

Tibboel, H., De Houwer, J., & Van Bockstaele, B. (2015). Implicit measures of "wanting" and "liking" in humans. *Neuroscience & Biobehavioral Reviews*, *57*, 350–364. https://doi.org/10.1016/j.neubiorev.2015.09.015.

Tindell, A. J., Berridge, K. C., Zhang, J., Pecina, S., & Aldridge, J. W. (2005). Ventral pallidal neurons code incentive motivation: Amplification by mesolimbic sensitization and amphetamine. *European Journal of Neuroscience*, *22*(10), 2617–2634. https://doi.org/10.1111/j.1460-9568.2005.04411.x.

Tindell, A. J., Smith, K. S., Berridge, K. C., & Aldridge, J. W. (2009). Dynamic computation of incentive salience: "Wanting" what was never "liked". *Journal of Neuroscience*, *29*(39), 12220–12228. https://doi.org/10.1523/JNEUROSCI.2499-09.2009.

Tindell, A. J., Smith, K. S., Pecina, S., Berridge, K. C., & Aldridge, J. W. (2006). Ventral pallidum firing codes hedonic reward: When a bad taste turns good. *Journal of Neurophysiology*, *96*(5), 2399–2409. https://doi.org/10.1152/jn.00576.2006.

Tobler, P. N., Dickinson, A., & Schultz, W. (2003). Coding of predicted reward omission by dopamine neurons in a conditioned inhibition paradigm. *Journal of Neuroscience*, *23*(32), 10402–10410. https://doi.org/10.1523/JNEUROSCI.23-32-10402.2003.

Tricomi, E., Balleine, B. W., & O'Doherty, J. P. (2009). A specific role for posterior dorsolateral striatum in human habit learning. *European Journal of Neuroscience*, *29*(11), 2225–2232. https://doi.org/10.1111/j.1460-9568.2009.06796.x.

Tunstall, B. J., & Kearns, D. N. (2016). Cocaine can generate a stronger conditioned reinforcer than food despite being a weaker primary reinforcer. *Addiction Biology*, *21*(2), 282–293. https://doi.org/10.1111/adb.12195.

Ungless, M. A., Whistler, J. L., Malenka, R. C., & Bonci, A. (2001). Single cocaine exposure in vivo induces long-term potentiation in dopamine neurons. *Nature*, *411*(6837), 583–587. https://doi.org/10.1038/35079077.

van den Bos, R., van der Harst, J., Vijftigschild, N., Spruijt, B., van Luijtelaar, G., & Maes, R. (2004). On the relationship between anticipatory behaviour in a pavlovian paradigm and pavlovian-to-instrumental transfer in rats (rattus norvegicus). *Behavioural Brain Research*, *153*(2), 397–408. https://doi.org/10.1016/j.bbr.2003.12.017.

Van den Oever, M. C., Goriounova, N. A., Wan Li, K., Van der Schors, R. C., Binnekade, R., Schoffelmeer, A. N. M., … De Vries, T. J. (2008). Prefrontal cortex ampa receptor plasticity is crucial for cue-induced relapse to heroin-seeking. *Nature Neuroscience*, *11*, 1053. https://doi.org/10.1038/nn.2165.

van Huijstee, A. N., & Mansvelder, H. D. (2014). Glutamatergic synaptic plasticity in the mesocorticolimbic system in addiction. *Frontiers in Cellular Neuroscience*, *8*, 466. https://doi.org/10.3389/fncel.2014.00466.

Volkow, N. D., & Baler, R. D. (2014). Addiction science: Uncovering neurobiological complexity. *Neuropharmacology*, *76*(Pt B), 235–249. https://doi.org/10.1016/j.neuropharm.2013.05.007.

Volkow, N. D., & Morales, M. (2015). The brain on drugs: From reward to addiction. *Cell*, *162*(4), 712–725. https://doi.org/10.1016/j.cell.2015.07.046.

Volkow, N. D., Wang, G. J., Fowler, J. S., Logan, J., Gatley, S. J., Wong, C., … Pappas, N. R. (1999). Reinforcing effects of psychostimulants in humans are associated with increases in brain dopamine and occupancy of d(2) receptors. *Journal of Pharmacology and Experimental Therapeutics*, *291*(1), 409–415.

Vollstadt-Klein, S., Wichert, S., Rabinstein, J., Buhler, M., Klein, O., Ende, G., … Mann, K. (2010). Initial, habitual and compulsive alcohol use is characterized by a shift of cue processing from ventral to dorsal striatum. *Addiction*, *105*(10), 1741–1749. https://doi.org/10.1111/j.1360-0443.2010.03022.x.

Voon, V., Derbyshire, K., Rück, C., Irvine, M. A., Worbe, Y., Enander, J., … Bullmore, E. T. (2014). Disorders of compulsivity: A common bias towards learning habits. *Molecular Psychiatry*, *20*(3), 345–352. https://doi.org/10.1038/mp.2014.44.

Waelti, P., Dickinson, A., & Schultz, W. (2001). Dopamine responses comply with basic assumptions of formal learning theory. *Nature*, *412*(6842), 43–48. https://doi.org/10.1038/35083500.

Wanat, M. J., Willuhn, I., Clark, J. J., & Phillips, P. E. M. (2009). Phasic dopamine release in appetitive behaviors and drug addiction. *Current Drug Abuse Reviews*, *2*(2), 195–213.

Warren, H. C. (1916). Mental association from plato to hume. *Psychological Review*, *23*(3), 208–230. https://doi.org/10.1037/h0073099.

Wassum, K. M., Ostlund, S. B., Loewinger, G. C., & Maidment, N. T. (2013). Phasic mesolimbic dopamine release tracks reward seeking during expression of pavlovian-to-instrumental transfer. *Biological Psychiatry*, *73*(8), 747–755. https://doi.org/10.1016/j.biopsych.2012.12.005.

Weiss, F., Lorang, M. T., Bloom, F. E., & Koob, G. F. (1993). Oral alcohol self-administration stimulates dopamine release in the rat nucleus accumbens: Genetic and motivational determinants. *Journal of Pharmacology and Experimental Therapeutics*, *267*(1), 250–258.

Whitesell, M., Bachand, A., Peel, J., & Brown, M. (2013). Familial, social, and individual factors contributing to risk for adolescent substance use. *Journal of Addiction*, (579310), 1–9. https://doi.org/10.1155/2013/579310.

Wikler, A. (1965). Conditioning factors in opiate addiction and relapse. In D. I. Willner & G. G. Kassenbaum (Eds.), *Narcotics* (pp. 7–21). New York, NY: McGraw-Hill.

Willner, P., James, D., & Morgan, M. (2005). Excessive alcohol consumption and dependence on amphetamine are associated with parallel increases in subjective ratings of both 'wanting' and 'liking'. *Addiction*, *100*(10), 1487–1495. https://doi.org/10.1111/j.1360-0443.2005.01222.x.

Willuhn, I., Burgeno, L. M., Groblewski, P. A., & Phillips, P. E. M. (2014). Excessive cocaine use results from decreased phasic dopamine signaling in the striatum. *Nature Neuroscience*, *17*(5), 704–709. https://doi.org/10.1038/nn.3694.

Wise, R. A., Leone, P., Rivest, R., & Leeb, K. (1995). Elevations of nucleus accumbens dopamine and dopac levels during intravenous heroin self-administration. *Synapse*, *21*(2), 140–148. https://doi.org/10.1002/syn.890210207.

Wunderlich, K., Dayan, P., & Dolan, R. J. (2012). Mapping value based planning and extensively trained choice in the human brain. *Nature Neuroscience*, *15*(5), 786–791. https://doi.org/10.1038/nn.3068.

Wyvell, C. L., & Berridge, K. C. (2000). Intra-accumbens amphetamine increases the conditioned incentive salience of sucrose reward: Enhancement of reward "wanting" without enhanced "liking" or response reinforcement. *Journal of Neuroscience*, *20*(21), 8122–8130. https://doi.org/10.1523/JNEUROSCI.20-21-08122.2000.

Wyvell, C. L., & Berridge, K. C. (2001). Incentive sensitization by previous amphetamine exposure: Increased cue-triggered "wanting" for sucrose reward. *Journal of Neuroscience*, *21*(19), 7831–7840. https://doi.org/10.1523/JNEUROSCI.21-19-07831.2001.

Yin, H. H., Knowlton, B. J., & Balleine, B. W. (2004). Lesions of dorsolateral striatum preserve outcome expectancy but disrupt habit formation in instrumental learning. *European Journal of Neuroscience*, *19*(1), 181–189. https://doi.org/10.1111/j.1460-9568.2004.03095.x.

Yin, H. H., Knowlton, B. J., & Balleine, B. W. (2005). Blockade of nmda receptors in the dorsomedial striatum prevents action–outcome learning in instrumental conditioning. *European Journal of Neuroscience*, *22*(2), 505–512. https://doi.org/10.1111/j.1460-9568.2005.04219.x.

Yin, H. H., Ostlund, S. B., Knowlton, B. J., & Balleine, B. W. (2005). The role of the dorsomedial striatum in instrumental conditioning. *European Journal of Neuroscience*, *22*(2), 513–523. https://doi.org/10.1111/j.1460-9568.2005.04218.x.

Zapata, A., Minney, V. L., & Shippenberg, T. S. (2010). Shift from goal-directed to habitual cocaine seeking after prolonged experience in rats. *The Journal of Neuroscience*, *30*(46), 15457–15463. https://doi.org/10.1523/jneurosci.4072-10.2010.

Zhang, J., Berridge, K. C., Tindell, A. J., Smith, K. S., & Aldridge, J. W. (2009). A neural computational model of incentive salience. *PLoS Computational Biology*, *5*(7), e1000437. https://doi.org/10.1371/journal.pcbi.1000437.

Zhang, W. H., Cao, K. X., Ding, Z. B., Yang, J. L., Pan, B. X., & Xue, Y. X. (2018). Role of prefrontal cortex in the extinction of drug memories. *Psychopharmacology*, 1–15. https://doi.org/10.1007/s00213-018-5069-3.

Zijlstra, F., Veltman, D. J., Booij, J., van den Brink, W., & Franken, I. H. A. (2009). Neurobiological substrates of cue-elicited craving and anhedonia in recently abstinent opioid-dependent males. *Drug and Alcohol Dependence*, *99*(1), 183–192. https://doi.org/10.1016/j.drugalcdep.2008.07.012.

Chapter 8

Episodic future thinking in drug addiction

Alejandro N. Morris[a], Mohamad El Haj[b,c,d], Ahmed A. Moustafa[e]
[a]*School of Social Sciences and Psychology, Western Sydney University, Sydney, NSW, Australia,* [b]*Psychology Laboratory of Pays de la Loire (EA 4638), University of Angers, Angers, France,* [c]*Geriatric Unit, Centre Hospitalier de Tourcoing, Tourcoing, France,* [d]*Institut Universitaire de France, Paris, France,* [e]*Marcs Institute for Brain, Behaviour, and Development and School of Psychology, Western Sydney University, Penrith, NSW, Australia*

Introduction

Episodic memory and episodic future thinking are two constructs that have been used in psychology to, respectively, assess individuals' ability to imagine their subjective self at two points in time: the past and future (Moustafa, Morris, & El Haj, 2018). Episodic memory is a category within autobiographical memory (AM) proposed by Tulving (1984, 2002) and Wheeler, Stuss, and Tulving (1997) as a memory system that allows individuals to re-experience events in a specific time and place (Prebble, Addis, & Tipett, 2013). Episodic future thinking is the ability to pre-experience and project oneself into specific future events (Atance & O'Neill, 2001). Both processes form part of autonoetic consciousness, which is defined by Wheeler et al. (1997) as the capacity to "mentally represent and to become aware of one's protracted existence across a subjective time and to focus directly on one's own subjective experience" (p. 335). Autonoeticity underpins important human social behaviors such as self-awareness of autonomy and responsibility for past and present behavior, as well as planning for and behaving in congruence with future goals (Sani, 2008). Research has focused on measuring patients' ability to remember and imagine future episodic events in various psychiatric disorders in an attempt to understand their etiology and maintenance factors. Compiling and assessing current literature on future-oriented thought in addiction disorders will improve conceptualization, measurement and possible treatment. The aim of the current chapter is to assess the method and task the studies employed to elucidate episodic thinking in addiction populations and whether this group exhibits deficits compared to healthy individuals.

Besides the future episodic memory paradigm, the intolerance of uncertainty scale (IUS) and the purpose in life test (PIL-R) (Newcomb & Bentler, 1987)

Cognitive, Clinical, and Neural Aspects of Drug Addiction. https://doi.org/10.1016/B978-0-12-816979-7.00008-X

were employed in several studies as well as ours to determine baseline measures of how chronic opiate and alcohol users perceive the future. The IUS aims at investigating how opiate users and alcohol dependent individuals deal with the uncertainty of the future. The PIL-R investigates life purpose, meaninglessness in life and suicidal thoughts. The UIS and PIL-R are psychologically sound instruments of future prospection. They both have adequate internal consistency (IUS: $\alpha=0.94$; PIL-R: $\alpha=0.81$). The IUS has temporal stability ($r=0.74$ over 5 weeks) and demonstrated cross-cultural validity (Sexton & Dugas, 2009).

Types of future thinking tasks used in the literature

Generally speaking, future thinking can be assessed by asking participants to imagine events that may reasonably happen in the future. Participants are typically asked to be precise and specific, that is, events have to last no more than a day, participants are also invited to provide details such as time and place at which events will occur, as well as to describe their feelings and emotions during those events. These instructions have been used in most of research reviewed in this paper. This research has also considered emotional aspects as future thinking, this by asking patients to imagine neutral, positive, or negative events. For instance, MacLeod and Cropley (1995) asked patients with dysphoria to imagine positive and negative events, this to observed that negative events were related with depression, while positive events were related with hopelessness. Emotional valence was also assessed by MacLeod, Tata, Kentish, and Jacobsen (1997) who found that anxious participants gave more negative responses but not fewer positive responses than controls.

Besides assessing emotional valence, research in this review was interested by phenomenological characteristics of future thinking. The latter thinking is believed to triggers a state of autonoetic consciousness by which the phenomenological (i.e., subjective) experience of future is relived thanks to mental time travel (Tulving, 2002). In the reviewed literature (e.g., D'Argembeau, Raffard, & Van der Linden, 2008; De Oliveira, Cuervo-Lombard, Salame, & Danion, 2009; Raffard et al., 2010; Raffard, Esposito, Boulenger, & Van der Linden, 2013). The phenomenological experience of future thinking has been by asking patients to fill scales of conscious experience. On these scales, participants were asked to rate amount of phenomenological characteristics such as visual details, amount of sounds, and amount of smell/taste that may characterize their future thinking, as well as clarity of location, clarity of the spatial arrangement of objects and people, and clarity of the time of day in their projection. Self-referential information was also assessed with items about representation of the patients' own behavior, representation of what she/he would say, and representation of what she/he would think. Participants are also invited to report the visual perspective they took in their future thinking depending on whether they "see" themselves in their memory or see the scene from their own perspective. Thanks to these demarches, studies observed that future thinking is less vivid in dysphoria (Anderson & Evans, 2014). Studies also reported poor visual imagery during future thinking in Schizophrenia (Raffard et al., 2010, 2013), depression (Morina, Deeprose, Pusowski, Schmid, & Holmes, 2011).

Episodic foresight in alcohol dependent populations

Chronic alcohol abuse is associated with poor performance in executive function tasks, including planning ability, rule detection and coordination of dual tasks (D'Argembeau, Van der Linden, Verbanck, & Noel, 2006). Alcohol dependent individuals are reported to have greater difficulty accessing past episodic events compared to healthy individuals. Nevertheless, when they did access specific past events, they reported similar event-specific details (sensory, contextual, self-referential, etc.) similar to healthy controls (D'Argembeau et al., 2006).

Research to date has focused on prospective memory (i.e., remembering to perform a task) in alcohol-induced individuals rather than clinically diagnosed alcohol dependent individuals (see Table 1). Results indicate that after participants were administered alcohol, they had more difficulty engaging in episodic future thinking (Leitz et al., 2009; Paraskevaides et al., 2010). However, further research indicated that engaging in episodic foresight improved performance on prospective memory tasks in individuals who had been intoxicated with alcohol (Paraskevaides et al., 2010). The clinical implications of these findings suggest future episodic event simulation could potentially reduce relapse rates

TABLE 1 A summary of future existing future thinking tasks in alcohol and drug addiction

	Clinical group	Future thinking task
Leitz, Morgan, Bisby, Rendell, and Curran (2009)	Alcohol induced individuals	Prospective task memory: virtual week
Paraskevaides et al. (2010)	Alcohol induced individuals	Virtual week
Bulley and Gullo (2017)	Alcohol purchase task (simulation)	Asked to engage in episodic foresight (events they looked forward to)
Snider, LaConte, and Bickel (2016)	Alcohol dependence	Generate future positive events
Mercuri et al. (2015)	Opiate abuse	Episodic foresight task with cue words
Mercuri et al. (2016)	Opiate abuse	Episodic foresight-cue sentences

(Paraskevaides et al., 2010). However, in order to gain greater clinical relevance, future research should focus on individuals who have been diagnosed with alcohol dependence.

Addiction

Table 1 shows summary of studies investigating future thinking in patients with alcohol and drug addiction.

Alcohol addiction

Leitz et al. (2009) studied the acute effects of alcohol upon prospective memory using the virtual week (having to remember to perform planned task in the future). Forty healthy volunteers were administered 0.6 g/kg ethanol or a matched placebo and were asked to complete the virtual week along prose recall and executive function tasks. It was found that alcohol produced impairments all prospective tasks. It also produced impairments of episodic memory, which positively correlated with perspective memory performance of irregular tasks.

Unlike prior study, Paraskevaides et al. (2010) investigated whether future simulation could attenuate the impairing effects of acute alcohol on prospective memory. Thirty-two healthy volunteers were administered 0.06 g/kg of ethanol. Prospective memory was assessed using a behavioral task (virtual week). The study replicated Leitz et al. (2009) findings that acute alcohol consumption impairs prospective memory for event-based tasks. Future event simulation significantly improved prospective memory performance.

Recently, Bulley and Gullo (2017) examined the effects of alcohol on delay discounting and alcohol related decision making. Forty-eight college students were administered both intertemporal choice and hypothetical alcohol purchase tasks during which personally relevant episodic future event cues or control imagery cues were presented. It was found that engaging in episodic foresight reduced rate of at which monetary rewards were discounted and initial alcohol demand relative to control.

Using different measures, Snider et al. (2016) examined episodic future thinking, delay discounting and alcohol purchasing in a group of alcohol-dependent individuals. Fifty participants with alcohol dependence were asked to generate positive future or recent past events for each of 5 time points. Episodic future thinking significantly increased the valuation of future monetary rewards, while decreasing initial consumption of alcoholic drinks.

Episodic future thinking in opiate dependent populations

The literature indicates that chronic opiate use is associated with neurocognitive impairment, specifically of the medial and prefrontal regions (Ersche, Clark, London, Robbins, & Sahakian, 2006; Liu et al., 2009). It is suggested that being able to project oneself into the future provides increased behavioral flexibility in present actions and increases survival chances in the future (Suddendorf & Corballis, 2007). A breakdown in episodic foresight may be part of the reason opiate users are associated with poor social and economic outcomes (Mercuri et al., 2015). In previous research, the assessment of EFT in opiate users has been conducted indirectly through prospective memory (i.e., remembering to perform a scheduled task) and decision-making. Opiate users have been shown to be less able to retrieve past episodic details (Ersche et al., 2005) and to make maladaptive decisions that result in long-term losses (Grant, Contoreggi, & London, 2000).

Mercuri et al. (2015) was the first to analyze episodic memory and foresight in opiate users using cue words such as "birthday" or "vacation". The results indicated that, relative to controls, episodic foresight was impaired in opiate users. Both controls and opiate users produced similar number of details, yet the latter generated more non-episodic details. The implications of these findings suggest that episodic foresight has an anticipatory element that allows individuals to work through various future hypothetical scenarios before executing goal-directed actions (Mercuri et al., 2015). If this ability is hindered, it may limit the frequency of behavioral contingencies that might lead to desired goals. In the case of opiate users, impaired episodic foresight can contribute to maladaptive decision-making as patients seek the fulfillment of current objectives (e.g., consuming drugs) over long-term goals that yield greater rewards (e.g., health maintenance) (Mercuri et al., 2015). However, there appeared to be no correlation between the Hayling inhibition task, which assesses the restraint function of inhibition, and episodic foresight. The implications being that therapeutic training involving episodic foresight could be hindered by deficient inhibitory control. Nevertheless, Mercuri et al. (2015) advised that further inhibition tasks were required to measure other aspects of control inhibition such as selective attention and emotion regulations. Enhancing episodic foresight in the context of relapse prevention has the potential to be an efficacious cognitive rehabilitation technique.

Mercuri et al. (2016) further extended previous research and asked participants to construct atemporal and future scenes. Opiate users did not differ with controls in the construction of atemporal situations, but instead underperformed in the creation of future scenarios. It suggests that opiate users have difficulty adopting an alternative perspective of the self, which would contribute to their reduced episodic foresight ability. Mercuri et al. (2016) findings indicate the potential value of constructing short-term treatment objectives (1–2 years) that rely on imagining the self in the nearer future. To engage in short-term episodic foresight (1–2 years) could have clinical value that would help opiate users manage exposure to relapse triggers and make them aware of the cost-benefit analysis of changing current drug-related behaviors.

Past research indicates that episodic foresight has the potential to increase self-regulation by stimulating future scenarios that involve planning, problem solving and decision-making. However, more research is required to understand which cognitive abilities can enhance episodic foresight and thereby increase the frequency of self-generated future simulations.

Opiate addiction

Mercuri et al. (2015) assessed how episodic foresight is affected in the context of opiate users, as well as the degree to which any deficits are related to difficulties with executive control and episodic memory. They employed an adaptation of the autobiographical memory task cue past and future events. Results showed that the clinical group exhibited significant impairment in episodic foresight but not episodic memory. In a follow-up study, Mercuri et al. (2016) aimed to better understand the circumstances in which chronic opiate users might expect to have problems with episodic foresight, by addressing whether deficits reflect compromised scene construction, self-projection, or narrative ability. There were deficits in self-projection. However, these deficits were independent of scene construction or narrative ability. Instead, it was due to a specific impairment in self-projection into the future.

To sum up, research into future prospection in alcohol induced individuals showed they had a diminished ability to remember future tasks. In addition, alcohol dependent individuals who engaged in episodic foresight showed greater delayed discounting and a reduced desire for immediate alcohol consumption. Episodic foresight in opiate dependent individuals was less specific compared to healthy controls, after controlling for narrative ability. The next important steps are to identify any deficits in components of episodic foresight such as semantic construction.

The literature indicates that chronic opiate use is associated with neurocognitive impairment, specifically of the medial and prefrontal regions (Ersche et al., 2006; Liu et al., 2009). It is suggested that being able to project oneself into the future provides increased behavioral flexibility in present actions and increases survival chances in the future (Suddendorf & Corballis, 2007). A breakdown in episodic foresight may be part of the reason opiate users are associated with poor social and economic outcomes (Mercuri et al., 2015). In previous research, the assessment of EFT in opiate users has been conducted indirectly through prospective memory (i.e., remembering to perform a scheduled task) and decision-making. Opiate users have been shown to be less able to retrieve past episodic details (Ersche et al., 2005) and to make maladaptive decisions that result in long-term losses (Grant et al., 2000).

Mercuri et al. (2015) was the first to analyze episodic memory and foresight in opiate users using cue words such as "birthday" or "vacation". The results indicated that, relative to controls, episodic foresight was impaired in opiate users. Both controls and opiate users produced similar number of details, yet the latter generated more non-episodic details. The implications of these findings suggest that episodic foresight has an anticipatory element that allows individuals to work through various future hypothetical scenarios before executing goal-directed actions (Mercuri et al., 2015). If this ability is hindered, it may limit the frequency of behavioral contingencies that might lead to desired goals. In the case of opiate users, impaired episodic foresight can contribute to maladaptive decision-making as patients seek the fulfillment of current objectives (e.g., consuming drugs) over long-term goals that yield greater rewards (e.g., health maintenance) (Mercuri et al., 2015). However, there appeared to be no correlation between the Hayling inhibition task, which assesses the restraint function of inhibition, and episodic foresight. The implications being that therapeutic training involving episodic foresight could be hindered by deficient inhibitory control. Nevertheless, Mercuri et al. (2015) advised that further inhibition tasks were required to measure other aspects of control inhibition such as selective attention and emotion regulations. Enhancing episodic foresight in the context of relapse prevention has the potential to be an efficacious cognitive rehabilitation technique.

Mercuri et al. (2016) further extended previous research and asked participants to construct atemporal and future scenes. Opiate users did not differ with controls in the construction of atemporal situations, but instead underperformed in the creation of future scenarios. It suggests that opiate users have difficulty adopting an alternative perspective of the self, which would contribute to their reduced episodic foresight ability. Mercuri et al. (2016) findings indicate the potential value of constructing short-term treatment objectives (1–2 years) that rely on imagining the self in the nearer future. To engage in short-term episodic foresight (1–2 years) could have clinical value that would help opiate users manage exposure to relapse triggers and make them aware of the cost-benefit analysis of changing current drug-related behaviors.

Past research indicates that episodic foresight has the potential to increase self-regulation by stimulating future scenarios that involve planning, problem solving and decision-making. However, more research is required to understand which cognitive abilities can enhance episodic foresight and thereby increase the frequency of self-generated future simulations.

Episodic foresight in alcohol dependent populations

Chronic alcohol abuse is associated with poor performance in executive function tasks, including planning ability, rule detection and coordination of dual tasks (D'Argembeau et al., 2006). Alcohol dependent individuals are reported to have greater difficulty accessing past episodic events compared to healthy individuals. Nevertheless, when they did access specific past events, they reported similar event-specific details (sensory, contextual, self-referential, etc.) similar to healthy controls (D'Argembeau et al., 2006).

Research to date has focused on prospective memory (i.e., remembering to perform a task) in alcohol-induced individuals rather than clinically diagnosed alcohol dependent individuals (see Table 1). Results indicate that after participants were administered alcohol, they had more difficulty engaging in episodic future thinking (Leitz et al., 2009; Paraskevaides et al., 2010). However, further research indicated that engaging in episodic foresight improved performance on prospective memory tasks in individuals who had been intoxicated with alcohol (Paraskevaides et al., 2010). The clinical implications of these findings suggest future episodic event simulation could potentially reduce relapse rates (Paraskevaides et al., 2010). However, in order to gain greater clinical relevance, future research should focus on individuals who have been diagnosed with alcohol dependence.

This study examined differences across the three groups in two questionnaires that measure people's belief about future uncertainty and purpose in life. Both questionnaires were employed to determine baseline measures of how chronic opiate and alcohol users perceive the future. The intolerance of uncertainty scale (IUS) (Sexton & Dugas, 2009) aims at investigating how opiate users and alcohol dependent individuals deal with the uncertainty of the future.

Intolerance to uncertainty scale (IUS) and purpose in life in alcohol and opiate populations

In addition to the episodic future thinking paradigms discussed above, intolerance of uncertainty and purpose in life are also related to future thinking. The intolerance to uncertainty scale (IUS) is a 27- item self-report research instrument used to measure negative beliefs about uncertainty and its perceived consequences (Sexton & Dugas, 2009). This instrument is used on individuals 18 and above. The 27 items are divided into two factors. Factor 1 encompasses 15 items, which assesses the participants' beliefs that being uncertain impairs behavior and reflects badly on an individual's character. Factor 2 items reflect the notion that the future should be predictable and that unpredictability is unfair and distressing (Sexton & Dugas, 2009). Participants are instructed to rate items using a 5-point likert scale (1 = not at all characteristic of me, 3 = somewhat characteristic of me, 5 = entirely characteristic of me). Individual item scores are summed to provide a total score out of 75 for factor 1 and out of 60 for factor 2. Higher scores on both factors indicate greater intolerance to uncertainty.

The purpose in life test (PIL-R) (Harlow, Newcomb, & Bentler, 1987) is a 20-item self-report instrument used to assess constructs of happiness, suicidality and meaninglessness in persons over the age of 18. Individuals are instructed to rate a 7-point likert scale (1 = strongly agree, 7 = strongly disagree). Eleven items are reverse scored due to their negative phrasing (e.g., "In life I have no goals or aims at all"). Scores range from 10 (low purpose in life) to 140 (high purpose in life). The UIS and PIL-R are psychologically sound instruments of future prospection. They both have adequate internal consistency (IUS: $\alpha=0.94$; PIL-R: $\alpha=0.81$). The IUS has temporal stability ($r=0.74$ over 5 weeks) and demonstrated cross-cultural validity (Sexton & Dugas, 2009).

We have conducted a study to compare IUS across three groups: patients with heroin addiction, patients with alcohol addiction, and healthy controls. A total of 76 participants (*M* age=49.6 years, *SD*=11.79 years, 30 females, 46 males) took part in the study. Thirty-two opiate users aged 22–59 were recruited purposively from the methadone clinic at the Royal Prince Alfred Hospital (RPAH). Opiate users were currently enrolled in an opiate substitution program (methadone $n=26$, suboxone $n=4$, buprenorphine $n=2$), with an average daily dose of 63 mg of methadone (*SD*=45.51), 25 mg (*SD*=8.25) of suboxone and 12 mg (*SD*=5.66) of buprenorphine. Subjects were approached at the methadone clinic and asked if they were willing to participate in a fourth year Honors addiction project in exchange for $20. Subjects who accepted were examined prior to their daily dose in order to reduce the effects of drugs on performance. They were recruited in accordance with the approval granted by the research development office at the RPAH and Western Sydney University (approval number: x12-0187). Data for one participant was discarded due to incomplete responses ($n=31$).

Twenty-one chronic alcohol users aged 37–71 were recruited from the baclofen trial database at RPAH. All alcohol patients met the criteria for alcohol dependence as described by the Diagnostic and Statistical Manual of Mental Disorders (4th ed. rev; *DSM-IV-TR*; American Psychiatric Association, 2000). Patients who had finished the trial were contacted via letter asking if they were willing to complete some follow-up tests in exchange for $20. Participants were recruited in accordance with the approval granted by the research development office at the RPAH and Western Sydney University (approval number: x11-0154).

The control group consisted of 23 adults aged 34–71 years with no reported history of alcohol, opiate, or other drug dependence. Participants were recruited via the healthy control database at the RPAH. All subjects were reimbursed $20.

All testing for the opiate and alcohol groups took place in interview rooms at the Drug and Health Services department at the King George V Memorial Hospital for Mothers and Babies. The control groups were tested either at a public library or the quiet of their respective homes. The three groups did not differ significantly on gender, $\chi^2(2, 76)=0.74$, $P=0.691$. Age, education and gender for all three groups is shown in Table 2. In addition, we summarize the frequency of participants engaging in current polysubstance abuse. (See Table 3.)

A one-way between groups analysis of variance (ANOVA) was used to investigate any significant age difference between the groups. Inspections of skewness, kurtosis and Shapiro-Wilk statistics indicated the assumption of normality was supported for all three groups. Levene's statistic was non-significant $F(2,72)=2.79$, $P=0.068$, and thus the assumption of homogeneity of variance was not violated. The ANOVA was significant, indicating that there were significant differences in age between the 3 groups, $F(2, 72)=12.55$, $P<0.001$. Post-hoc analysis using Tukey's HSD (using an α of 0.05) revealed that the opiate group had a significantly lower mean age than the alcohol and control groups.

TABLE 2 Background characteristics of the controls, opiate and alcohol user groups

	Control group		Opiate group		Alcohol group	
	n=23		*n*=31		*n*=21	
	M	*SD*	*M*	*SD*	*M*	*SD*
Age (in years)	53.04	12.30	42.65	8.83	56.09	9.88
Education (in years)	16.22	1.95	10.58	2.67	13.10	2.84
Proportion of males (%)	60.9%		54.8%		66.7%	
Years of problematic drinking[a]					17.24	6.98
Standard drinks[b]					12.21	4.69

[a] *Drinking significantly affecting one's family and work life.*
[b] *Number of standard drinks per day one month prior to beginning baclofen trial.*

TABLE 3 Frequency of participants engaging in current substance use

	Opiate group	Alcohol group
Substance	*n*=31	*n*=21
Alcohol (no:yes)[a]	24:7	10:11
Cannabis (no:yes)[b]	25:6	21:0
Methamphetamine (no:yes)[c]	16:15	21:0
Amphetamine (no:yes)[d]	30:1	20:1
Cocaine (no:yes)	27:4	20:1
Benzodiazepines (no:yes)	28:3	21:0
Heroin (no:yes)[e]	5:26	20:1
Oxycodone (no:yes)	29:2	21:0
Barbiturates (no:yes)	31:0	20:1

[a] *Daily frequency of alcohol reported for both groups.*
[b] *Daily frequency reported by opiate group.*
[c] *Weekly frequency reported by opiate group.*
[d] *Less than once a month reported by both groups.*
[e] *Daily frequency reported by both groups.*

A second ANOVA was used to assess any differences in education between the 3 groups. Skewness and kurtosis were not violated across the three groups. However, Shapiro-Wilk revealed normality was violated for alcohol and opiate groups, but not for controls. Levene's statistic was significant, $F(2,72)=4.01$, $P=0.02$, indicating the assumption of homogeneity was violated. Consequently, the results should be interpreted with caution. The ANOVA was significant, indicating years of education were significantly different across the groups, $F(2,72)=32.90$, $P<0.001$. Post-hoc analysis using Tukey's HSD (α set at 0.05) revealed that the opiate group was significantly less educated than the alcohol and control group. The alcohol group was significantly less educated than the control group.

Three factorial ANCOVAs were used to measure mean differences in IUS Factors 1 and 2. For IUS Factor 1, there were minor departures from normality as indicated by the Shapiro-Wilk. Therefore, results should be interpreted with caution. Scatterplots indicated that the relationship between the covariates and the dependent variable (IUS Factor 1) was linear. The assumptions of homogeneity of regression slopes and homogeneity of variances were supported by the absence of a significant IV-by-covariates interaction, $F(3,67)=2.08$, $P=0.111$ and a non-significant Levene's test, $F(2,72)=1.51$, $P=0.227$. The ANCOVA indicated that, after accounting for age and levels of education, there was a non-significant effect of group membership on levels of IUS Factor 1, $F(2,70)=0.02$, $P=0.982$, partial $\eta^2=0.001$.

The assumptions of normality were met for IUS Factor 2. The assumption of linearity was met as indicated in the scatterplots. The assumptions of homogeneity of regression slopes and homogeneity of variances were supported by the absence of a significant IV-by-covariates interaction, $F(3,67)=2.15$, $P=0.102$, and a non-significant Levene's test, $F(2,72)=0.29$, $P=0.748$. The ANCOVA indicated that, after controlling for age and level of education, there was a non-significant effect of group membership on levels of IUS Factor 2, $F(2,70)=0.31$, $P=0.737$, partial $\eta^2=0.009$.

This study sought to explore whether the addiction groups displayed greater intolerance to future uncertainty compared to controls. No significant differences were found in IUS Factors 1 and 2 across the three groups. Uncertainty about the future has been linked with vulnerability to addiction, suggesting that addicts seek familiar options over the unknown (Radell, Myers, Beck, Moustafa, & Allen, 2016). Radell et al. (2016) tested uncertainty towards the future using conditioned preference tasks, while the current study employed a self-report measure. Items were undisguised and participants may have been reluctant to rate statements such as "Being uncertain means that I lack confidence", as it would reflect on their own personality. Social desirability bias may have distorted the current findings. In addition, IUS questionnaire items have been associated with anxiety (Sexton & Dugas, 2009) and the current study did not test levels of anxiety. It may be that worry, irrespective of addiction, significantly contributed to IUS ratings. Further research should explore levels of anxiety in alcohol-dependent individuals and opiate users and its effect on future thinking.

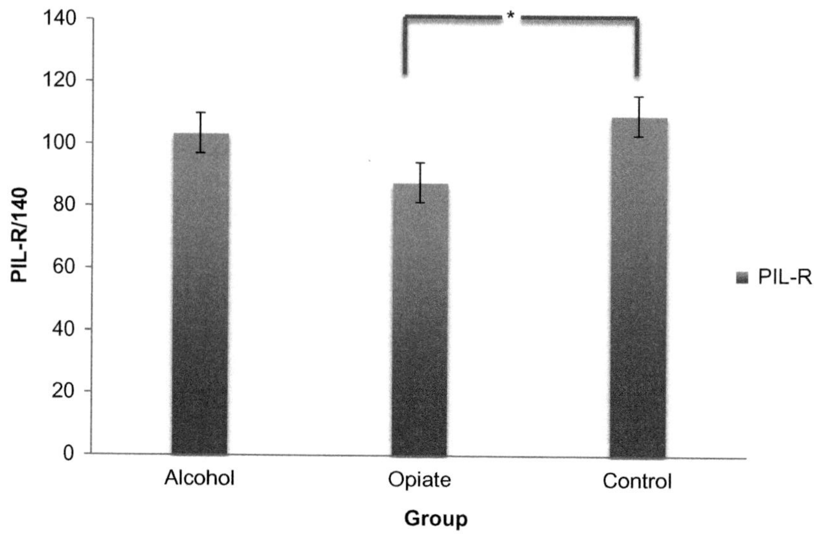

FIG. 1 Difference in average PIL-R score across the three groups (*$P<0.05+SE$). Post-Hoc comparisons showed a significant difference between opiate and control groups at $\alpha=0.05$, yet after the Bonferroni correction, the difference was non-significant.

For PIL-R, the assumptions of normality were met. The assumption of linearity was met for both covariates and the IV as indicated in the scatterplots. The assumptions of homogeneity of regression slopes and homogeneity of variances were supported by the non-significant IV-by-covariates interaction, $F(3,67)=0.62$, $P=0.604$, and a non-significant Levene's test, $F(2,72)=0.34$, $P=0.713$. The ANCOVA showed that, after controlling for age and levels of education, there was a statistically significant effect of group membership on levels of PIL-R, $F(2,70)=3.29$, $P=0.043$, partial $\eta^2=0.086$ (see Fig. 1). Post-Hoc testing using Bonferroni adjustments for multiple tests 0.033 (0.10/3) revealed that participants in the control group reported significantly higher levels of PIL-R than the opiate group ($M_{diff}=16.731$, $P=0.044$, 95% CI [0.320, 33.142]), yet after the alpha adjustment to 0.033, this difference was no longer significant.

Discussion

Studies that analyzed alcohol patients' ability to envision future episodic events (Bulley & Gullo, 2017; Leitz et al., 2009; Paraskevaides et al., 2010; Snider et al., 2016) evidenced that there is a difficulty for patients to engage in episodic foresight. Similarly, Mercuri et al. (2015, 2016) found that opiate users showed difficulty to visualize themselves in the future. However, the latest study performed at the Royal Prince Alfred Hospital (Sydney) (Moustafa et al., 2018) found opiate users did significantly worse at episodic future thinking compared to alcohol dependent patients and healthy individuals. This suggested different levels of ability of envisioning future events in different addiction populations.

Opiate users tend to engage in more poly-drug abuse and are exposed to opiate substitution programs that may hinder their ability of episodic foresight. Langleben et al. (2008) found that individuals who were opiate users had a diminished anterior cingulate cortex, which is understood to be the region responsible for anticipating future events. This region has not yet been investigated in alcohol dependent individuals. Further research is required to investigate if this link exists in this addiction population.

The PIL-R was expected to be higher in the controls compared to the addiction groups, and this was not supported. Opiate users did worse (albeit not significantly) than controls, indicating they had higher levels of meaninglessness and were more open to suicidality. Higher purpose in life is viewed as a coping mechanism that can be protective against high-risk situations (e.g. stress) from drug use (Martin, MacKinnon, Johnson, & Rohsenow, 2011). This may partially explain why opiate users are linked to a 59% relapse rate (Smyth, Barry, Keenan, & Ducray, 2010). However, purpose in life scores could be mediated by depression, which is prevalent in opiate populations (Martin et al., 2011) and was not measured in this study.

Alcohol-dependent individuals and controls performed similarly on PIL-R, indicating they both had comparable levels of purpose in life. This is contrary to Martin et al. (2011)'s findings of drug users having lower purpose in life scores. A potential explanation may be that the alcohol group had finalized their baclofen trial and were more hopeful about the future, as seen in other post-treatment groups (Dale, Adair, & Humphris, 2010). Nevertheless, the generalizability of these conclusions is limited, as no post-treatment questions were administered regarding their views on their future quality of life and mental wellbeing.

This study sought to explore whether the addiction groups displayed greater intolerance to future uncertainty compared to controls. No significant differences were found in IUS Factors 1 and 2 across the three groups. Uncertainty about the future has been linked with vulnerability to addiction, suggesting that addicts seek familiar options over the unknown (Radell et al., 2016). Radell et al. (2016) tested uncertainty towards the future using conditioned preference tasks, while the current study employed a self-report measure. Items were undisguised and participants may have been reluctant to rate statements such as "Being uncertain means that I lack confidence", as it would reflect on their own personality. Social desirability bias may have distorted the current findings. In addition, IUS questionnaire items have been associated with anxiety (Sexton & Dugas, 2009) and the current study did not test levels of anxiety. It may be that worry, irrespective of addiction, significantly contributed to IUS ratings. Further research should explore levels of anxiety in alcohol-dependent individuals and opiate users and its effect on future thinking.

Limitations and future studies

The current studies have established the existence of a tendency to overgeneralize future episodic events in individuals diagnosed with psychiatric conditions such as MDD, schizophrenia and addiction populations. The current studies have limitations. Firstly, only one study (Mercuri et al., 2016) sought to understand and explain what factors maintained individuals' difficulty at envisaging future episodic events. Related to this idea, more longitudinal studies are required to trace the development and associations between an individual's specific condition and their ability to imagine past and future episodic events.

References

Anderson, R. J., & Evans, G. L. (2014). Mental time travel in Dysphoria: Differences in the content and subjective experience of past and future episodes. *Consciousness and Cognition*, *37*, 237–248. https://doi.org/10.1016/j.concog.2014.05.006.

Atance, C. M., & O'Neill, D. K. (2001). Episodic future thinking. *Trends in Cognitive Sciences*, *5*(12), 533–539.

Bulley, A., & Gullo, M. J. (2017). The influence of episodic foresight on delay discounting and demand for alcohol. *Addictive Behaviors*, *66*, 1–6. https://doi.org/10.1016/j.addbeh.2016.11.003.

Dale, H. L., Adair, P. M., & Humphris, G. M. (2010). Systematic review of post-treatment psychosocial and behaviour change interventions for men with cancer. *Psycho-Oncology*, *19*(3), 227–237. https://doi.org/10.1002/pon.1598.

D'Argembeau, A., Raffard, S., & Van der Linden, M. (2008). Remembering the past and imagining the future in schizophrenia. *Journal of Abnormal Psychology*, *117*(1), 247–251. https://doi.org/10.1037/0021-843X.117.1.247.

D'Argembeau, A., Van der Linden, M., Verbanck, P., & Noel, X. (2006). Autobiographical memory in non-amnesiac alcohol-dependent patients. *Psychological Medicine*, *36*(12), 1707–1715. https://doi.org/10.1017/S0033291706008798.

De Oliveira, H., Cuervo-Lombard, C., Salame, P., & Danion, J. M. (2009). Autonoetic awareness associated with the projection of the self into the future: An investigation in schizophrenia. *Psychiatry Research*, *169*(1), 86–87. https://doi.org/10.1016/j.psychres.2008.07.003.

Ersche, K. D., Clark, L., London, M., Robbins, T. W., & Sahakian, B. J. (2006). Profile of executive and memory function associated with amphetamine and opiate dependence. *Neuropsychopharmacogology*, *31*(5), 1036–1047. https://doi.org/10.1038/sj.npp.1300889.

Ersche, K. D., Roiser, J. P., Clark, L., London, M., Robbins, T. W., & Sahakian, B. J. (2005). Punishment induces risk risky decision-making in methadone-maintained opiate users but not in heroin users or healthy volunteers. *Neuropsychopharmacology*, *30*(11), 2115–2124. https://doi.org/10.1038/sj.npp.1300812.

Grant, S., Contoreggi, C., & London, E. D. (2000). Drug abusers show impaired performance in a laboratory test of decision making. *Neuropsychologia*, *38*(8), 1180–1187.

Harlow, L. L., Newcomb, M. D., & Bentler, P. M. (1987). Purpose in Life Test assessment using latent variable methods. *British Journal of Clinical Psychology*, *26*(3), 235–236. https://doi.org/10.1111/j.2044-8260.1987.tb01355.x.

Langleben, D. D., Ruparel, K., Elman, I., Busch-Winokur, S., Pratiwadi, R., Loughead, J., et al. (2008). Acute effect of methadone maintenance dose on brain FMRI response to heroin-related cues. *American Journal of Psychiatry*, *165*(3), 390–394.

Leitz, J. R., Morgan, C. J., Bisby, J. A., Rendell, P. G., & Curran, H. V. (2009). Global impairment of prospective memory following acute alcohol. *Psychopharmachology*, *205*(3), 379–387. https://doi.org/10.1007/s00213-009-1546-z.

Liu, J., Liang, J., Qin, W., Tian, J., Yuan, K., Bai, L., et al. (2009). Dysfunctional connectivity patterns in chronic heroin users: An fMRI study. *Neuroscience Letters*, *460*(1), 72–77. https://doi.org/10.1016/j.neulet.2009.05.038.

MacLeod, A. K., & Cropley, M. L. (1995). Depressive future thinking: The role of valence and specificity. *Cognitive Therapy and Research*, *19*(1), 35–50. https://doi.org/10.1007/BF02229675.

MacLeod, A. K., Tata, P., Kentish, J., & Jacobsen, H. (1997). Retrospective and prospective cognitions in anxiety and depression. *Cognition & Emotion*, *11*(4), 467–479. https://doi.org/10.1080/026999397379881.

Martin, R. A., MacKinnon, S., Johnson, J., & Rohsenow, D. J. (2011). Purpose in life predicts treatment outcome among adult cocaine abusers in treatment. *Journal of Substance Abuse Treatment*, *40*(2), 183–188. https://doi.org/10.1016/j.jsat.2010.10.002.

Mercuri, K., Terrett, G., Henry, J. D., Bailey, P. E., Curran, H. V., & Rendell, P. G. (2015). Episodic foresight deficits in long-term opiate users. *Psychopharmacology*, *232*(7), 1337–1345. https://doi.org/10.1007/s00213-014-3772-2.

Mercuri, K., Terrett, G., Bailey, P. E., Henry, J. D., Curran, H. V., & Rendell, P. G. (2016). Deconstructing the nature of episodic foresight deficits associated with chronic opiate use. *British Journal of Clinical Psychology*, *55*(4), 401–413. https://doi.org/10.1111/bjc.12110.

Morina, N., Deeprose, C., Pusowski, C., Schmid, M., & Holmes, E. A. (2011). Prospective mental imagery in patients with major depressive disorder or anxiety disorders. *Journal of Anxiety Disorders*, *25*(8), 1032–1037. https://doi.org/10.1016/j.janxdis.2011.06.012.

Moustafa, A. A., Morris, A., & El Haj, M. (2018). A review on future episodic thinking in mood and anxiety disorders. *Reviews in the Neurosciences*.

Moustafa, A. A., Morris, A. N., Nandrino, J. L., Misiak, B., Szewczuk-Bogusławska, M., Frydecka, D., et al. (2018). Not all drugs are created equal: Impaired future thinking in opiate, but not alcohol, users. *Experimental Brain Research*, 1–11.

Newcomb, M. D., & Bentler, P. M. (1987). Changes in drug use from high school to young adulthood: Effects of living arrangement and current life pursuit. *Journal of Applied Developmental Psychology*, *8*(3), 221–246. https://doi.org/10.1016/0193-3973(87)90001-3.

Paraskevaides, T., Morgan, C. J., Leitz, J. R., Bisby, J. A., Rendell, P. G., & Curran, H. V. (2010). Drinking and future thinking: Acute effects of alcohol on prospective memory and future simulation. *Psychopharmacology*, *208*(2), 301–308. https://doi.org/10.1007/s00213-009-1731-0.

Prebble, S. C., Addis, D. R., & Tipett, L. J. (2013). Autobiographical memory and sense of self. *Psychological Bulletin*, *139*(4), 815–840. https://doi.org/10.1037/a0030146.

Radell, M. L., Myers, C. E., Beck, K. D., Moustafa, A. A., & Allen, M. T. (2016). The personality trait of intolerance to uncertainty affects behaviour in a novel computer-based conditioned place preference task. *Frontiers in Psychology*, *7*, 1175. https://doi.org/10.3389/fpsyg.2016.01175.

Raffard, S., D'Argembeau, A., Lardi, C., Bayard, S., Boulenger, J. P., & Van der Linden, M. (2010). Narrative identity in schizophrenia. *Consciousness and Cognition*, *19*(1), 328–340. https://doi.org/10.1016/j.concog.2009.10.005.

Raffard, S., Esposito, F., Boulenger, J. P., & Van der Linden, M. (2013). Impaired ability to imagine future pleasant events is associated with apathy in schizophrenia. *Psychiatry Research*, *209*(3), 393–400. https://doi.org/10.1016/j.psychres.2013.04.016.

Sani, F. (2008). *Self-continuity: Individual and collective perspectives*. East Sussex: Psychology Press. https://doi.org/10.4324/9780203888513.

Sexton, K. A., & Dugas, M. J. (2009). Defining distinct negative beliefs about uncertainty: Validating the factor structure of the Intolerance of Uncertainty Scale. *Psychological Assessment*, *21*(2), 176–186. https://doi.org/10.1037/a0015827.

Smyth, B. P., Barry, J., Keenan, E., & Ducray, K. (2010). Lapse and relapse following inpatient treatment of opiate dependence. *Irish Medical Journal, 103*(6), 176–179.

Snider, S. E., LaConte, S. M., & Bickel, W. K. (2016). Episodic future thinking: Expansion of the temporal window in individuals with alcohol dependence. *Alcoholism, Clinical and Experimental Research, 40*(7), 1558–1566. https://doi.org/10.1111/acer.13112.

Suddendorf, T., & Corballis, M. C. (2007). The evolution of foresight: What is mental time travel, and is it unique to humans? *The Behavioural and Brain Sciences, 30*(3), 313–351. https://doi.org/10.1017/S0140525X07001975.

Tulving, E. (1984). Relations among components of memory and processes of memory. *Behavioural Brain Science, 7*, 257–268.

Tulving, E. (2002). Episodic memory: From mind to brain. *Annual Review Psychology, 53*, 1–25. https://doi.org/10.1146/annurev.psych.53.100901.135114.

Wheeler, M. A., Stuss, D. T., & Tulving, E. (1997). Toward a theory of episodic memory: The frontal lobes and autonoetic consciousness. *Psychological Bulletin, 121*, 331–354.

Further reading

Addis, D. R., Hach, S., & Tippett, L. J. (2016). Do strategic processes contribute to the specificity of future simulation in depression? *British Journal of Clinical Psychology*, https://doi.org/10.1111/bjc.12103.

Andersen, S. M., & Limpert, C. (2001). Future-event schemas: Automaticity and rumination in major depression. *Cognitve Therapy and Research, 25*(3), 311–333. https://doi.org/10.1023/A:1026447600924.

Anderson, R. J., Boland, J., & Garner, S. R. (2016). General past and future thinking in Dysphoria: The role of emotional cues and cueing methodology. *Memory, 24*(5), 708–719. https://doi.org/10.1080/09658211.2015.1046134.

Belcher, J., & Kangas, M. (2013). Reduced goal specificity is associated with reduced memory specificity in depressed adults. *Cognition and Emotion, 28*(1), 163–171. https://doi.org/10.1080/02699931.2013.807776.

Brown, A. D., Root, J. C., Romano, T. A., Chang, L. J., Bryant, R. A., & Hirst, W. (2013). Overgeneralized autobiographical memory and future thinking in combat veterans with posttraumatic stress disorder. *Journal of Behavior Therapy and Experimental Psychiatry, 44*(1), 129–134.

Crane, C., Winder, R., Hargus, E., Amaransighe, M., & Barnhofer, T. (2012). Effects of mindfulness-based cognitive therapy on specificity of life goals. *Cognitive Therapy and Research, 26*(3), 182–189.

Cropley, M., & McLeod, A. K. (2003). Dysphoria, attributional reasoning and future event probability. *Clinical Psychology and Psychotherapy, 10*(4), 220–227. https://doi.org/10.1002/cpp.360.

D'Argembeau, A., Lardi, C., & Van der Linden, M. (2012). Self-defining future projections: Exploring the identity function of thinking about the future. *Memory, 20*(2), 110–120.

D'Argembeau, A., & Mathy, A. (2011). Tracking the construction of episodic future thoughts. *Journal of Experimental Psychology: General, 140*(2), 258–271. https://doi.org/10.1037/a0022581.

D'Argembeau, A., & Van der Linden, M. (2004). Phenomenal characteristics associated with projecting oneself back into the past and forward into the future: Influence of valence and temporal distance. *Consciousness and Cognition, 13*(4), 844–858. https://doi.org/10.1016/j.concog.2004.07.007.

Dickson, J. M., & Bates, G. W. (2006). Autobiographical memories and views of the future: In relation to Dysphoria. *International Journal of Psychology, 41*(02), 107–116. https://doi.org/10.1080/00207590500188025.

Dickson, J. M., Moberly, J. M., & Kinderman, N. J. (2011). Depressed people are not less motivated by personal goals but are more pessimistic about attaining them. *Journal of Abnormal Psychology, 120*(4), 975–980. https://doi.org/10.1037/a0023665.

El Haj, M., Antoine, P., & Kapogiannis, D. (2015a). Flexibility decline contributes to similarity of past and future thinking in Alzheimer's disease. *Hippocampus, 25*(11), 1447–1455. https://doi.org/10.1002/hipo.22465.

El Haj, M., Antoine, P., & Kapogiannis, D. (2015b). Similarity between remembering the past and imagining the future in Alzheimer's disease: Implication of episodic memory. *Neuropsychologia, 66*(0), 119–125. https://doi.org/10.1016/j.neuropsychologia.2014.11.015.

Eysenck, M., Payne, S., & Santos, R. (2006). Anxiety and depression: Past, present, and future events. *Cognition & Emotion, 20*(2), 274–294.

Hach, S., Tippett, L. J., & Addis, D. R. (2014). Neural changes associated with the generation of specific past and future events in depression. *Neuropsychologia, 65*, 41–55. https://doi.org/10.1016/j.neuropsychologia.2014.10.003.

Hassabis, D., Kumaran, D., Vann, S. D., & Maguire, E. A. (2007). Patients with hippocampal amnesia cannot imagine new experiences. *Proceedings of the National Academy of Science, 104*(5), 1726–1731. https://doi.org/10.1073/pnas.0610561104.

Hoerger, M., Quirk, S. W., Chapman, B. P., & Duberstein, P. R. (2012). Affective forecasting and self-rated symptoms of depression, anxiety, and hypomania: Evidence for Dysphoric forecasting Bias. *Cognition and Emotion, 26*(6), 1098–1106. https://doi.org/10.1080/02699931.2011.631985.

Holmes, E. A., Lang, T. J., Moulds, M. L., & Steele, A. M. (2008). Prospective and positive mental imagery deficits in dysphoria. *Behaviour Research and Therapy, 46*(8), 976–981. https://doi.org/10.1016/j.brat.2008.04.009.

Irush, M., & Piguet, O. (2013). The pivotal role of semantic memory in remembering the past and imagining the future. *Frontiers in Behavioral Neuroscience, 7*, 27.

Ji, J. L., Holmes, E. A., & Blackwell, S. E. (2017). Seeing light at the end of the tunnel: Positive prospective mental imagery and optimism in depression. *Psychiatry Research, 247*, 155–162. https://doi.org/10.1016/j.psychres.2016.11.025.

King, M. J., Williams, L. A., MacDougall, A. G., Ferris, S., Smith, J. R., Ziolkowski, N., et al. (2011). Patients with bipolar disorder show a selective deficit in the episodic simulation of future events. *Consciousness and Cognition, 20*(4), 1801–1807. https://doi.org/10.1016/j.concog.2011.05.005.

Kleim, B., Graham, B., Fihosy, S., Stott, R., & Ehlers, A. (2014). Reduced specificity in episodic future thinking in posttraumatic stress disorder. *Clinical Psychological Science, 2*(2), 165–173. https://doi.org/10.1177/2167702613495199.

Klein, S. B. (2013). The complex act of projecting oneself into the future. *Wiley Interdisciplinary Reviews: Cognitive Science, 4*, 63–79.

Kwan, D., Kurczek, J., & Rosenbaum, R. S. (2016). Specific, personally meaningful cues can benefit episodic prospection in medial temporal lobe amnesia. *British Journal of Clinical Psychology, 55*(2), 137–153. https://doi.org/10.1111/bjc.12095.

Lavender, A., & Watkins, E. (2004). Rumination and future thinking in depression. *British Journal of Clinical Psychology, 43*(2), 129–142. https://doi.org/10.1348/014466504323088015.

Maccallum, F., & Bryant, R. A. (2011). Imagining the future in complicated grief. *Depression and Anxiety, 28*(8), 658–665. https://doi.org/10.1002/da.20866.

MacLeod, A. K., & Byrne, A. (1996). Anxiety, depression, and the anticipation of future positive and negative experiences. *Journal of Abnormal Psychology, 105*(2), 286–289.

Miles, H., MacLeod, A. K., & Pote, H. (2004). Retrospective and prospective cognitions in adolescents: Anxiety, depression, and positive and negative affect. *Journal of Adolescence*, *27*(6), 691–701. https://doi.org/10.1016/j.adolescence.2004.04.001.

Miloyan, B., Bulley, A., & Suddendorf, T. (2016). Episodic foresight and anxiety: Proximate and ultimate perspectives. *The British Journal of Clinical Psychology*, *55*(1), 4–22. https://doi.org/10.1111/bjc.12080.

Okuda, J., Fuji, T., Ohtake, H., Tsukiura, T., Tanji, K., Suzuki, K., et al. (2003). Thinking of the future and past: The roles of the frontal pole and the medial temporal lobes. *NeuroImage*, *19*(4), 1369–1380.

Robinaugh, D. J., & McNally, R. J. (2013). Remembering the past and envisioning the future in bereaved adults with and without complicated grief. *Clinical Psychological Science*, *1*(3), 290–300. https://doi.org/10.1177/2167702613476027.

Roepke, A. M., & Seligman, M. E. (2016). Depression and prospection. *British Journal of Clinical Psychology*, *55*(1), 23–48. https://doi.org/10.1111/bjc.12087.

Sarkohi, A., Bjarehed, J., & Andersson, G. (2011). Links between future thinking and autobiographical memory specificity in major depression. *Psychology*, *2*(3), 261–265. https://doi.org/10.4236/psych.2011.23041.

Starcevic, V. (2007). Dysphoric about dysphoria: Towards a greater conceptual clarity of the term. *Australasian Psychiatry*, *15*(1), 9–13.

Stöber, J. (2000). Prospective cognitions in anxiety and depression: Replication and methodological extension. *Cognition & Emotion*, *14*(5), 725–729. https://doi.org/10.1080/02699930050117693.

Thimm, J. C., Holte, A., Brennen, T., & Wang, C. E. (2013). Hope and expectancies for future event in depression. *Frontiers in Psychology*, (40), 24. https://doi.org/10.3389/fpsyg.2013.00470.

Thomsen, D. K., Schnieber, A., & Olesen, M. H. (2011). Rumination is associated with the phenomenal characteristics of autobiographical memories and future scenarios. *Memory*, *19*(6), 574–584. https://doi.org/10.1080/09658211.2011.591533.

Vihauer, J. S., Young, S., Kealoha, S., Borrmann, J., Ishak, W. W., Rapaport, M. H., et al. (2012). Treating major depression by creating positive expectations for the future: A pilot study for the effectiveness of future-directed therapy (FDT) on symptom severity and quality of life. *CNS Neuroscience & Therapeutics*, *18*(2), 102–109. https://doi.org/10.1111/j.1755-5949.2011.00235.x.

Vincent, P. J., Boddana, P., & MacLeod, A. K. (2004). Positive life goals and plans in parasuicide. *Clinical Psychology and Psychotherapy*, *11*, 90–99. https://doi.org/10.1002/cpp.394.

Williams, J. M. G., & Broadbent, K. (1986). Autobiographical memory in suicide attempters. *Journal of Abnormal Psychology*, *95*(2), 144–149.

Wu, J. Q., Szpunar, K. K., Godovich, S. A., Schacter, D. L., & Hofmann, S. G. (2015). Episodic future thinking in generalized anxiety disorder. *Journal of Anxiety Disorders*, *36*, 1–8. https://doi.org/10.1016/j.janxdis.2015.09.005.

Chapter 9

Intolerance of uncertainty and addiction

Belinda Favaloro[a], Ahmed A. Moustafa[b]
[a]*School of Social Sciences and Psychology, Western Sydney University, Sydney, NSW, Australia,*
[b]*Marcs Institute for Brain, Behaviour, and Development and School of Psychology, Western Sydney University, Penrith, NSW, Australia*

Introduction

Substance use (drug addiction) involves a disruption of reward processing causing the drug user to be dominated by the pursuit and use of a substance, regardless of associated negative consequences (Clark & Robbins, 2002; Wise & Bozarth, 1987). A prolonged exposure to drug use causes neurological changes that mark the transition to drug abuse and subsequently addiction (Everitt et al., 2008; Everitt & Robbins, 2005; Wise & Bozarth, 1987). Drug addiction has been described as poor decision making that results from associative learning to a maladaptive stimulus (Myers et al., 2016). Associative learning is when an association between two stimuli is learned, for example, we may associate our bed with sleep, or a discarded syringe with heroin use. Therefore, understanding the decision-making process and vulnerabilities that increase susceptibility to making poor choices is useful for the treatment of drug addiction.

Intolerance of uncertainty in addiction

Not all individuals develop an addiction to drugs; some individuals are more vulnerable than others. Vulnerability factors may include certain personality traits, such as behavioral inhibition (BI) and intolerance to uncertainty (IU), that cause biases in decision-making, such as pursuing a familiar source of reward rather than exploring other sources. BI is the tendency to withdraw from or avoid novel, threatening or uncertain situations; reactivity to these situations is often enhanced (Radell, Myers, Beck, Moustafa, & Allen, 2016). BI may predispose an individual to interpret ambiguous information more negatively, therefore contributing to their avoidant response (Radell et al., 2016). Behaviourally inhibited individuals exhibit enhanced associative learning (Allen, Myers, & Servatius, 2016; Holloway, Allen, Myers, & Servatius, 2014) and increased

Cognitive, Clinical, and Neural Aspects of Drug Addiction. https://doi.org/10.1016/B978-0-12-816979-7.00009-1

avoidance learning (Sheynin et al., 2014), in other words learning to avoid an aversive situation. This type of learning has been reported in several addiction populations (Sheynin et al., 2016), as well as in individuals with anxiety disorders (Radell et al., 2016).

IU is a personality trait characterized by the tendency to perceive uncertain situations negatively and make attempts to avoid them (Nelson, Kessel, Jackson, & Hajcak, 2016). This perception may include beliefs that uncertainty is unfair, has unfavorable consequences, as well as other beliefs regarding the capacity to cope with the unpredictable and the necessity of certainty (Buhr & Dugas, 2002). In addition, it often involves an underlying fear of the unknown (Carleton, 2012).

IU can be split into two factors, which are prospective IU and inhibitory IU. Prospective IU refers to the desire for predictability and actively seeking out certainty, often exhibited through cognitive distress. Inhibitory IU describes the paralysis of both cognitions and actions when facing uncertain situations (Tanovic, Hajcak, & Joormann, 2018). To measure intolerance of uncertainty, the Intolerance of Uncertainty Scale (IUS) (Buhr & Dugas, 2002) and the Intolerance of Uncertainty Index (IUI) (Carleton, Gosselin, & Asmundson, 2010) are often used. These are self-report measures that assess reactions to uncertain and ambiguous situations. The IUS measures both cognitive distress (prospective IU) and behavioral inhibition (inhibitory IU) elicited by uncertainty (Nelson et al., 2016).

Almost every decision made requires an individual to consider some type of uncertainty or unknown information. The capacity to tolerate uncertainty can affect an individuals' decision-making process, including taking additional time to seek more information before making a choice or choosing more immediate options in order to reduce the amount of time spent waiting. Waiting for an uncertain outcome or reward may be considered aversive and often elicits negative emotions, and the desire to end the period of uncertainty outweighs the desire to wait for a greater gain (Ladouceur, Blais, Freeston, & Dugas, 1998). Individuals with higher IU often find uncertainty to be negative and stressful, and believe that it should be avoided (Buhr & Dugas, 2002). They also tend to use ineffective problem solving strategies in uncertain situations and interpret ambiguity as threatening (Ladouceur, Talbot, & Dugas, 1997).

A common finding is that individuals with higher IU tend to choose low-probability immediate rewards over high-probability delayed rewards; this impulsive behavior is known as delay discounting (Luhmann, Ishida, & Hajcak, 2011). In general, most people tend to perceive future rewards as less certain (Patak & Reynolds, 2007). The value of rewards that are uncertain is reduced when the probability of receiving that reward is unknown. A reduced tolerance for uncertainty may strengthen this belief and result in more immediate, or less thought-out, choices to be made (i.e. delay discounting). Worries about not receiving the future reward could reduce the subjective value of the delayed reward enough to make the immediate reward more attractive, even if it is not (Ladouceur et al., 1997).

Tanovic, Gee, and Joormann (2018) and Tanovic, Hajcak, et al. (2018) found that people who found it difficult to wait in a state of uncertainty were characterized by higher levels of neuroticism, inhibitory IU, and worry. These individuals were more likely to engage in delay discounting behavior, suggesting that inhibitory IU may influence decision-making in this way. Another possible explanation for delay discounting may be that these individuals are not willing to inhibit their desire to choose immediate rewards for long enough to wait for the delayed reward (Tanovic, Gee, et al., 2018; Tanovic, Hajcak, et al., 2018). This would suggest that restructuring cognitions regarding the value of waiting might be effective in reducing impulsive decision-making.

In different forms of drug addictions, deficits in processing reward-related information and managing reward-driven behavior are common. The decision-making process is altered, which influences the development, maintenance of addiction and risk of relapse. Drug use is often a coping mechanism for distress or negative affect; subsequently individuals with this coping style are at an increased risk of drug dependence (Hasin & Carpenter, 1999; Kraemer, McLeish, & O'Bryan, 2015; Park & Levenson, 2002). As previously mentioned, IU is associated with the tendency to make quick decisions to alleviate distress and to cope or avoid negative emotions. This suggests that individuals with greater IU may be at a higher risk of using drugs to cope with the negative affect associated with uncertainty, it therefore may increase vulnerability to addiction (Kraemer et al., 2015).

Individuals with substance use disorders exhibit delay discounting behavior, as they tend to choose immediate small rewards over larger uncertain rewards (Clark & Robbins, 2002; Hefner & Starr, 2017; MacKillop et al., 2011). This pattern of behavior appears to be greater in drug users than healthy individuals (Carleton, Norton, & Asmundson, 2007; Garami et al., 2017; Hefner & Starr, 2017; Nelson et al., 2016; Radell et al., 2018). When IU was explored in individuals with substance use disorders, specifically with opioid use disorder (Carleton et al., 2007; Garami et al., 2017) and individuals on methadone maintenance therapy (Radell et al., 2018), several studies found that substance users possessed higher levels of IU than healthy individuals. Therefore, greater levels of IU in individuals appear to be associated with increased delay discounting behavior, which suggests an association with IU and impulsive behavior or decision-making.

IU and impulsivity in addiction

Impulsivity can be understood by a preference for immediate reward over delayed rewards, despite the delayed reward being greater or more valuable than the immediate (Xia, Gu, Zhang, & Luo, 2017). Subsequently, delay discounting is the experimental paradigm often used to examine impulsivity. Much of research has established that anxiety felt in the face of uncertainty can result in maladaptive behaviors such as impulsive decision-making. Individuals with anxiety disorders have been shown to have greater trait and state impulsivity than individuals without an anxiety disorder (Del Carlo et al., 2012).

There are two current ideas about impulsivity and its association with anxiety. Firstly, that anxiety and impulsivity are two independent biologically based dimensions of personality, which correspond to avoidance and approach motivations. Approach motivation refers to behavior that is instigated by a desirable outcome, whereas in avoidance motivation, behavior is instigated by an undesirable outcome (Elliot & Thrash, 2002). Behavioral inhibition and safety-seeking are features of anxiety that are consistent with impulsivity. The alternative concept is that impulsive behavior may be caused by increased arousal and reduced cognitive efficiency. Despite these mixed findings, many studies have reported a link between anxiety and impulsivity (Garami et al., 2017).

Many anxiety disorders involve the fear of uncertainty, which has been associated with greater psychophysiological reactivity and exaggerated startle responses in anticipation of an unpredictable negative event (Nelson et al., 2016). Therefore, anxious individuals are more likely to underestimate the value of delayed rewards. At a behavioral level, individuals with high trait anxiety are more impulsive than those with low trait anxiety. This is evident through their immediate preference for immediate rewards over delayed, as well as, the increased speed at which they choose the immediate option, in comparison to individuals with low trait anxiety (Xia et al., 2017). This positive relationship between trait anxiety and impulsivity appears to be a consequence of enhanced attentional orientation to immediate options, overevaluation of immediate rewards and greater levels of motivations linked to immediacy (Xia et al., 2017).

Overall, intolerance of uncertainty may actually promote impulsivity, especially in substance abusers. Impulsive behaviors are often found in individuals who have substance abuse disorders, and there is evidence for a link between uncertainty and increased striatal activation in the brain, specifically the left caudate, which is related to impulsivity in drug users (Garami et al., 2017). Anxiety mediates the relationship between IU and impulsivity, which indicates that negative beliefs about uncertainty may not play a unique role in impulsivity. When impulsivity was examined in opioid-dependent individuals, a significant positive relationship between impulsivity and IU was found. Therefore, there is some association between IU and impulsivity within substance abusers. Since both anxiety and impulsivity are risk factors for addiction, understanding the association is important (Garami et al., 2017).

Intolerance of uncertainty and anxiety

Uncertain situations are more likely to be perceived as aversive by individuals with greater IU. Subsequently, these individuals have stronger emotional responses, such as increased stress, when in those conditions. They also require further information before making a decision but since they are often behaviorally inhibited in uncertain situations, obtaining further information is challenging and stress-provoking (Ladouceur et al., 1997). It is, therefore, unsurprising that IU is found in many anxiety disorders, both inhibitory and prospective IU (Tanovic, Gee, et al., 2018; Tanovic, Hajcak, et al., 2018).

In clinical anxiety, the focus is often on perceived or unrealized threat and involves many uncertainties. High levels of IU has been associated with a number of anxiety disorders including generalized anxiety disorder (GAD), obsessive compulsive disorder (OCD) (Tolin, Abramowitz, Brigidi, & Foa, 2003), social anxiety disorder (Carleton, Collimore, & Asmundson, 2010) and panic disorder (Carleton et al., 2014; McEvoy & Mahoney, 2012). IU partially mediates the relationship between neuroticism and symptoms of anxiety disorders and different aspects of IU are related to different disorders. McEvoy and Mahoney (2012) found that prospective IU was associated with GAD (i.e. worry) and OCD, while inhibitory IU was associated with social anxiety disorder, panic disorder, agoraphobia and even depression. IU's association with a variety of these disorders has lead to the suggestion that IU may be a trait-like predisposition to developing these disorders (Carleton, 2012).

Early research shows that IU plays a causal role in the development and maintenance of worry and anxiety disorders. IU has been associated with worry (Luhmann et al., 2011), rumination (Liao & Wei, 2011), post-event processing (i.e. repetitive negative thought of an event; Shiktani, Antony, Cassin, & Kuo, 2016), and increased sensitivity to anxiety (fear of physiological symptoms associated with anxiety), which all maintain a heightened level of anxiety. Theses repetitive cognitive behaviors are validated because the possibility of potential threat remains (Dugas, Gosselin, & Ladouceur, 2001). Thus, the ability to tolerate uncertainty regarding the possibility of being exposed to the potential threat will impact the frequency of worry and therefore, a person's anxiety (Laugesen, Dugas, & Bukowski, 2003).

Moreover, high IU leads to impairment in problem-solving skills and is associated with inaction or avoidance of uncertain situations (Carleton et al., 2007). This avoidance occurs through checking, reassurance-seeking and hypervigilance (Tanovic, Gee, et al., 2018). OCD patients with checking compulsions showed greater difficulties tolerating uncertainty (Tolin et al., 2003). These behaviors, along with the aforementioned cognitive behaviors, may have been developed to control for the unknown (Dugas, Buhr, & Ladouceur, 2004). Delayed discounting, is another behavior observed in individuals with anxiety disorders (Xia et al., 2017). Greater delayed discounting is associated with less willingness to wait in uncertainty and the more anxiety an individual experienced, the more compromised their decisions tend to be (Tanovic, Hajcak, et al., 2018). Similarly to addiction, anxiety related to uncertainty and unpredictability disrupts the functioning of the reward system (Radell et al., 2016) and may result in impulsive behavior.

Uncertainty itself may activate the fight or flight response, due to potential threats associated with it (Thayer, Ahs, Fredrikson, Sollers, & Wager, 2012), and the behavioral inhibition system (Gray & McNaughton, 2003). The fear of the unknown or uncertainty increases anxiety in anticipation of an unpredictable negative event (Nelson et al., 2016); this may be perceived as threatening and influence decisions by emphasizing short- over long-term benefits (Gray, 1999). Thus, trying to alleviate this anxiety is often attempted through making hasty or impulsive decisions. Consequently, these decisions lead to maladaptive coping mechanisms, such as cognitive or behavioral avoidance, including engaging in drug use. These coping mechanisms are developed in an attempt escape or to control the unknown, gain more certainty and a sense of control (Carleton, 2012; Shihata, McEvoy, & Mullan, 2017).

Increased psychological stress, a manifestation of anxiety disorders, is associated with greater levels of IU (Oglesby, Albanese, Chavarria, & Schmidt, 2014). When exposure to stress is repeated or chronic, as it is in an anxiety disorder, the risk of drug abuse is greater (Covington & Miczek 2001; Wills, McNamara, Vaccaro, & Hirky, 1996). Drug abuse in the context of stress and anxiety appears to be a greater motivator in comparison to social or enhancement reasons (Sinha, 2001), which suggests that IU may play a significant role in decision-making.

There is a high comorbidity rate between addiction and anxiety disorders (Radell et al., 2016). Anxiety disorders and addiction share common mechanisms, neural processes, maintenance cycles and maladaptive decision-making (Clark & Robbins, 2002; Koob & Volkow, 2010; Packard, 2009). In both, anxiety and addiction, a pattern of maladaptive behavior is often continued despite long-term negative consequences (Radell et al., 2018). Additionally, anxiety and stress increase the risk of drug use and relapse (Jacobsen, Southwick, & Kosten, 2001; Sinha, 2001), especially since drugs are often used as a method of coping with distressing emotions or experiences.

The risk of relapse is greater with increased levels of stress and IU (Gray, 1999; Kosten, Miserendino, & Kehoe, 2000; Radell et al., 2018). Stress increases cravings for drugs such as cocaine, and decreases any positive emotions. Drug cravings, frequency of drug use and decisions to chase immediate rewards increase with the severity of the emotional response to stress (Covington & Miczek, 2001; Sinha, 2001). Individuals suffering from substance use also report stress to be a trigger for drug use after a period of not using (Jacobsen et al., 2001) and also report greater levels of IU.

Overall, stress, acute or chronic, appears to increase vulnerability to substance use and therefore addiction. As similar to greater IU, stress causes hasty choices for smaller rewards over larger rewards in future. Greater levels of IU and stress also influence drug reinstatement, or relapse, after a period of abstinence.

IU and contextual cues in addiction

Contextual or environmental stimuli associated with drug use can play a role in continued drug use and drug relapse. These may be cues associated with the rewarding effects of the drug, cues associated with drug withdrawal symptoms or contextual cues previously associated with the drug (Lu et al., 2005). Stimuli associated with drugs elicit cravings for the drug; therefore contextual cues associated with drug use may contribute to these cravings and can help explain drug-seeking behavior and relapse. Research using rats has found context-induced reinstatement of heroin, cocaine, alcohol and nicotine (Crombag, Bossert, Koya, & Shaham, 2008). Contexts associated with drug use tend to gain an incentive value, and as extinction often occurs in another context; these drug-associated contexts retain their incentive value and consequently increase motivation for using drugs again (Mayo et al., 2013).

When this effect was explored in humans, Childs and de Wit (2009) found that places associated with drugs, whether it was purchase or use, influenced the resumption of drugs. This could explain why many addicted individuals relapse when leaving rehabilitation programs, as returning to environments that were associated with drug use can influence drug reuptake (Carter & Tiffany, 1999). In addition, since IU is associated with impulsive and avoidant behaviors, individuals with high IU may be more vulnerable to relapse. When combined with contextual cues that are associated with drugs, the risk of relapse may be greater. Future research should investigate this speculation.

Conditioned place preference (CPP) is a paradigm that compares preference for contexts paired with reward. It has frequently been used to examine the motivation for drug and non-drug rewards. Many animal models have been tested with this paradigm in an attempt to understand the reinforcing effects of rewards, including drugs, food, and access to a mate (Bardo & Bevins, 2000; Tzschentke, 2007). In these models, animal subjects (typically rodents) explore an apparatus that contains two distinct interconnected compartments, and initial preference is measured by comparing time spent in each compartment (context). In conditioning sessions, the animal is injected with a drug and confined to one context, whilst the other context is paired with saline. Subsequently, the animal subjects, now drug free, are allowed to choose between the two contexts and preference is assessed.

The advantage of this paradigm is its ability to measure drug reward through a single context-drug unconditioned stimulus pairing and advantages in determining factors contributing to relapse. Many studies have found that animals spend more time in the drug-paired condition than the saline-paired condition for a variety of drugs, including opioids and psychomotor stimulants (Bardo, Rowlett, & Harris, 1995). The CPP paradigm has recently been adapted for human participants, into a computer-based design that requires an individual to guide an avatar into either a room containing frequent reward or one containing less frequent reward. This design allows for individuals to make a choice or decision, which is an important part of humans initially taking drugs and continued use of drugs. Obtaining reward, in this computer CPP task, was contingent on operant responding by participants (Radell et al., 2018). When given a choice, individuals with higher levels of IU return to the location that was associated with higher chance of reward. Individuals with lower levels of IU do not seem to show a preference for either, a location associated with low-reward or high-reward frequency (Radell et al., 2018). Thus, IU may influence decisions to return to a preferred location that may be associated with rewards, specifically more certain rewards.

Many studies have shown that individuals have a preference for locations associated with rewards (Astur et al., 2016; Astur, Carew, & Deaton, 2014; Childs & de Wit, 2009); these preferences are greater when individuals have been deprived of the rewards (Astur et al., 2014, 2015; Childs, Astur, & de Wit, 2017). Individuals that prefer familiarity over unpredictability, show a preference for locations associated with high probabilities of reward (Radell et al., 2016). In comparison, Radell et al. (2018) found no preference was shown when testing a population of Opioid Use Disorder patients that were undergoing methadone maintenance therapy. This may be due to chronic drug use altering reward thresholds and decreasing reward sensitivity (Volkow et al., 2010).

Chronic drug use can result in reduction of dopamine production; this would explain the lower sensitivity to natural rewards (Volkow, Fowler, Wang, & Goldstein, 2002). Another possible explanation for not finding the CPP effect may be that Opioid Use disorder patients were using a different search strategy, such as foraging. This involves choosing where and how to look for reward by weighing a number of options associated with different reward contingencies (Radell et al., 2016). There is a cost associated with leaving a location to search in a new location, including loss of time and energy traveling to another location as well as having to assess the value of possible rewards (Radell et al., 2018). Belin, Belin-Rauscent, Murray, and Everitt (2013) suggested that this search strategy is emphasized in drug addicts when under the control of stimuli that act as reinforcers.

Although there is sufficient evidence to suggest that intolerance of uncertainty (IU) plays a role in decision-making, an association between IU and conditioned place preference is not clear. However, Radell et al. (2018) did find that individuals with greater levels of IU engaged in more exploration efforts, suggesting greater motivation to acquire reward. In addition, it is consistent with research that suggests individuals with high IU require more information and engage in greater levels of information-seeking under uncertain conditions (Ladouceur et al., 1998).

The impact of context on addiction can be understood through associative learning principles; cues from the context are paired with prior drug use and gain incentive value (Bardo et al., 1995; Childs & de Wit, 2009, 2013; Tzschentke, 2007; Radell et al., 2018). That is, an initially neutral stimulus gained positive salience or motivation properties, through its association with the purchase or use of a drug, which is rewarding. Context appears to influence reward learning, which may create a bias in decision-making regarding returning to a previous location associated with reward and to resume collecting that particular reward. This effect may be emphasized for individuals with a greater intolerance for uncertainty (Radell et al., 2016). Thus, a behavioral treatment involving extinction targeting drug-associated cues, such as contextual cues, and beliefs of uncertainty may help to reduce relapse and therefore improve long-term drug abstinence.

Research using rats has found context-induced reinstatement of heroin, cocaine, alcohol and nicotine (Crombag et al., 2008). Drug extinction often occurs in another context; hospitals, in-patient units, an environment that is often considered "safe" due to the absence of cues related to drug-seeking or drug-taking behaviors. However, when patients are released back into their original environments, they are exposed to drug-associated cues that have retained their incentive value, consequently resulting in increased motivation for using drugs again (Mayo et al., 2013). When combined with exposure to stressors, motivation for drug use appears to be enhanced further (Biala et al., 2018; Radell et al., 2018). Therefore it may be that stressors that involve a degree of uncertainty may influence an individual, who may have a high intolerance to uncertainty, to choose to engage in behaviors that provide an immediate outcome over one that may result in a more appealing long-term outcome e.g. using substances for immediate relief as opposed to tolerating anxiety.

Treatments that involve reducing or inhibiting the conditioned place preference, or reducing the incentive value of drug-associated cues, may be an effective objective to preventing drug relapse. One such therapy, Cue Exposure Therapy (CET), involves exposing individuals to drug cues to extinguish conditioned responses. The aim of this therapy is to extinguish learned responses to drug cues through repeated non-reinforced exposure. However, recent research has shown controversial results for the effectiveness of CET therapy, suggesting that further improvements are required (Mellentin et al., 2017) or perhaps testing the efficacy of CET when combined with other treatments.

Research in the area of conditioned place preference also has implications for pharmacological treatment of drug addictions. Disrupting the CPP effect or the mechanisms underlying the effect may be the aim of treatment. In fact, targeting certain neurotransmitters involved in the CPP effect has shown to prevent reinstated drug seeking in animals (Do Couto, Aguilar, Manzanedo, Rodriguez-Arias, & Minarro, 2005). Mehri, Sajjadi, Tabatabai, and Hosseinzadeh (2018) found that by modulating the neurotransmitter pathways, specifically the glutamate neurotransmitter, involved in morphine induced CPP; they were able to inhibit acquisition and reinstatement effects. Medications to prevent or inhibit the expression of place preference may be effective treatment option to augment cognitive-behavioral treatments. If contextual cues emphasize the effect that stress and uncertainty have on decision-making, then by finding treatments to reduce the effect of contextual cues we may be able to reduce rate of relapse for individuals with greater IU.

IU and working memory in addiction

Working memory (WM) is the part of the cognitive system that is used to retain a limited amount of information for a short period of time. It is usually associated with the frontal lobe; damage to this area of the brain often results in deficits specific to attentional allocation and information manipulation (Shallice & Burgess, 1991). Research has shown that impulsive behavior, such as delay discounting, is related to WM function. More specifically, an increase in WM load results in a decline in the value of delayed monetary reward and produces more impulsive decision-making (Hinson, Jameson, & Whitney, 2003). An increase in WM load can prevent the development of affective markers and produce more impulsive decision-making. This suggests that even when emotional or motivational values of reward and punishment are involved, WM processes can still have an impact on decision-making.

WM limitations, such as low WM capacity, can exacerbate the poor impulse control resulting from excessive consumption of alcohol (Finn, Justus, Mazas, & Steinmetz, 1999). This is based on the presumption that impulsivity increases as the executive control of behavioral inhibition systems decreases. If an individual attempts to hold more information than their WM capacity allows, they may become overloaded. A way of coping with this overload may be to make a quick decision based on the simplest or earliest information, in order to reduce any potential anxiety and confusion. Part of what may overload working memory may be stress and intolerance of uncertainty, therefore we can speculate that individuals with greater IU may be at greater risk of overloading their working memory, which may result in more impulsive choices to reduce negative affect e.g. engaging in substance use. Individuals with greater working memory are predicted to be more able to resist strong impulses, such as those associated with drug reward, and focus on more distant consequences (Khurana, Romer, Betancourt, & Hurt, 2017). These speculations should be tested experimentally in future research to confirm or disconfirm our hypothesis regarding the relationship between IU and WM.

Substance users may share a similar dysfunction in WM, which may influence their decision-making process. Working memory deficits and excessive delay discounting have been found in alcohol-, cocaine-, methamphetamine- and opioid-dependent groups (Bickel, Moody, & Quisenberry, 2014). When examined in adolescents, those with a weaker WM had less control over impulsive urges, which put them at risk for later developing substance use disorders (Khurana et al., 2017). As discussed previously, delay discounting is associated with greater intolerance to uncertainty and impulsive behavior. This can also explain an indirect relationship between IU and WM, as well.

Research has identified benefits of training working memory, including improvement in impulse control disorders. A study of working memory training conducted with alcohol dependent individuals reported reduced alcohol consumption particularly in those more impulsive (Houben, Wiers, & Jansen, 2011). These changes also lasted for a month after training was ceased. Another study examined the effects of working memory training on stimulant dependent individuals, and found that delay discounting behavior was reduced significantly, suggesting increased self-regulation and valuation of delayed rewards

(Bickel, Christensen, & Marsch, 2011). These findings suggests that by training working memory, aspects of self-regulation may improve, which may result in reduced impulsive decision-making. WM training may be an adjunct intervention to help reduce impulsive behavior associated with psychiatric illnesses, including substance abuse disorders. It is also important to note that IU may interact with all of these processes, and should be taken into account in clinical practice.

Summary and future research

Due to the chronic nature of substance use disorders, relapse is a common experience for those undergoing treatment. There is an increased need to understand the mechanisms underlying decision-making in substance use disorders, and how those influence the risk of relapse. Evidence suggests that contextual cues associated with drug use influences the resumption of drug use (Crombag et al., 2008), and this is often seen when individuals return to their original environments after leaving rehab (Carter & Tiffany, 1999). When combined with intolerance to uncertainty, the decision to choose more immediate rewards over longer, more valuable rewards is evident in patients with drug addiction (Radell et al., 2018). These rewards may not necessarily be of monetary value, but may include immediate relief from distress, anxiety and stress (Clark & Robbins, 2002). Thus, contextual cues and intolerance to uncertainty appear to influence decision-making in addiction, including the decision to use drugs again after a period of abstinence.

Future research may consider modifying the CPP paradigm to include different reward types to examine whether this affects preference in addicted and non-addicted individuals. Examining punishment learning or using losses, in addition to rewards may be another modification applied to the CPP paradigm. Loss and threat are particularly relevant to intolerance of uncertainty for some individuals (Tanovic, Hajcak, et al., 2018). By including losses, there is greater risk involved, which may allow for the exploration of the influence different levels of risk may have on preference. These modifications would allow for further understanding of the impact of contextual cues on drug use and relapse.

In much of the aforementioned research, IU is often measured through self-report measures. Future studies may consider administering multiple measures of IU to examine whether they converge in their association to neural or physiological measures. Using self-report measures means the individual needs to have sufficient insight of their attitude and beliefs of uncertainty. In addition, taking a closer look at inhibitory and prospective IU may also shed light on whether one influences more drug abuse than the other. Overall, further research is needed to understand these specific mechanisms and the influence of environmental cues. This is necessary to improve treatments for substance use disorders and may also help with the prevention of addiction development.

In summary, intolerance of uncertainty appears to play a major role in increased avoidance behavior in anxiety and substance use disorders (Carleton, 2012; Nelson et al., 2016). Intolerance of uncertainty increases both perceived (Nelson et al., 2016) and physiological stress (Sinha, 2001), and substance use is resorted to as a way of coping with increased stress. Intolerance of uncertainty appears to influence individuals to chase short-term rewards rather than long-term rewards (Radel et al., 2018) possibly due to thinking that long-term rewards may never eventuate. Consequently, this results in less thought-out and quicker choices made to avoid uncertainty and the associated negative affect. Impulsivity appears to be a risk factor for substance use disorders; therefore being able to reduce this type of behavior may prevent development of addictions. Individuals high in IU may use drugs as a way of reducing distress due to anxiety, stress and worry about the future. Consequently, risk of relapse is greater when stress levels are increased, and drug-associated contextual cues further influence this decision to use drugs again.

References

Allen, M. T., Myers, C. E., & Servatius, R. J. (2016). Uncertainty of trial timing enhances acquisition of conditioned eyeblinks in anxiety vulnerable individuals. *Behavioural Brain Research, 304*, 86–91. https://doi.org/10.1016/j.bbr.2016.02.007.

Astur, R. S., Carew, A. W., & Deaton, B. E. (2014). Conditioned place preferences in humans using virtual reality. *Behavioural Brain Research, 267*, 173–177.

Astur, R. S., Palmisano, A. N., Carew, A. W., Deaton, B. E., Kuhney, F. S., Niezrecki, R. N., et al. (2016). Conditioned place preferences in humans using secondary reinforcers. *Behavioural Brain Research, 297*, 15–19.

Astur, R. S., Palmisano, A. N., Hudd, E. C., Carew, A. W., Deaton, B. E., Kuhney, F. S., et al. (2015). Pavlovian conditioning to food reward as a function of eating disorder risk. *Behavioural Brain Research, 291*, 277–282.

Bardo, M. T., & Bevins, R. A. (2000). Conditioned place preference: What does it add to our preclinical understanding of drug reward? *Psychopharmacology, 153*(1), 31–43. https://doi.org/10.1007/s002130000569.

Bardo, M. T., Rowlett, J. K., & Harris, M. J. (1995). Conditioned place preference using opiate and stimulant drugs: A meta-analysis. *Neuroscience & Biobehavioral Reviews, 19*(1), 39–51. https://doi.org/10.1016/0149-7634(94)00021-R.

Belin, D., Belin-Rauscent, A., Murray, J. E., & Everitt, B. J. (2013). Addiction: Failure of control over maladaptive incentive habits. *Current Opinion in Neurobiology, 23*, 564–572.

Biala, G., Pekala, K., Boguszewska-Czubara, A., Michalak, A., Kruk-Slomka, M., Grot, K., et al. (2018). Behavioral and biochemical impact of chronic unpredictable mild stress on the acquisition of nicotine conditioned place preference in rats. *Molecular Neurobiology, 55*(4), 3270–3289. https://doi.org/10.1007/s12035-017-0585-4.

Bickel, W. K., Christensen, D. R., & Marsch, L. A. (2011). A review of computer-based interventions used in the assessment, treatment, and research of drug addiction. *Substance Use & Misuse, 46*(1), 4–9.

Bickel, W. K., Moody, L., & Quisenberry, A. (2014). Computerized working-memory training as a candidate adjunctive treatment for addiction. *Alcohol Research: Current Reviews, 36*(1), 123.

Buhr, K., & Dugas, M. J. (2002). The intolerance of uncertainty scale: Psychometric properties of the English version. *Behaviour Research and Therapy, 40*, 931–946.

Carleton, R. N. (2012). The intolerance of uncertainty construct in the context of anxiety disorders: Theoretical and practical perspectives. *Expert Review of Neurotherapeutics, 12*(8), 937–947. https://doi.org/10.1586/ern.12.82.

Carleton, R. N., Collimore, K. C., & Asmundson, G. J. (2010). "It's not just the judgements—It's that I don't know": Intolerance of uncertainty as a predictor of social anxiety. *Journal of Anxiety Disorders, 24*(2), 189–195.

Carleton, R. N., Duranceau, S., Freeston, M. H., Boelen, P. A., McCabe, R. E., & Antony, M. M. (2014). "But it might be a heart attack": Intolerance of uncertainty and panic disorder symptoms. *Journal of Anxiety Disorders, 28*(5), 463–470.

Carleton, R. N., Gosselin, P., & Asmundson, G. J. (2010). The intolerance of uncertainty index: Replication and extension with an English sample. *Psychological Assessment, 22*(2), 396.

Carleton, R. N., Norton, M.A.P.J., & Asmundson, G. J. G. (2007). Fearing the unknown: A short version of the intolerance of uncertainty scale. *Journal of Anxiety Disorders, 21*, 105–117.

Carter, B. L., & Tiffany, S. T. (1999). Meta-analysis of cue-reactivity in addiction research. *Addiction, 94*, 327–340. https://doi.org/10.1046/j.1360-0443.1999.9433273.x.

Childs, E., Astur, R. S., & de Wit, H. (2017). Virtual reality conditioned place preference using monetary reward. *Behavioural Brain Research, 322*, 110–114.

Childs, E. L., & de Wit, H. (2009). Amphetamine-induced place preference in humans. *Biological Psychiatry, 65*(10), 900–904. https://doi.org/10.1016/j.biopsych.2008.11.016.

Childs, E., & de Wit, H. (2013). Contextual conditioning enhances the psychostimulant and incentive properties of d-amphetamine in humans. *Addiction Biology, 18*(6), 985–992.

Clark, L., & Robbins, T. W. (2002). Decision-making deficits in drug addiction. *Trends in Cognitive Sciences, 6*, 361–363.

Covington, H. E., & Miczek, K. A. (2001). Repeated social-defeat stress, cocaine or morphine. *Psychopharmacology, 158*(4), 388–398.

Crombag, H. S., Bossert, J. M., Koya, E., & Shaham, Y. (2008). Context-induced relapse to drug seeking: A review. *Philosophical Transactions of the Royal Society of London B: Biological Sciences, 363*(1507), 3233–3243. https://doi.org/10.1098/rstb.2008.0090.

Del Carlo, A., Benvenuti, M., Fornaro, M., Toni, C., Rizzato, S., & Swann, A. C. (2012). Different measures of impulsivity in patients with anxiety disorders: A case control study. *Psychiatry Research, 197*(3), 231–236.

Do Couto, B. R., Aguilar, M. A., Manzanedo, C., Rodriguez-Arias, M., & Minarro, J. (2005). NMDA glutamate but not dopamine antagonists blocks drug-induced reinstatement of morphine place preference. *Brain Research Bulletin, 64*, 493–503.

Dugas, M. J., Buhr, K., & Ladouceur, R. (2004). *The role of intolerance of uncertainty in etiology and maintenance*.

Dugas, M. J., Gosselin, P., & Ladouceur, R. (2001). Intolerance of uncertainty and worry: Investigating specificity in a nonclinical sample. *Cognitive Therapy and Research, 25*, 551–558.

Elliot, A. J., & Thrash, T. M. (2002). Approach-avoidance motivation in personality: Approach and avoidance temperaments and goals. *Journal of Personality and Social Psychology, 82*(5), 804–818.

Everitt, B. J., Belin, D., Economidou, D., Pelloux, Y., Dalley, J. W., & Robbins, T. W. (2008). Neural mechanisms underlying the vulnerability to develop compulsive drug-seeking habits and addiction. *Philosophical Transactions of the Royal Society of London Series B, Biological Sciences, 363*, 3125–3135. Review.

Everitt, B. J., & Robbins, T. W. (2005). Neural systems of reinforcement for drug addiction: From actions to habits to compulsion. *Nature Neuroscience*, *8*, 1481–1489.

Finn, P. R., Justus, A., Mazas, C., & Steinmetz, J. E. (1999). Working memory, executive processes and the effects of alcohol on Go/No-Go learning: Testing a model of behavioral regulation and impulsivity. *Psychopharmacology*, *146*(4), 465–472.

Garami, J., Haber, P., Myers, C. E., Allen, M. T., Misiak, B., Frydecka, D., et al. (2017). Intolerance of uncertainty in opioid dependency—Relationship with trait anxiety and impulsivity. *PLoS ONE*, *12*, 1–12.

Gray, J. R. (1999). A bias toward short-term thinking in threat-related negative emotional states. *Personality and Social Psychology Bulletin*, *25*(1), 65–75. https://doi.org/10.1177/0146167299025001006.

Gray, J. A., & McNaughton, N. (2003). *The neuropsychology of anxiety: An enquiry into the function of the septo-hippocampal system (No. 33)*. Oxford University Press.

Hasin, D. S., & Carpenter, K. M. (1999). Drinking to cope with negative affect and DSM-IV alcohol use disorders: A test of three alternative explanations. *Journal of Studies on Alcohol*, *60*(5), 694–704. https://doi.org/10.15288/jsa.1999.60.694.

Hefner, K. R., & Starr, M. J. (2017). Altered subjective reward valuation among female heavy marijuana users. *Experimental and Clinical Psychopharmacology*, *25*(1), 1–12. https://doi.org/10.1037/pha0000101.

Hinson, J. M., Jameson, T. L., & Whitney, P. (2003). Impulsive decision making and working memory. *Journal of Experimental Psychology: Learning, Memory, and Cognition*, *29*(2), 298.

Holloway, J., Allen, M. T., Myers, C. E., & Servatius, R. J. (2014). Behaviorally inhibited individuals demonstrate significantly enhanced conditioned response acquisition under non-optimal learning conditions. *Behavioural Brain Research*, *261*(15), 49–55. https://doi.org/10.1016/j.bbr.2013.10.041.

Houben, K., Wiers, R. W., & Jansen, A. (2011). Getting a grip on drinking behavior: Training working memory to reduce alcohol abuse. *Psychological Science*, *22*(7), 968–975.

Jacobsen, L. K., Southwick, S. M., & Kosten, T. R. (2001). Substance use disorders in patients with posttraumatic stress disorder: A review of the literature. *American Journal of Psychiatry*, https://doi.org/10.1176/appi.ajp.158.8.1184.

Khurana, A., Romer, D., Betancourt, L. M., & Hurt, H. (2017). Working memory ability and early drug use progression as predictors of adolescent substance use disorders. *Addiction*, *11*(7), https://doi.org/10.1111/add.13792.

Koob, G. F., & Volkow, N. D. (2010). Neurocircuitry of addiction. *Neuropsychopharmacology*, *35*(1), 217–238. https://doi.org/10.1038/npp.2009.110.

Kosten, T. A., Miserendino, M. J., & Kehoe, P. (2000). Enhanced acquisition of cocaine self-administration in adult rats with neonatal isolation stress experience. *Brain Research*, *875*(1–2), 44–50.

Kraemer, K. M., McLeish, A. C., & O'Bryan, E. M. (2015). The role of intolerance of uncertainty in terms of alcohol use motives among college students. *Addictive Behaviors*, *42*, 162–166. https://doi.org/10.1016/j.addbeh.2014.11.033.

Ladouceur, R., Blais, F., Freeston, M. H., & Dugas, M. J. (1998). Problem solving and problem orientation in generalized anxiety disorder. *Journal of Anxiety Disorders*, *12*, 139–152.

Ladouceur, R., Talbot, F., & Dugas, M. J. (1997). Behavioral expressions of intolerance of uncertainty in worry. Experimental findings. *Behavior Modification*, *21*, 355–371.

Laugesen, N., Dugas, M. J., & Bukowski, W. M. (2003). Understanding adolescent worry: The application of a cognitive model. *Journal of Abnormal Child Psychology*, *31*, 55–64.

Liao, K. Y. H., & Wei, M. (2011). Intolerance of uncertainty, depression, and anxiety: The moderating and mediating roles of rumination. *Journal of Clinical Psychology*, *67*(12), 1220–1239.

Lu, L., Chen, H., Su, W., Ge, X., Yue, W., Su, F., et al. (2005). Role of withdrawal in reinstatement of morphine-conditioned place preference. *Psychopharmacology*, *181*(1), 90–100.

Luhmann, C. C., Ishida, K., & Hajcak, G. (2011). Intolerance of uncertainty and decisions about delayed, probabilistic rewards. *Behavior Therapy*, *42*, 378–386.

MacKillop, J., Amlung, M. T., Few, L. R., Ray, L. A., Sweet, L. H., & Munafò, M. R. (2011). Delayed reward discounting and addictive behavior: A meta-analysis. *Psychopharmacology*, *216*, 305–321.

Mayo, L. M., Fraser, D., Childs, E., Momenan, R., Hommer, D. W., de Wit, H., et al. (2013). Conditioned preference to a methamphetamine-associated contextual cue in humans. *Neuropsychopharmacology*, *38*(6), 921–929. https://doi.org/10.1038/npp.2013.3.

McEvoy, P. M., & Mahoney, A. E. J. (2012). To be sure, to be sure: Intolerance of un-certainty mediates symptoms of various anxiety disorders and depression. *Behavior Therapy*, *43*(3), 533–545.

Mehri, S., Sajjadi, S. S., Tabatabai, S. M., & Hosseinzadeh, H. (2018). Effects of clavulanic acid on the acquisition and reinstatement following morphine-induced conditioned place preference in mice. *Basic and Clinical Neuroscience*, *9*(4), 289.

Mellentin, A. I., Skot, L., Nielsen, B., Schippers, G. M., Nielsen, A. S., Stenager, E., et al. (2017). Cue exposure therapy for the treatment of alcohol use disorders: A meta-analytic review. *Clinical Psychology Review*, *57*, 195–207. https://doi.org/10.1016/j.cpr.2017.07.006.

Myers, C. E., Sheynin, J., Balsdon, T., Luzardo, A., Beck, K. D., Hogarth, L., et al. (2016). Probabilistic reward- and punishment-based learning in opioid addiction: Experimental and computational data. *Behavioural Brain Research*, *296*, 240–248.

Nelson, B. D., Kessel, E. M., Jackson, F., & Hajcak, G. (2016). The impact of an unpredictable context and intolerance of uncertainty on the electrocortical response to monetary gains and losses. *Cognitive, Affective, & Behavioral Neuroscience*, *16*, 153–163.

Oglesby, M. E., Albanese, B. J., Chavarria, J., & Schmidt, N. B. (2014). Intolerance of uncertainty in relation to motives for alcohol use. *Cognitive Therapy and Research*, *39*, 356–365. https://doi.org/10.1007/s10608-014-9665-1.

Packard, M. G. (2009). Anxiety, cognition, and habit: A multiple memory systems perspective. *Brain Research*, *1293*, 121–128. https://doi.org/10.1016/j.brainres.2009.03.029.

Park, C. L., & Levenson, M. R. (2002). Drinking to cop among college students: Prevalence, problems and coping processes. *Journal of Studies on Alcohol*, *63*(4), 486–497. https://doi.org/10.15288/jsa.2002.63.486.

Patak, M., & Reynolds, B. (2007). Question- based assessments of delay discounting: Do respondents spontaneously incorporate uncertainty into their valuations for delayed rewards? *Addictive Behaviors*, *32*(2).

Radell, M. L., Allen, M. T., Favaloro, B., Myers, C. E., Haber, P., Morley, K., et al. (2018). Intolerance of uncertainty and conditioned place preference in opioid addiction. *PeerJ*, *6*, e4775. https://doi.org/10.7717/peerj.4775.

Radell, M. L., Myers, C. E., Beck, K. D., Moustafa, A. A., & Allen, M. T. (2016). The personality trait of intolerance to uncertainty affects behavior in a novel computer-based conditioned place preference task. *Frontiers in psychology*, *7*, 1175.

Shallice, T., & Burgess, P. W. (1991). Deficits in strategy application following frontal lobe damage in man. *Brain*, *114*, 727–741.

Sheynin, J., Beck, K. D., Pang, K. C., Servatius, R. J., Shikari, S., Ostovich, J., et al. (2014). Behaviourally inhibited temperament and female sex, two vulnerability factors for anxiety disorders, facilitate conditioned avoidance (also) in humans. *Behavioural Processes*, *103*, 228–235. https://doi.org/10.1016/j.beproc.2014.01.003.

Sheynin, J., Moustafa, A. A., Beck, K. D., Servatius, R. J., Casbolt, P. A., Haber, P., et al. (2016). Exaggerated acquisition and resistance to extinction of avoidance behavior in treated heroin-dependent men. *The Journal of Clinical Psychiatry*, *77*(3), 386–394. https://doi.org/10.4088/JCP.14m09284.

Shihata, S., McEvoy, P. M., & Mullan, B. A. (2017). Pathways from uncertainty to anxiety: An evaluation of a hierarchical model of trait and disorder-specific intolerance of uncertainty on anxiety disorder symptoms. *Journal of Anxiety Disorders*, *45*, 72–79.

Shiktani, B., Antony, M. M., Cassin, S. E., & Kuo, J. R. (2016). Examining the role of perfectionism and intolerance of uncertainty in postevent processing in social anxiety disorder. *Journal of Psychopathology and Behavioral Assessment*, *38*(2), 297–306.

Sinha, R. (2001). How does stress increase risk of drug abuse and relapse? *Psychopharmacology*, *158*(4), 343–359. https://doi.org/10.1007/s002130100917.

Tanovic, E., Gee, D. G., & Joormann, J. (2018). Intolerance of uncertainty: Neural and psychophysiological correlates of the perception of uncertainty as threatening. *Clinical Psychology Review*, *60*, 87–99.

Tanovic, E., Hajcak, G., & Joormann, J. (2018). Hating waking: Individual differences in willingness to wait in uncertainty. *Journal of Experimental Psychopathology*, *9*(1), https://doi.org/10.1177/2043808718778982.

Thayer, J. F., Ahs, F., Fredrikson, M., Sollers, J. J., & Wager, T. D. (2012). A meta-analysis of heart rate variability and neuroimaging studies: Implications for heart rate variability as a marker of stress and health. *Neuroscience and Biobehavioral Reviews*, *36*, 747–756.

Tolin, D. F., Abramowitz, J. S., Brigidi, B. D., & Foa, E. B. (2003). Intolerance of uncertainty in obsessive-compulsive disorder. *Journal of Anxiety Disorders*, *17*(2), 233–242.

Tzschentke, T. M. (2007). Review on CPP: Measuring reward with the conditioned place preference (CPP) paradigm: Update of the last decade. *Addiction Biology*, *12*(3–4), 227–462. https://doi.org/10.1111/j.1369-1600.2007.00070.x.

Volkow, N. D., Fowler, J. S., Wang, G.-J., & Goldstein, R. Z. (2002). Role of dopamine, the frontal cortex and memory circuits in drug addiction: Insights from imaging studies. *Neurobiology of Learning and Memory*, *78*, 610–624.

Volkow, N. D., Wang, G.-J., Fowler, J. S., Tomasi, D., Telang, F., & Baler, R. (2010). Addiction: Decreased reward sensitivity and increased expectation sensitivity conspire to overwhelm the brain's control circuit. *BioEssays*, *32*, 748–755.

Wills, T. A., McNamara, G., Vaccaro, D., & Hirky, A. E. (1996). Escalated substance use: A longitudinal grouping analysis from early to middle adolescence. *Journal of Abnormal Psychology*, *105*(2), 166.

Wise, R. A., & Bozarth, M. A. (1987). A psychomotor stimulant theory of addiction. *Psychological Review*, *94*, 469–492.

Xia, L., Gu, R., Zhang, D., & Luo, Y. (2017). Anxious individuals are impulsive decision-makers in the delay discounting task: An ERP study. *Frontiers in Behavioral Neuroscience*, *11*, 5. https://doi.org/10.3389/fnbeh.2017.00005.

Further Reading

Blum, K., Braverman, E. R., Holder, J. M., Lubar, J. F., Monastra, V. J., Miller, D., et al. (2000). Reward deficiency syndrome: A biogenetic model for the diagnosis and treatment of impulsive, addictive, and compulsive behaviors. *Journal of Psychoactive Drugs*, *32*, 1–112. Suppl. i–iv.

Chapter 10

Social cognition impairment in different types of drug addictions

Jean-Louis Nandrino[a], Mohamad El Haj[b,c,d], Ahmed A. Moustafa[e]
[a]Univ. Lille, CNRS, CHU Lille, UMR 9193—SCALab—Cognitive and Affective Sciences Lab, Lille, France, [b]Laboratoire de Psychologie des Pays de la Loire (EA 4638), University of Nantes, Nantes, France, [c]Geriatric Unit, Centre Hospitalier de Tourcoing, Tourcoing, France, [d]Institut Universitaire de France, Paris, France, [e]Marcs Institute for Brain, Behaviour, and Development and School of Psychology, Western Sydney University, Penrith, NSW, Australia

Introduction

The presence of emotional and social disturbances are key factors underlying the onset of substance use disorders (Baker, Piper, McCarthy, Majeskie, & Fiore, 2004; Sinha et al., 2009) as well as the continuation of addictive behaviors and relapse (Kornreich et al., 2002). The first observations of an inability to identify and verbalize emotions in substance users have been part of the premise of the study of emotion processes in substance users (Khantzian, 1997; Taylor, Bagby, & Parker, 1997). Since these first findings, the use of direct measures in a more ecological way, tends today to suggest the presence of a more global disruption of emotional information processing. Along these lines, a set of studies have extended the investigations of such emotional disturbances to social abilities and corresponding to what researchers called "social cognition" (Frith, 2008). This generic term refers to how people process, store, and apply information about other individuals and the understanding of their social situations. Thus, a set of processes involved in social impairment observed in individuals suffering from SUD were identified including cognitive and affective processes. Importantly, recent studies show that while some substances of abuse may temporarily enhance social cognitive processes (e.g., alcohol), others may impact these processes, such as heroin and cocaine (for a review, see Miller, Bershad, & de Wit, 2015).

However, the heterogeneity of the concepts considered, and the clinical diversity of the term "substance use disorder" make it difficult to identify clear

Cognitive, Clinical, and Neural Aspects of Drug Addiction. https://doi.org/10.1016/B978-0-12-816979-7.00010-8

results characterizing these processes in these various drug using populations. The purpose of this chapter is to present empirical studies on social cognition disorders in different types of substance users (alcohol, opiate, cocaine, or methamphetamine). For each of these populations, we will examine different psychological processes involved in social cognition: emotional facial expression (EFE) recognition, theory of mind (ToM) including empathic ability and finally prosodic impairment.

Emotional facial expression (EFE) and substance abuse

In the scientific literature, there are significant observations depicting difficulties in emotional facial recognition in different populations of patients with SUD. Classically, alcoholic individuals demonstrate deficits in decoding basic and complex emotional facial expressions (Castellano et al., 2015; D'Hondt, Campanella, Kornreich, Philippot, & Maurage, 2014; Donadon & de Lima Osório, 2014; Kornreich et al., 2001, 2002; Maurage et al., 2009; Maurage, Campanella, Philippot, Martin, & De Timary, 2008; Maurage et al., 2011b; Philippot et al., 1999; Uekermann & Daum, 2008). In particular, the recognition of negative affect such as disgust and anger are especially impaired in individuals with alcohol use disorders (AUD) (for a review, see Bora & Zorlu, 2017). Similar effects were also reported for recognizing sad faces (Attwood, Ohlson, Benton, Penton-Voak, & Munafó, 2009; Craig, Attwood, Benton, Penton-Voak, & Munafò, 2009; Kamboj et al., 2013). This impairment may contribute to interpersonal problems in everyday life, for example, misperceiving a facial grimace for aggression (Kornreich et al., 2002). In addition, longer duration of alcohol misuse and more depressive symptoms were associated with more severe deficits in recognition of facial emotions alcoholic individuals. Unlike negative emotional states, drug use may enhance the rating of positive facial expressions. For example, in one study, it was reported that the amount of alcohol used positively correlates with the rating of attractiveness of individuals (Attwood, Penton-Voak, Goodwin, & Munafó, 2012; Parker, Penton-Voak, Attwood, & Munafó, 2008), especially those of the opposite sex (Jones, Jones, Thomas, & Piper, 2003). From a neuroscientific perspective, it has been shown that in comparison to healthy controls, individuals with AUD had a hypoactivation of the affective (rostral) division of the anterior cingulate (ACC) during the processing of facial expressions signaling flight or avoidance response (fear, disgust or sadness), while they had normal activation in response to anger (Salloum et al., 2007).

This impairment in EFE is also noted in polydrug users (Fernández-Serrano, Lozano, Pérez-García, & Verdejo-García, 2010; Verdejo-García, Rivas-Pérez, Vilar-López, & Pérez-García, 2007), individuals with opiate use disorders (Foisy et al., 2005; Kornreich et al., 2003), cocaine addicts (Ersche et al., 2015; Kemmis, Hall, Kingston, & Morgan, 2007), ecstasy users (Yip & Lee, 2006), methampethamine users (Potvin et al., 2018) and methadone patients

(Kornreich et al., 2003). This difficulty in discriminating facial expressions seems to be more widely observable especially regarding the different negative expressions such as fear (Bell, Bryson, & Lysaker, 1997; Kemmis et al., 2007; Verdejo-García et al., 2007), anger (Bell et al., 1997), sadness and disgust (Bell et al., 1997; Yip & Lee, 2005). For example, Ersche et al. (2015) tested patients with cocaine use disorder as well as healthy controls while performing a face recognition task and observed that the patient group has impairment recognizing fear as well as faces expressing anger. Interestingly, this effect was related to intelligence level in the patient group.

In addition, in one study, Kornreich et al. (2003) compared recently detoxified alcoholics, detoxified participants with both alcohol and opiate dependence antecedents, opiate addicts under methadone maintenance treatment, and detoxified opiate addicts. Results showed that opiate dependence is also associated with an impaired EFE decoding but less than in alcoholism. As suggested by the authors, alcohol and opiate chronic consumption could both produce a negative effect on brain areas responsible for EFE, but chronic alcohol use would have a more deleterious effect than opiates on neuropsychological functions.

However, two studies did not find such deficits. Payer et al. (2008) found that methamphetamine users did not experience difficulty in recognizing facial expressions, but observed a particular cortical activity in these patients during the performance of the task, perhaps suggesting the existence of socio-emotional deficits. Similarly, this lack of disruption in the ability of consumers to correctly discriminate facial expressions was described by Martin et al. (2006) in opioid users, despite slow recognition of facial expressions, which the authors attribute this finding to the sedative effect of methadone.

In addition, a main issue that remains to be explored concerns the link between emotional facial recognition deficits and the toxic effects of chronic use of psychoactive substances. Some researchers argue that a set of difficulties in the processing of emotional information can occur before the onset of substance use disorders and could be expressed through these EFE disorders (D'Hondt et al., 2014; Kornreich et al., 2003). Other researchers argue that these emotional facial recognition deficits are only one of the harmful effects of chronic substances use. Thus, Kornreich et al. (2001) observed that deficits in decoding accuracy for anger and disgust, and to a lesser degree sadness, persist with abstinence of 2 months and beyond. Moreover, Foisy et al. (2005) compared the performance of opioid-dependent patients who had been abstinent for less than 2 months to that of patients who had been abstinent for more than 2 months in a facial expression decoding task. They did not observe any difference between the two groups and suggest that the duration of abstinence does not improve the ability to discriminate emotions. In addition, in this study, Foisy et al. (2005) also demonstrated that a history of alcohol dependence increased emotional facial recognition impairment in opioid users.

Importantly, it should also be noted that when other emotional or social indices are used, the capacity for emotional recognition remains disrupted or even more degraded. In cross-modal emotional context (e.g., simultaneous appearance of a voice and a face presenting congruent emotional state), individuals with AUD have higher difficulties compared to unimodal situations, that is appearance of either voice or face (Maurage et al., 2013; Maurage, Campanella, Philippot, Pham, & Joassin, 2007) suggesting an inability to integrate these two information streams in a coherent percept. Similarly, when multiple indices are coupled with facial expressions, such as prosody and upper body movements, individuals with SUD also have greater difficulty in detecting the nature of the emotion expressed by others, particularly for negative emotions (Bell et al., 1997).

As argued by D'Hondt et al. (2014), the contrasting results reported by some literature reviews (for example, Donadon & de Lima Osório, 2014) might be partly explained by the nature and the sensitivity of the emotional task used to evaluate EFE recognition. D'Hondt et al. (2014) argue that individuals with SUD and especially with alcohol consumption difficulties would show preserved performance in tasks with low complexity (e.g., classifying the intensity of EFE as high or low). These types of tasks require superficial EFE processing supposed to be low cost in terms of cognitive resources for the person. Conversely, when complex and high-level emotional processing is required (e.g., multiple choices between a large range of emotional labels, emotional intensity manipulation, use of mixed emotions), impairments are systematically reported in individuals with SUD. This "task emotional complexity" hypothesis (for a discussion, see D'Hondt et al., 2014) could also be applied to positive EFE. Maurage et al. (2011a) have shown that when various positive emotional EFE are used, recognition deficit is also observed in patients with AUD. The absence of a deficit usually observed in the previous studies and classically interpreted as a preservation of positive emotions recognition, could rather correspond to a complexity effect, as happiness is usually the only positive EFE assessed.

Theory of mind (ToM) abilities in individuals with SUD

Understanding social situations is essential for the ability to quickly and accurately judge people's mental states, such as their beliefs, intentions or desires. This ability to understand mental states in others as well as oneself has been termed a theory of mind (ToM; Premack & Woodruff, 1978). ToM is first developed around the age of 4 years old, when children realize that people can have beliefs about the world that are false or different from theirs (Perner, Leekam, & Wimmer, 1987). According to Leslie (1994), the capacity for mentalization depends on the ability to form meta-representations of reality. When an individual evaluates the behavior of others, they integrate the specific objectives of their social partners who guide their behavior and beliefs about the world, considering the fact that these may not correspond to reality or to our own vision of the world. Accordingly, we must both consider how individuals perceive the world, as well as the reality of the world from our perspective, be able to distinguish the two and compare them. The field of social neuroscience has recently begun to address the neural underpinnings of TOM and empathy. Different brain regions were shown to play a role in TOM as reported in many different studies of social cognition, including the medial prefrontal cortex, superior temporal sulcus, temporo-parietal junction, temporal poles, amygdala and insula. Kennedy and Adolphs (2012) and more recently Happé and Frith (2014) in a developmental perspective suggested four major social processing networks identified by function/anatomy, which they labeled the amygdala, mentalizing, empathy and mirror networks. The amygdala network corresponds to the amygdala and orbitofrontal regions involved in threat detection, emotional evaluation and emotion regulation; the mentalizing network corresponds to medial prefrontal and superior temporal regions implicated in the automatic attribution of mental states; The empathy network refers to the insula and amygdala regions involved in automatically detecting and responding emotionally to others' distress and finally; the mirror network involves parietal and prefrontal regions containing neurons that are responsive both to observed and executed actions.

Two recent meta-analysis studies performed by Bora and Zorlu (2017) and Onuoha, Quintana, Lyvers, and Guastella (2016) suggest that Alcohol use disorder (AUD) may be associated with impaired understanding of others' intentions and emotions, which can leave an individual vulnerable to misinterpreting social cues. Bora and Zorlu mentioned that AUD was associated with deficits in ToM (d=0.58, 95% CI=0.36–0.81) in tasks measuring both decoding (d=0.46, 95% CI=0.19–0.73) and reasoning (d=0.72, 95% CI=0.37–1.06) aspects of ToM. In addition, results suggest that the percentage of males in the sample had a significant impact on the effect size, while intelligence (IQ) and level of education did not the results (Onuoha et al., 2016).

Concerning other types of consumed substances, one meta-analysis performed by Potvin et al. (2018) showed that individuals with methamphetamine use disorders (MUD) presented prominent deficits in social cognition (e.g. facial emotion recognition and theory of mind ability), which are even more important than deficits in traditional cognitive constructs such as executive functions and working memory. In particular, severe MUD was related to ToM impairments in long-term abstinence (Henry, Mazur, & Rendell, 2009), while the abuse of methamphetamine was found to produce impairments in short- to medium- term abstinence since last drug use. These results are in line with the proposal of Homer et al. (2008), which suggests that methamphetamine use produces deleterious effects on the socio-cognitive functioning of individuals caused by impairments in cathecholaminergic neurotransmission on the functioning of the frontal lobe. Moreover, concerning opiate dependent patients, the results of a recent study by Gandolphe et al. (2018), confirmed a deficit in ToM. Opiate dependent patients demonstrated an impaired performance on both a basic task of intention recognition (V-LIS, Bazin et al., 2009) and a measure of different levels of complexity in ToM (Th.o.m.a.s scale; Bosco et al., 2009). The combination of perspective taking and reflecting on others' mental states was particularly difficult for opiate dependent patients. Conversely, ToM impairments were not found in recreational users of cannabis or cocaine (Platt, Kamboj, Morgan, & Curran, 2010).

It is important to note in these studies that ToM deficits in individuals with SUD are reported either using different methods:

(1) self-reported measures as with scales such as the Empathy Quotient (EQ, Baron-Cohen & Wheelwright, 2004), the Theory of Mind Assessment Scale (Th.o.m.a.s., (Bosco et al., 2009), the Interpersonal Reactivity Index (IRI, Davis, 1983),
(2) intention and mental state identification tasks, such as with the false belief tasks (Maurage, de Timary, Tecco, Lechantre, & Samson, 2015), the "faux-pas" test (Thoma, Winter, Juckel, & Roser, 2013), the reading the mind in the eyes test (RME-test, revised version of Baron-Cohen, Wheelwright, Hill, Raste, & Plumb, 2001),
(3) ecological video-based tasks such as with the Versailles-situational intention reading (V-SIR; Bazin et al., 2009), the Multifaceted Empathy Test (MET-Core; Edele, Dziobek, & Keller, 2013) or the Movie for Assessment of Social Cognition (MASC; Dziobek et al., 2006).

As for self-reports, Martinotti, Nicola, Tedeschi, Cundari, and Janiri (2009) and Ferrari, Smeraldi, Bottero, and Politi (2014) reported that recently detoxified alcohol-dependent patients scored significantly lower than healthy controls in the EQ total score (Baron-Cohen & Wheelwright, 2004), but also in factor 2 related to emotional empathy (Ferrari et al., 2014). In addition, Maurage et al. (2011b) used both IRI and EQ to distinguish cognitive empathy and emotional empathy subcomponents. Recently detoxified alcohol-dependent participants showed a dissociated pattern of impaired emotional empathy and preserved cognitive empathy relative to healthy controls. Furthermore, impaired trait emotional empathy correlated with the severity of self-reported interpersonal problems and alexithymia. This result supports the idea of a specific impairment of the so-called "affect processing system" in alcohol dependence. This finding is supported by the data of Ferrari et al. (2014), although this was not confirmed by Thoma, Winter, et al. (2013). In this latter study, the patient group tended to show higher emotional empathy sub-scores and increased dimension of personal distress of the IRI scores in particular, which might partly be explained by the inclusion of alcohol-dependent patients with comorbid depression (Thoma, Friedmann, & Suchan, 2013). Low IRI score was also observed by Maurage et al. (2015) but the authors added 2 false belief tasks (nonverbal and video based) to assess ability to infer other people's mental states and to predict their behavior. When considering the whole group, the authors observed that individuals with AUD had significantly lower scores on both ToM (false belief) tasks in comparison to control participants. However, using single-case statistical analyses, different sub-groups emerge: some patients presented severe deficits on the ToM tasks that can be compared to what is observed in brain-damaged individuals while other participants with AUD did not differ from controls. Interestingly, individuals who presented deficits in the ToM tasks had a longer history of alcoholism and used to consume a higher number of alcohol units.

This distinction between cognitive and affective systems in processing ToM tasks, was completed with the use of more ecological and complex tasks of mental states identification. In the "reading the mind in the eyes-test" (RME-test, Baron-Cohen et al., 2001), 36 photographs of the eye area were presented to participants who have to determine the emotion corresponding to the eyes presented on these photographs from a choice of four proposals. The emotions expressed in the photographs correspond to so-called "complex" mental states, implying the attribution of a belief or an intention to the individual (e.g., optimistic, offended), as different to 'basic' emotions that can be recognized without needing to access the mental states of others (e.g., happy, sad). This test implies that participants have the emotional lexicon in their memory and that they are able to apply the appropriate vocabulary to each picture (Baron-Cohen et al., 2001). Although Kornreich et al. (2011) did not find any deficits in alcohol-dependent patients, Maurage et al. (2011b), Thoma, Winter, et al. (2013) and Nandrino et al. (2014) reported an overall reduced response accuracy. More detailed analyses by Maurage et al. (2011b) revealed significant impairments only for emotional (positive and negative) mental states.

Using both a cognitive ToM task (the Versailles-Situational Intention Reading, (V-SIR) that identifies the intentions and actions of characters from video) and an affective ToM task (RME-Test), Nandrino et al. (2014) showed that patients with AUD maintained a good performance in the cognitive task but had a poorer performance in the affective task. Moreover, a similar result is reported in the study by Maurage et al. (2016) where alcohol-dependent individuals (in their third week of abstinence) performed the Movie for Assessment of Social Cognition (MASC). This is a multiple-choice task that requires the identification of the emotions, thoughts, and intentions expressed in 45 short video sequences depicting real-life social interactions. Individuals with AUD showed a global ToM impairment, indexed by a reduced MASC global score. However, the investigation of ToM's subcomponents showed that the overall deficit was driven by a massive reduction in affective ToM, with the cognitive subcomponent preserved.

These results based on ecological ToM evaluation suggest that deficits in individuals with AUD are not related to a generalized ToM deficit but rather to a dissociation between a preserved cognitive subcomponent and an impaired affective one. In agreement with these studies, Grynberg, Maurage, and Nandrino (2017) asked individuals with AUD to complete the Multifaceted Empathy Test (MET-Core; Edele et al., 2013), in which they were instructed to identify complex emotional states expressed by individuals in contextual scenes and to state to what extent they shared them. Compared to healthy individuals, individuals with AUD showed impairment identifying negative and positive emotional states but, conversely, presented intact abilities in sharing others' emotional states.

This study shows that alcohol dependence is characterized by an impaired ability to decode complex emotional states (both positive and negative), despite the presence of complementary contextual cues, but by preserved emotion-sharing.

Regarding the consumption of substances other than alcohol, Kim, Kwon, and Chang (2011) reported difficulties of emotional recognition especially for expressions of fear and a degraded score at the "reading "the mind in the eyes-test. Results also especially showed that impairments in cognitive flexibility were correlated with impairments in facial emotion recognition and ToM in methamphetamine abusers. Deficits in global facial expression recognition and theory of mind would persist after 6 months of abstinence and could be explained in terms of the deleterious effects of the substance on cognitive functioning (Henry et al., 2009). In cocaine users, no difficulty in recognizing complex emotions has been demonstrated with the RME test, whereas with a traditional facial recognition task, a deficit in the recognition of fear expressions was observed in regular drug users (Kemmis et al., 2007). These results highlight the complementarity of these two types of evaluations. In agreement with these reports, Platt et al. (2010) combined the RME-test with a Dynamic Emotional Expression Recognition Task (DEER-T) to measure the ability of cannabis users to apprehend the emotional state of others. This dynamic facial recognition task is in the form of movie sequences constructed from pictures of emotional facial expressions, in which a face progressively changes from a neutral expression to an emotional expression. Participants should identify as soon as possible the emotional expression that appears. In this study, drug users demonstrate no difficulty in identifying the expressed emotional expression, either with the DEER-T or the RME-test, however the latency of emotional identification during the DEER-T is longer for cannabis consumers in comparison to the control individuals.

Finally, it is important to confirm difficulties associated with the complexity levels of ToM tasks, which are especially observed by using self-reported tools manipulating the ability to change perspective. This ability was examined especially by Uekermann, Channon, Winkel, Schlebusch, and Daum (2007) on the ability to identify humoristic messages. After the presentation of each joke, the mentalizing ability was assessed by questions about the perspective of the first and second protagonists. Reduced mentalizing ability was associated with humor-processing deficits, both being dependent upon the ability to take the perspectives of the protagonists to comprehend the correct punchlines.

Some searchers have also used specific tools (such as the Th.o.m.a.s scale; Bosco et al., 2009) to investigate different complex aspects of the ToM. This tool is a semi-structured interview composed of 39 open-ended questions through which the participant is directly asked to express how they understands their own mental states and those of others. In each question, the participant is invited to illustrate their answer with examples from everyday life. Bosco, Capozzi, Colle, Marostica, and Tirassa (2013) observed that persons with AUD performed worse at third-person than at first-person ToM, and at the allocentric than at the egocentric perspective.

Using the same tool with opiate dependent patients, Gandolphe et al. (2018) identified difficulties in understanding own and others' mental states, including positive and negative emotions, volitional states and epistemic states. The opiate dependent patients also performed worse on both first-order and second-order ToM, first-person and third-person ToM, and the egocentric and allocentric perspectives. These findings support the hypothesis that the ability to understand and describe mental states is impaired in individuals with SUD but is also associated to the difficulty of the construction of the mental states.

To sum up, there are several important observations. On the one hand, there seem to be different effects depending on the affective or cognitive dimension of the ToM task. This distinction seems to be rather characteristic of individuals with alcohol use disorders. These findings suggest that the emotional component of ToM (i.e., experiencing and sharing emotional states of another person) can be particularly impaired in alcoholic patients, while the cognitive component of empathy (i.e., understanding mental states of another person) can be relatively preserved in this patient group (Maurage et al., 2011a; Maurage et al., 2016; Nandrino et al., 2014).

Moreover, there appears to be a more global degradation in opioid users, affecting both the cognitive and affective aspects but also the self-evaluative skills of individuals (Gandolphe et al., 2018). On the other hand, in occasional cocaine users or cannabis users, the decrease in performing ToM tasks is not observed (Platt et al., 2010), but rather a slowing down in the performance of the task. Nevertheless, studies of populations of individuals with heavy drug use (e.g., heroin, cocaine, and methamphetamine) are much fewer than those conducted with individuals with AUD. There is today a real need to continue the investigations of ToM in ecological contexts and in particular with different populations of substance users.

Prosody and SUD

Prosody, which is a non-linguistic aspect of language, involves acoustic features such as stress, timing, intonation patterns, pitch, rhythm and differential pausing. Prosody enables the transfer of information beyond the information which is expressed by word choice or grammar.

Impairment in decoding affective prosody has also been observed in individuals with AUD (Maurage et al., 2009; Monnot, Lovallo, Nixon, & Ross, 2002; Monnot, Nixon, Lovallo, & Ross, 2001). In a study by Monnot et al. (2001), the Aprosodia Battery was administered to individuals with AUD. They were asked to assess whether two sentences are spoken with the same or a different emotional tone (presented in a happy, angry, sad, surprised, disinterested, bored and neutral voice). The performance of these patients was impaired in comparison to healthy controls and was correlated significantly with clinical variables related to alcohol consumption (e.g. age at first consumption, alcohol abuse chronicity, abuse onset age, and alcohol use of mother). In addition, individuals with AUD appear to be especially impaired in the ability to understand the emotional valence in the speech of others, which results in errors of judgment that may impair social interactions.

Moreover, emotional prosody deficits may be exacerbated when the affective prosody does not match the semantic content in sentences or when trying to match affective prosody to facial expressions (Uekermann, Daum, Schlebusch, & Trenckmann, 2005). As alcoholic patients also had problems in identifying emotional prosody for neutral semantic content, inhibitory control difficulties alone cannot explain the prosody perception deficits. Thus, as also suggested for the processing of EFE, it seems that alcoholic patients do not benefit from the cross-modal processing facilitation effect (Maurage et al., 2007, 2013).

These observations have also been confirmed in other drug users. For example, among ecstasy users, Yip and Lee (2006) have shown that emotional discrimination is affected in both facial and prosodic recognition tasks. In cocaine users, Hulka, Preller, Vonmoos, Broicher, and Quednow (2013) showed that cocaine users had significantly lower scores than those in the quotient scales of "emotion recognition" and "prosody recognition" and the subtests "conflicting prosody/meaning—attend to prosody" and "match emotional prosody to emotional face" either requiring to attend to prosody or to integrate cross-modal information. Cumulative cocaine doses and duration of cocaine use correlated negatively with emotion processing. This study highlights that cocaine users as observed in individuals with chronic alcohol use, show impaired cross-modal integration of different emotion processing channels particularly with regard to prosody, whereas more basic aspects of emotion processing such as facial affect perception are comparable to the performance of healthy controls.

Conclusion

When considering the studies on social cognition in different types of substance use, several important elements emerge:

1. There appears to be an effect of the type of substances consumed: indeed, a more deleterious effect is observed in chronic abusers of alcohol or opiates.
2. It seems to be an effect of the emotional nature of the task of ToM: performance seems more impaired when the task is emotional especially for chronic alcohol users. One can think that this dissociation between the affective or cognitive domain corresponds to affective deficit related to specific changes in stress responses (for example blunted stress-induced cortisol) (Sinha et al., 2009) and compromises of strategies of emotion regulation (Berking et al., 2011; Claisse et al., 2017).
3. Another observed effect seems associated with the complexity levels of ToM tasks to produce. Patients with SUD performed worse on both first-order and second-order ToM, first-person and third-person ToM, and the egocentric and allocentric perspectives.
4. There would also be an impairment of cross-modal integration of different emotion processing channels both in AUD.

One way to explain some of these deficits is to consider them part of a global alteration of the mechanisms involved in self-awareness. As argued by Moeller and Goldstein (2014), self-awareness impairment corresponds to the failure to ascribe personal relevance or significance to internal or external stimuli or events that have implications for the self. Such information may be environmental cues/feedback, interoceptive sensations, or ongoing behavior. Impaired self-awareness in drug addiction may reflect dysfunction of the rostral anterior cingulate (rACC) and ventromedial prefrontal cortex (vmPFC), thought to underlie the attribution of personal relevance to behavior or stimuli in one's environment that have implications for the self.

The investigation of such deficits in social cognition and emotional processing, along with lack of awareness of these deficits, are especially important because they have the potential to contribute to family, social, and career-related difficulties exhibited by many alcoholics (Kornreich et al., 2002; Philippot et al., 1999). Individuals with SUD can misunderstand their relatives' state of mind, leading to family strife, or having trouble recognizing negative affect such as anger from relatives regarding their drinking behavior (Maurage et al., 2009). Nevertheless, a set of important questions remain to be investigated: it will be important to check if the impairment of ToM remains permanent or not after abstinence. This question is central to verify if these socio-affective disorders are already present before substance use disorders and thus defend a neurodevelopmental approach or on the contrary defend that these deficits are rather associated with the deleterious neurotoxic effects of chronic consumption. It seems also important to investigate the level of complexity of ToM tasks. Only two studies, one with AUD patients and the other with opiate-dependent individuals, investigated first- or third-person perspectives in ego or allocentric positions and showed increased difficulties with complexity levels.

Further studies are also needed to confirm more pronounced deleterious effects depending on the type of substance consumed; indeed, most of the work concerns alcohol-related disorders and studies on other drugs are essential, especially to evaluate performance in ecological situations. These previous studies underline the importance of ecological measures to examine subcomponents of social cognition in SUD individuals.

Finally, an important question also concerns the links with more general cognitive deficits and more particularly with the executive processes or memory function modifications especially in encoding and accessing experiences during consumption periods (Nandrino, Gandolphe, & El Haj, 2017).

References

Attwood, A., Ohlson, C., Benton, C., Penton-Voak, I., & Munafó, M. (2009). Effects of acute alcohol consumption on processing of perceptual cues of emotional expression. *Journal of Psychopharmacology*, *23*, 23–30.

Attwood, A. S., Penton-Voak, I. S., Goodwin, C., & Munafó, M. R. (2012). Effects of acute nicotine and alcohol on the rating of attractiveness in social smokers and alcohol drinkers. *Drug and Alcohol Dependence*, *125*, 43–48.

Baker, T. B., Piper, M. E., McCarthy, D. E., Majeskie, M. R., & Fiore, M. C. (2004). Addiction motivation reformulated: An affective processing model of negative reinforcement. *Psychological Review, 111*(1), 33.

Baron-Cohen, S., & Wheelwright, S. (2004). The empathy quotient: An investigation of adults with Asperger syndrome or high functioning autism, and normal sex differences. *Journal of Autism and Developmental Disorders, 34*(2), 163–175.

Baron-Cohen, S., Wheelwright, S., Hill, J., Raste, Y., & Plumb, I. (2001). The "Reading the Mind in the Eyes" test revised version: A study with normal adults, and adults with Asperger syndrome or high-functioning autism. *Journal of Child Psychology and Psychiatry, 42*(2), 241–251.

Bazin, N., Brunet-Gouet, E., Bourdet, C., Kayser, N., Falissard, B., Hardy-Baylé, M. C., & Passerieux, C. (2009). Quantitative assessment of attribution of intentions to others in schizophrenia using an ecological video-based task: A comparison with manic and depressed patients. *Psychiatry Research, 167*(1–2), 28–35.

Bell, M., Bryson, G., & Lysaker, P. (1997). Positive and negative affect recognition in schizophrenia: A comparison with substance abuse and normal control subjects. *Psychiatry Research, 73*(1–2), 73–82.

Berking, M., et al. (2011). Deficits in emotion-regulation skills predict alcohol use during and after cognitive–behavioral therapy for alcohol dependence. *Journal of Consulting and Clinical Psychology, 79*(3), 307–318.

Bora, E., & Zorlu, N. (2017). Social cognition in alcohol use disorder: A meta-analysis. *Addiction, 112*(1), 40–48.

Bosco, F. M., Capozzi, F., Colle, L., Marostica, P., & Tirassa, M. (2013). Theory of mind deficit in subjects with alcohol use disorder: An analysis of mindreading processes. *Alcohol and Alcoholism, 49*(3), 299–307.

Bosco, F. M., Colle, L., De Fazio, S., Bono, A., Ruberti, S., & Tirassa, M. (2009). Thomas: An exploratory assessment of theory of mind in schizophrenic subjects. *Consciousness and Cognition, 18*(1), 306–319.

Castellano, F., Bartoli, F., Crocamo, C., Gamba, G., Tremolada, M., Santambrogio, J., … Carrà, G. (2015). Facial emotion recognition in alcohol and substance use disorders: A meta-analysis. *Neuroscience & Biobehavioral Reviews, 59*, 147–154.

Claisse, C., Cottencin, O., Ott, L., Berna, G., Danel, T., & Nandrino, J. L. (2017). Heart rate variability changes and emotion regulation abilities in short-and long-term abstinent alcoholic individuals. *Drug and Alcohol Dependence, 175*, 237–245.

Craig, L. C., Attwood, A. S., Benton, C. P., Penton-Voak, I. S., & Munafò, M. R. (2009). Effects of acute alcohol consumption and alcohol expectancy on processing of perceptual cues of emotional expression. *Journal of Psychopharmacology, 23*, 258–265.

D'Hondt, F., Campanella, S., Kornreich, C., Philippot, P., & Maurage, P. (2014). Below and beyond the recognition of emotional facial expressions in alcohol dependence: From basic perception to social cognition. *Neuropsychiatric Disease and Treatment, 10*, 2177.

Davis, M. H. (1983). Measuring individual differences in empathy: Evidence for a multidimensional approach. *Journal of Personality and Social Psychology, 44*(1), 113–117.

Donadon, M. F., & de Lima Osório, F. (2014). Recognition of facial expressions by alcoholic patients: A systematic literature review. *Neuropsychiatric Disease and Treatment, 10*, 1655.

Dziobek, I., Fleck, S., Kalbe, E., Rogers, K., Hassenstab, J., Brand, M., … Convit, A. (2006). Introducing MASC: A movie for the assessment of social cognition. *Journal of Autism and Developmental Disorders, 36*(5), 623–636.

Edele, A., Dziobek, I., & Keller, M. (2013). Explaining altruistic sharing in the dictator game: The role of affective empathy, cognitive empathy, and justice sensitivity. *Learning and Individual Differences*, *24*, 96–102.

Ersche, K. D., Hagan, C. C., Smith, D. G., Jones, P. S., Calder, A. J., & Williams, G. B. (2015). In the face of threat: Neural and endocrine correlates of impaired facial emotion recognition in cocaine dependence. *Translational Psychiatry*, *5*, e570.

Fernández-Serrano, M. J., Lozano, Ó., Pérez-García, M., & Verdejo-García, A. (2010). Impact of severity of drug use on discrete emotions recognition in polysubstance abusers. *Drug and Alcohol Dependence*, *109*(1–3), 57–64.

Ferrari, V., Smeraldi, E., Bottero, G., & Politi, E. (2014). Addiction and empathy: A preliminary analysis. *Neurological Sciences*, *35*(6), 855–859.

Foisy, M. L., Philippot, P., Verbanck, P., Pelc, I., Van Der Straten, G., & Kornreich, C. (2005). Emotional facial expression decoding impairment in persons dependent on multiple substances: Impact of a history of alcohol dependence. *Journal of Studies on Alcohol*, *66*(5), 673–681.

Frith, C. D. (2008). Social cognition. *Philosophical Transactions of the Royal Society of London B: Biological Sciences*, *363*(1499), 2033–2039.

Gandolphe, M. C., Lecluyse, B., Triquet, C., Brunelle, E., Duparcq, J. P., & Nandrino, J. L. (2018). Mind reading abilities in opiate-dependent patients: An exploratory study. *Comprehensive Psychiatry*, *83*, 46–52.

Grynberg, D., Maurage, P., & Nandrino, J. L. (2017). Preserved affective sharing but impaired decoding of contextual complex emotions in alcohol dependence. *Alcoholism: Clinical and Experimental Research*, *41*(4), 779–785.

Happé, F., & Frith, U. (2014). Annual research review: Towards a developmental neuroscience of atypical social cognition. *Journal of Child Psychology and Psychiatry*, *55*(6), 553–577.

Henry, J. D., Mazur, M., & Rendell, P. G. (2009). Social-cognitive difficulties in former users of methamphetamine. *British Journal of Clinical Psychology*, *48*(3), 323–327.

Homer, B. D., Solomon, T. M., Moeller, R. W., Mascia, A., DeRaleau, L., & Halkitis, P. N. (2008). Methamphetamine abuse and impairment of social functioning: A review of the underlying neurophysiological causes and behavioral implications. *Psychological Bulletin*, *134*, 301–310.

Hulka, L. M., Preller, K. H., Vonmoos, M., Broicher, S. D., & Quednow, B. B. (2013). Cocaine users manifest impaired prosodic and cross-modal emotion processing. *Frontiers in Psychiatry*, *4*, 98.

Jones, B. T., Jones, B. C., Thomas, A. P., & Piper, J. (2003). Alcohol consumption increases attractiveness ratings of opposite-sex faces: A possible third route to risky sex. *Addiction*, *98*, 1069–1075.

Kamboj, S. K., Joye, A., Bisby, J. A., Das, R. K., Platt, B., & Curran, H. V. (2013). Processing of facial affect in social drinkers: A dose–response study of alcohol using dynamic emotion expressions. *Psychopharmacology*, *227*, 31–39.

Kemmis, L., Hall, J. K., Kingston, R., & Morgan, M. J. (2007). Impaired fear recognition in regular recreational cocaine users. *Psychopharmacology*, *194*(2), 151–159.

Kennedy, D. P., & Adolphs, R. (2012). The social brain in psychiatric and neurological disorders. *Trends in Cognitive Sciences*, *16*(11), 559–572.

Khantzian, E. J. (1997). The self-medication hypothesis of substance use disorders: A reconsideration and recent applications. *Harvard Review of Psychiatry*, *4*(5), 231–244.

Kim, Y. T., Kwon, D. H., & Chang, Y. (2011). Impairments of facial emotion recognition and theory of mind in methamphetamine abusers. *Psychiatry Research*, *186*, 80–84.

Kornreich, C., Blairy, S., Philippot, P.I.E.R.R.E., Hess, U., Noël, X., Streel, E., … Verbanck, P. (2001). Deficits in recognition of emotional facial expression are still present in alcoholics after mid-to long-term abstinence. *Journal of Studies on Alcohol*, *62*(4), 533–542.

Kornreich, C., Delle-Vigne, D., Knittel, J., Nerincx, A., Campanella, S., Noel, X., … Ermer, E. (2011). Impaired conditional reasoning in alcoholics: A negative impact on social interactions and risky behaviors? *Addiction*, *106*(5), 951–959.

Kornreich, C., Foisy, M. L., Philippot, P., Dan, B., Tecco, J., Noel, X., … Verbanck, P. (2003). Impaired emotional facial expression recognition in alcoholics, opiate dependence subjects, methadone maintained subjects and mixed alcohol-opiate antecedents subjects compared with normal controls *Psychiatry Research*, *119*(3), 251–260.

Kornreich, C., Philippot, P., Foisy, M. L., Blairy, S., Raynaud, E., Dan, B., … Verbanck, P. (2002). Impaired emotional facial expression recognition is associated with interpersonal problems in alcoholism. *Alcohol and Alcoholism*, *37*(4), 394–400.

Leslie, A. M. (1994). Pretending and believing: Issues in the theory of ToMM. *Cognition*, *50*(1–3), 211–238.

Martin, L., Clair, J., Davis, P., O'Ryan, D., Hoshi, R., & Curran, H. V. (2006). Enhanced recognition of facial expressions of disgust in opiate users receiving maintenance treatment. *Addiction*, *101*(11), 1598–1605.

Martinotti, G., Nicola, M. D., Tedeschi, D., Cundari, S., & Janiri, L. (2009). Empathy ability is impaired in alcohol-dependent patients. *The American Journal on Addictions*, *18*(2), 157–161.

Maurage, P., Campanella, S., Philippot, P., Charest, I., Martin, S., & de Timary, P. (2009). Impaired emotional facial expression decoding in alcoholism is also present for emotional prosody and body postures. *Alcohol & Alcoholism*, *44*(5), 476–485.

Maurage, P., Campanella, S., Philippot, P., Martin, S., & De Timary, P. (2008). Face processing in chronic alcoholism: A specific deficit for emotional features. *Alcoholism: Clinical and Experimental Research*, *32*(4), 600–606.

Maurage, P., Campanella, S., Philippot, P., Pham, T. H., & Joassin, F. (2007). The crossmodal facilitation effect is disrupted in alcoholism: A study with emotional stimuli. *Alcohol & Alcoholism*, *42*(6), 552–559.

Maurage, F., de Timary, P., Tecco, J. M., Lechantre, S., & Samson, D. (2015). Theory of mind difficulties in patients with alcohol dependence: Beyond the prefrontal cortex dysfunction hypothesis. *Alcoholism: Clinical and Experimental Research*, *39*(6), 980–988.

Maurage, P., D'hondt, F., de Timary, P., Mary, C., Franck, N., & Peyroux, E. (2016). Dissociating affective and cognitive theory of mind in recently detoxified alcohol-dependent individuals. *Alcoholism: Clinical and Experimental Research*, *40*(9), 1926–1934.

Maurage, P., Grynberg, D., Noël, X., Joassin, F., Hanak, C., Verbanck, P., … Philippot, P. (2011a). The "Reading the Mind in the Eyes" test as a new way to explore complex emotions decoding in alcohol dependence. *Psychiatry Research*, *190*(2–3), 375–378.

Maurage, P., Grynberg, D., Noël, X., Joassin, F., Philippot, P., Hanak, C., … Campanella, S. (2011b). Dissociation between affective and cognitive empathy in alcoholism: A specific deficit for the emotional dimension. *Alcoholism: Clinical and Experimental Research*, *35*(9), 1662–1668.

Maurage, P., Joassin, F., Pesenti, M., Grandin, C., Heeren, A., Philippot, P., & de Timary, P. (2013). The neural network sustaining crossmodal integration is impaired in alcohol-dependence: An fMRI study. *Cortex*, *49*(6), 1610–1626.

Miller, M. A., Bershad, A. K., & de Wit, H. (2015). Drug effects on responses to emotional facial expressions: Recent findings. *Behavioural Pharmacology*, *26*(6), 571–579.

Moeller, S. J., & Goldstein, R. Z. (2014). Impaired self-awareness in human addiction: Deficient attribution of personal relevance. *Trends in Cognitive Sciences*, *18*(12), 635–641.

Monnot, M., Lovallo, W. R., Nixon, S. J., & Ross, E. (2002). Neurological basis of deficits in affective prosody comprehension among alcoholics and fetal alcohol–exposed adults. *Journal of Neuropsychiatry and Clinical Neurosciences*, *14*(3), 321–328.

Monnot, M., Nixon, S., Lovallo, W., & Ross, E. (2001). Altered emotional perception in alcoholics: Deficits in affective prosody comprehension. *Alcoholism: Clinical and Experimental Research*, *25*(3), 362–369.

Nandrino, J. L., Gandolphe, M. C., Alexandre, C., Kmiecik, E., Yguel, J., & Urso, L. (2014). Cognitive and affective theory of mind abilities in alcohol-dependent patients: The role of autobiographical memory. *Drug and Alcohol Dependence*, *143*, 65–73.

Nandrino, J. L., Gandolphe, M. C., & El Haj, M. (2017). Autobiographical memory compromise in individuals with alcohol use disorders: Towards implications for psychotherapy research. *Drug and Alcohol Dependence*, *179*, 61–70.

Onuoha, R. C., Quintana, D. S., Lyvers, M., & Guastella, A. J. (2016). A meta-analysis of theory of mind in alcohol use disorders. *Alcohol and Alcoholism*, *51*(4), 410–415.

Parker, L. L., Penton-Voak, I. S., Attwood, A. S., & Munafó, M. R. (2008). Effects of acute alcohol consumption on ratings of attractiveness of facial stimuli: Evidence of long-term encoding. *Alcohol and Alcoholism*, *43*, 636–640.

Payer, D. E., Lieberman, M. D., Monterosso, J. R., Xu, J., Fong, T. W., & London, E. D. (2008). Differences in cortical activity between methamphetamine-dependent and healthy individuals performing a facial affect matching task. *Drug and Alcohol Dependence*, *93*(1–2), 93–102.

Perner, J., Leekam, S. R., & Wimmer, H. (1987). Three-year-olds' difficulty with false belief: The case for a conceptual deficit. *British Journal of Developmental Psychology*, *5*(2), 125–137.

Philippot, P., Kornreich, C., Blairy, S., Baert, I., Dulk, A. D., Bon, O. L., … Verbanck, P. (1999). Alcoholics' deficits in the decoding of emotional facial expression. *Alcoholism: Clinical and Experimental Research*, *23*(6), 1031–1038.

Platt, B., Kamboj, S., Morgan, C. J., & Curran, H. V. (2010). Processing dynamic facial affect in frequent cannabis-users: Evidence of deficits in the speed of identifying emotional expressions. *Drug and Alcohol Dependence*, *112*(1–2), 27–32.

Potvin, S., Pelletier, J., Grot, S., Hébert, C., Barr, A., & Lecomte, T. (2018). Cognitive deficits in individuals with methamphetamine use disorder: A meta-analysis. *Addictive Behaviors*, *80*, 154–160.

Premack, D., & Woodruff, G. (1978). Does the chimpanzee have a theory of mind? *Behavioral and Brain Sciences*, *1*(4), 515–526.

Salloum, J. B., Ramchandani, V. A., Bodurka, J., Rawlings, R., Momenan, R., George, D., & Hommer, D. W. (2007). Blunted rostral anterior cingulate response during a simplified decoding task of negative emotional facial expressions in alcoholic patients. *Alcoholism: Clinical and Experimental Research*, *31*(9), 1490–1504.

Sinha, R., Fox, H. C., Hong, K. A., Bergquist, K., Bhagwagar, Z., & Siedlarz, K. M. (2009). Enhanced negative emotion and alcohol craving, and altered physiological responses following stress and cue exposure in alcohol dependent individuals. *Neuropsychopharmacology*, *34*(5), 1198.

Taylor, G. J., Bagby, R. M., & Parker, J. D. A. (1997). *Disorders of affect regulation: Alexithymia in medical and psychiatric illness*. Cambridge: Cambridge University Press.

Thoma, P., Friedmann, C., & Suchan, B. (2013). Empathy and social problem solving in alcohol dependence, mood disorders and selected personality disorders. *Neuroscience & Biobehavioral Reviews*, *37*(3), 448–470.

Thoma, P., Winter, N., Juckel, G., & Roser, P. (2013). Mental state decoding and mental state reasoning in recently detoxified alcohol-dependent individuals. *Psychiatry Research*, *205*(3), 232–240.

Uekermann, J., Channon, S., Winkel, K., Schlebusch, P., & Daum, I. (2007). Theory of mind, humour processing and executive functioning in alcoholism. *Addiction*, *102*(2), 232–240.

Uekermann, J., & Daum, I. (2008). Social cognition in alcoholism: A link to prefrontal cortex dysfunction? *Addiction*, *103*(5), 726–735.

Uekermann, J., Daum, I., Schlebusch, P., & Trenckmann, U. (2005). Processing of affective stimuli in alcoholism. *Cortex*, *41*(2), 189–194.

Verdejo-García, A., Rivas-Pérez, C., Vilar-López, R., & Pérez-García, M. (2007). Strategic self-regulation, decision-making and emotion processing in poly-substance abusers in their first year of abstinence. *Drug and Alcohol Dependence*, *86*(2–3), 139–146.

Yip, J. T., & Lee, T. M. (2005). Effect of ecstasy use on neuropsychological function: A study in Hong Kong. *Psychopharmacology*, *179*(3), 620–628.

Yip, J. T., & Lee, T. M. (2006). Selective impairment of sadness and disgust recognition in abstinent ecstasy users. *Neuropsychologia*, *44*(6), 959–965.

Further reading

Homer, B. D., Halkitis, P. N., Moeller, R. W., & Solomon, T. M. (2013). Methamphetamine use and HIV in relation to impaired social cognition. *Journal of Health Psychology*, *18*, 900–910.

Part II

Clinical and treatment aspects of drug addiction

Chapter 11

Causes and clinical characteristics of drug abuse

Ahmed A. Moustafa[a], Julia Garami[b], Alaa Solieman[c], Abeer M. Eissa[c], Mohamed Shakir[d]

[a]*Marcs Institute for Brain, Behaviour, and Development and School of Psychology, Western Sydney University, Penrith, NSW, Australia,* [b]*School of Social Sciences and Psychology, Western Sydney University, Sydney, NSW, Australia,* [c]*Psychogeriatric Research Center, Institute of Psychiatry, Faculty of Medicine, Ain Shams University, Cairo, Egypt,* [d]*Al-Mamoura Mental Hospital, Alexandria, Egypt*

Introduction

Drug addiction

Addiction is defined as compulsive drug use despite negative consequences. The goals of the addicted person are focused on obtaining and using drugs, regardless of the occurrence of medical illnesses, risk of incarceration, and other problems. There are three essential features of dependence on substances: (1) the development of tolerance; (2) the presence of withdrawal symptoms when the drug dose is lowered or drug use is stopped; and (3) compulsive use of a substance (Hser, Hoffman, Grella, & Anglin, 2001; Hser, Huang, Chou, & Anglin, 2007). Compulsive use of a substance refers to an individual's failure to abstain from using the substance, in spite of the negative consequences of that use. Not only can the individual not stop using the drug, he or she cannot even regulate the amount consumed on a given occasion (Rhee et al., 2003). Tolerance occurs when an addict consumes an increasing amount of a drug to obtain the effects originally produced by lower doses. Tolerance can also be psychological. An example of psychological tolerance is when a marijuana smoker or heroin user no longer experiences a "high" after the initial dose of the substance. Tolerance does not develop uniformly to all actions of drugs of abuse. Tolerance to some actions of drugs can be so high that a hundredfold increase in dose is required to produce the original effect (Merikangas et al., 1998). For example, terminally ill cancer patients may need 200–300 mg a day of morphine, whereas a dose of 60 mg can easily be fatal to a healthy person (Kendler, Prescott, Myers, & Neale, 2003).

Cognitive, Clinical, and Neural Aspects of Drug Addiction. https://doi.org/10.1016/B978-0-12-816979-7.00011-X

Withdrawal is characterized by a group of symptoms that occur when a person stops taking a drug or lowers the dose. Some common minor withdrawal symptoms are restlessness, fatigue, and an inability to sleep. More serious withdrawal symptoms include hallucinations and depression (Merikangas et al., 1998). Some drugs, such as marijuana and hallucinogens, do not typically produce a withdrawal syndrome after use is stopped. However, withdrawal from other drugs such as opioids may cause seizures or hallucinations. In addition to physical withdrawal symptoms, anxiety and depression are also common. The symptoms of withdrawal do not often appear unless a person has been using drugs for a long time or when cessation is particularly abrupt, as occurs functionally when an opioid antagonist is given (Merikangas et al., 1998). The long-term use of drugs results in changes in the number and sensitivity of various kinds of receptors, which mediate at least some of the effects of tolerance and withdrawal. Long-term use of drugs is associated with increased sensitivity of the dopaminergic, cholinergic, serotonergic, and noradrenergic neurons (Kendler et al., 2003). Short-term use of drugs of abuse decreases the activity of the noradrenergic neurons in the locus coeruleus; long-term use activates a compensatory homeostatic mechanism within the neurons and withdrawal results in rebound hyperactivity (Kendler et al., 2003).

Potential causes of drug abuse

Prior research has shown that there are many motives for the initiation and continuation of drug use. For example, the main motives for drug use in males were found to be peer pressure, seeking pleasure, to elevate mood, improve sex and to show masculinity (El-Sawy, Abdel Hay, & Badawy, 2010). Peer pressure was found to be significantly higher among younger addicts (<20 years) compared to older addicts (El-Sawy et al., 2010). One of the most important factors for the initiation of substance use in different studies is the peer effect. For example, Abdel Wahab, Ely Ray, Erfan, and Fathy (2006) stated that 65% of substance abuser consumed drugs under the effect of friends. The same result was also found by Ahmed (2005) who reported that 88.6% of drug users were influenced by peer effect.

Studies also found that the availability of drugs had positive association with substance abuse (Ahmed, 2005). Erickson and Cheung (1999) found that unemployment and poverty can be a contributing factor of drug trading and abuse in different societies. It is also known that numerous socioeconomic and psychological factors all play an important role in the etiology of drug abuse (Tafesh, 2013). These conceptualizations are not mutually exclusive; rather they are facets of the multifactorial etiology of drug abuse. The most robust evidence highlights peer drug use, availability of drugs and also elements of family interaction, including parental discipline and family cohesion, as a significant risk factors for drug misuse (Frisher, Crome, Macleod, Bloor, & Hickman, 2007).

Abdel-Azim (2001) studied 154 patients with substance abuse and reported that 22.7% of patient related substance abuse to sexual practice. In regards to religious beliefs and attitudes, there was a negative correlation between religiosity and substance abuse (Abdel-Azim, 2001). On the other hand, Hashem (2002) studied female secondary school students and found that 70% of the sample reported using illicit drugs due to stressful situations while 30% of the sample consumed drugs due to somatic complaints. In the same study, 75% of the sample reported drug use when being in company with friend while 25% of students reported use of drugs in response to psychological troubles. Regarding parents' level of education, it was found that consumption of alcohol was more prevalent among those whose parents were highly educated. Marijuana and inhalants use was more prevalent among those whose parents were educated till primary school (Ahmed, 2005). In females, the most common motives for the initiation of drug abuse has found to be family troubles and sexual abuse (31%), followed by mood improvement (29%) followed by seeking pleasure and curiosity (13%) (El-Sawy et al., 2010). Multivariate regression analysis revealed that being younger, unmarried, male and earning an unsatisfactory income were significant risk factors for substance abuse and dependence (Amr, El Gilany, El Mogy, & Fathi, 2014).

The present study is a descriptive cross-sectional study investigating causes and problems associated with drug abuse. We study the causes of drug abuse in an Egyptian sample (e.g., anxiety and depression, peer pressure, fatigue, loneliness, insomnia, sexual problems, or curiosity), as well as the most common problems associated with drug abuse (e.g., legal, financial, or psychiatric).

Methods

Subjects

We recruited patients attending Al-Mamoura outpatient addiction treatment clinics over the course of 4 months. The clinic provides services for male patients 3 days a week and once a week) and once a week for female patients. All eligible and consenting subjects are included in this study during the 4 months of the study. Most of the patients in the current study had a long history of drug abuse and sought medical advice from the clinics as they suffered from intense pain and insomnia while trying to abstain. The inclusion criteria were as follows: (a) fulfilling the criteria of substance abuse according to DSM-IV and (b) aged between 18 and 65 years old. Our sample included 516 patients, the majority of which. Were polysubstance abusers.

Method

After approval of ethical and scientific committee of Neuro-Psychiatric Department, Ain Shams University, all participants provided written informed consent and were notified that refusal to participate would not affect their medical treatment A structured questionnaire was devised to record socio-demographic data; including age, gender, education, residency, occupation, marital status and socioeconomic status. Socioeconomic status was evaluated using the Upgraded Classification Scale (El-Gilany, El-Wehady, & El-Wasify, 2012). Substance use information was recorded; including type of substance, duration of intake, recent (last week) amount taken daily, route of intake, reasons for intake, intoxication symptoms, withdrawal symptoms, current and histories of physical complications related to drug abuse, current and histories of substance abuse-related psychiatric disorders, and drug related financial problems. Drug-related medico-legal problems were also assessed, such as violence, accidents, penalties for possession or dealing of illicit substances.

In addition, we utilized the following scales:

1. The Upgraded Social Classification Scale (El-Gilany et al., 2012):
   It includes 7 domains with a total score of 84, with a higher score indicating better SES: Education and cultural, Occupation, Family, Family possessions, Economic, Home sanitation, Health care. Scores between 0 and 21, 22–42, 43–62 and 63–84 are classified as very low, low, middle, high socioeconomic class respectively.
2. The Structured Clinical Interview for DSM-IV Axis II Personality Disorders (SCID-II; (First, Spitzer, Gibbon, & Williams, 1996); Arabic version (Hatata, Khalil, Asaad, Abo Zeid, & Okasha, 2004)
   The SCID-II is a clinician-administered semi-structured interview for diagnosing Axis II personality disorders of the Diagnostic and Statistical Manual of Mental Disorders (3rd ed., rev.), in addition to the Appendix category of self-defeating personality disorder. The SCID-II is unique in that it was designed with the primary goal of providing a rapid clinical assessment of personality disorders without sacrificing reliability or validity. It can be used in conjunction with a self-report personality questionnaire, which allows the interview to focus only on the items corresponding to positively endorsed questions on the questionnaire, thus shortening the administration time of the interview (First, Spitzer, Gibbon, & Williams, 1995).
3. The Structured Clinical Interview for DSM-IV Axis I Disorders (SCID-I; First, Spitzer, Gibbon, Williams, & Benjamin, 1994); Arabic version (Ahmed, 2005)
   The SCID-I is a semistructured interview that is used to diagnose major DSM-IV Axis I disorders, which is designed to be administered by a clinician but can also be administered by non-clinician researchers who have extensive experience with the study population in question (Ventura, Liberman, Green, Shaner, & Mintz, 1998).

Results

The majority of patients were male (90.5%) and the average participant age was 27.62 years (±6.4), with a mean age of substance abuse onset of 24.12 years (±6.88). The sociodemographic characteristics of the 516 studied patients are presented in Table 1.

TABLE 1 Distribution of the studied cases according to demographic data

	No.	%
Sex		
Male	467	90.5
Female	49	9.5
Age		
Mean ± SD	27.62 ± 6.4	
Range	18–57	
Age of onset of substance abuse		
Mean ± SD	24.12 ± 6.88	
Range	14–55	
Educational level		
Illiterate	138	26.74
Read and write	144	27.9
Technical	162	31.39
University	72	13.95
Employment		
Employed	222	43.0
Unemployed	294	56.97
Marital status		
Single	267	51.74
Married	167	32.36
Divorced	82	15.89
Residence		
Urban	330	63.95
Rural	186	36.04
Socioeconomic level		
High (I)	76	14.72
Medium (II)	182	35.27
Low (III)	194	37.6
Very low (IV)	64	12.4

TABLE 2 The distribution of the studied cases according to causes of substance abuse.

	No.	%
Causes of intake		
Anxiety and depression	126	24.4
Peer pressure	303	58.72
Fatigue	41	7.94
Loneliness	57	11.04
Insomnia	26	5.03
Sexual problems	91	17.63
Curiosity	173	33.52
Others	31	6.0

TABLE 3 The distribution of the studied cases according to negative consequences of substance abuse.

	Present		Absent	
	No.	%	No.	%
Legal problems	84	16.27	432	83.72
Financial problems	137	26.55	379	73.45
Somatic problems	182	35.27	334	64.72
Psychiatric problems	276	53.48	240	46.5

Based on the distribution of the studied cases according to the causes of substance abuse, peer pressure was found to be the most common cause of drug intake (58.72% of cases), while the second cause was curiosity (33.52% of cases). Insomnia was the least cited reason for drug use (5.03% of cases). The full account of reported causes of drug intake can be found in Table 2.

In regards to negative consequences of substance abuse, the most prevalent was psychiatric problems (53.48% of cases) and somatic complications (35.27% of cases) (Table 3).

When withdrawal symptoms were assessed, it was found that 92.63% of participants reported psychiatric symptoms, predominantly insomnia & depression. Central nervous system symptoms, such as pain and seizures, were reported by

TABLE 4 The distribution of the studied cases according to substance withdrawal symptoms

	No.	%
Withdrawal symptoms		
CNS	450	87.2
CVS	39	7.56
GIT	376	72.86
Autonomic	418	81.0
Psychiatric	478	92.63
Others	57	11.04

Abbreviation: CNS, central nervous system; *CVS*, cardiovascular system; *GIT*, gastrointestinal tract.

87.2% of cases, such as pain and seizures. The least prevalent withdrawal symptom among patients was cardiovascular problems (7.56%). (Table 4).

Discussion

This is a cross-sectional study exploring the psychosocial and physiological factors associated with drug addiction in an Egyptian sample. The main findings of our study show that peer pressure is one of the most predominant motivating factor for drug use in adults, followed by curiosity and anxiety/depression. We also found that over half of the sample suffered from psychiatric complaints as a consequence of drug use, followed by somatic symptoms.

Gender ratio in drug addiction

The finding that drug addiction was more common in males than females in our sample is in agreement with studies in other cultures. However, it is important to note that there were more days of treatment offered to males to females, which could have affected the data. Similar findings were reported with an Iranian population (Jadidi & Nakhaee, 2014). Similarly, Tafesh (2013) showed that 89.0% from their drug abuse sample are male; these researchers attributed this male dominance to the fact that women in Palestinian communities are afraid of the stigma, especially knowing that the community is considered a conservative community. El-Sawy et al. (2010) found that males started drug use earlier than females and had longer duration of dependence compared to females. Also, the results of their study suggest that males abuse drugs more commonly than females. In the Egyptian culture, males have more opportunities to abuse drugs than females at an earlier age due to earlier work careers and more freedom given to males. In the USA, the National Institute on Drug Abuse reported that men are more likely than women to have opportunities to use drugs, but men and women given an opportunity to use drugs for the first time are equally likely to do so and to progress from initial use to addiction. However, women and men appear to differ in their vulnerability to some drugs (see report on Principles of Drug Addiction Treatment: A Research-Based Guide). However, the disproportionate number of males versus females in our sample may have been an artifact of the sampling method used, as the drug treatment centre provided more days of therapy for men than women.

Peer pressure in drug addiction

The most striking finding is that peer pressure was the strongest factor associated with drug use in our sample. It has been widely observed that peer pressure is a strong motivating factor for drug use in adolescents (Rumpold et al., 2006; Simons-Morton, Haynie, Crump, Eitel, & Saylor, 2001), including in Egyptian adolescents (Loffredo et al., 2015). Peer pressure motivations for drug use initiation and continuing abuse has been found in numerous culturally diverse groups, across a variety of substances. Peer pressure has been found as an important predictive factor for alcohol and cannabis use in Swiss samples (Haug, Nunez, Becker, Gmel, & Schaub, 2014; Jadidi & Nakhaee, 2014; Studer et al., 2014), an etiological factor in the development of opioid addiction in Bosnia (Niškanović, 2013), and is a common barrier to substance abuse treatment for Asian American Pacific Islanders (Masson et al., 2013), and Taiwanese users of betel quid (Lin et al., 2017). It would appear that peer pressure is a common motivating for drug abuse in Egypt as has been found in many other countries around the world.

Similar associations between peer pressure and substance use have also been reported in culturally similar countries to Egypt, such as India and Bangladesh, in which peer pressure is listed as one of the most common reasons for initial use of alcohol, cigarettes, and other drugs of abuse (Bilques, Bashir, Munshi, Sheikh, & Firdosi, 2015; Gupta, 2018; Maruf, Khan, & Jahan, 2016; Nithin et al., 2013; Prabhu, Srinivas, Vishwanathan, & Raavi, 2014). Peer pressure has also shown to contribute to drug relapse in Bangladesh (Rahman, Rahaman, Hamadani, Mustafa, & Shariful Islam, 2016) and India (Sau, Mukherjee, Manna, & Sanyal, 2013), and the transition to intravenous drug use in Iran (Mosayeb Yarmohamadi, Mahmoud, Fatemeh, Ali, & Mohamad Hasan, 2015). The current study finds that peer pressure is a motivating factor for ongoing substance abuse and is not limited to the initiation of drug use. The current findings also show that peer pressure is a motivating factor for ongoing drug use in Egypt.

Attitudes of peers towards drug use can play a pivotal role in the onset of substance use, as social norms towards prescription opioid use in Iranian adolescents has shown strong associations with positive attitudes towards use and future abuse (Bashirian, Barati, & Fathi, 2014). It is widely accepted that social norms shape human behavior (Ajzen, 1991), so it is not surprising that peer pressure motivations for drug use were so commonly reported in our sample. Our findings may also be partly explained by peer pressure's effect on risk-taking behaviors, in which groups tend to increase impulsivity and increase the likelihood of antisocial behaviors (see Balogh, Mayes, & Potenza, 2013 for a review).

It is interesting that anxiety and depression were reported as only the third most common motivating factors for drug abuse in the current study, and at half the rate of peer pressure, given that drugs are used to increase mood, decrease negative affect and/or reduce stress (Khantzian, 1985; Sinha, 2001). It would appear that peer pressure has a very significant association with continued drug use, over and beyond personal needs to medicate anxiety and negative affect.

Psychiatric symptoms associated with drug addiction

Other findings of the occurrence of psychiatric disorder in drug abusers are in agreement with other studies. For example, Youssef, Fahmy, and Haggag (2005) found that anxiety, insomnia, social dysfunction and severe depression were prevalent among substance user. Also, Abdel Wahab et al. (2006) revealed that 60% of substance users had co-morbid psychiatric disorders. Amr et al. (2014) found that anxiety disorders particularly generalized anxiety disorder was prevalent in abuse group (60.7%,) whereas mood disorders and schizophrenia (25% each) were more encountered in the dependence group. Individuals with co-occurring substance use and mental disorders are often more difficult to treat and experience worse outcomes than other patients. Persons with these co-occurring disorders also have higher rates of emergency department use than those with mental disorders alone (Amr et al., 2014).

In summary, our results highlight important correlates and symptoms associated with drug addiction. As in other cultures, we found that peer pressure is the strongest factor underlying drug addiction. Future research should explain the details underlying peer pressure, and whether it impacts initiation and/or continuation of drug use. Importantly, clinical treatments should take into account peer pressure effect on drug use, and suggest alternative treatments that minimize it effects. Further, in agreement with studies in other countries, we found that psychiatric diseases are common among patients with addiction. Pharmacological and psychological treatments should also consider these psychiatric co-morbidities as they interfere with the course of treatment and quality of life of the patients.

References

Abdel Wahab, M., Ely Ray, L., Erfan, S., & Fathy, H. (2006). Substance dependence in a sample of Egyptian Adolescents. *Egyptian Journal of Psychiatry*, *25*(2), 7–14.

Abdel-Azim, K. A. (2001). *Psychosocial correlates of substance abuse: A study in an Egyptian sample*. Cairo: Ain Shams University.

Ahmed, A. (2005). *Substance use disorders among male general secondary school students in Cairo*. Cairo: Ain Shams University.

Ajzen, I. (1991). The theory of planned behavior. *Organizational Behavior and Human Decision Processes*, *50*(2), 179–211. https://doi.org/10.1016/0749-5978(91)90020-T.

Amr, M., El Gilany, A. H., El Mogy, A., & Fathi, W. (2014). Substance abuse and dependence among patients attending an emergency hospital in eastern Nile delta, Egypt. *Journal of Psychiatry*, *17*, 532–537.

Balogh, K. N., Mayes, L. C., & Potenza, M. N. (2013). Risk-taking and decision-making in youth: Relationships to addiction vulnerability. *Journal of Behavioral Addictions*, *2*(1), https://doi.org/10.1556/JBA.2.2013.1.1.

Bashirian, S., Barati, M., & Fathi, Y. (2014). Prevalence and factors associated with tramadol abuse among college students in west of Iran: An application of the theory of planned behavior. *Avicenna Journal of Neuro Psycho Physiology*, *1*(1), 1–5.

Bilques, S., Bashir, N., Munshi, I., Sheikh, A., & Firdosi, M. (2015). Socio-demographic correlates of substance use disorder in patients seeking de-addiction services in Kashmir India—a cross sectional study. *European Psychiatry*, *30*, 511.

El-Gilany, A., El-Wehady, A., & El-Wasify, M. (2012). Updating and validation of the socioeconomic status scale for health research in Egypt. *Eastern Mediterranean Health Journal*, *18*(9), 962–968.

El-Sawy, H., Abdel Hay, M., & Badawy, A. (2010). Gender differences in risks and pattern of drug abuse in Egypt. *Egyptian Journal of Neurology, Psychiatry and Neurosurgery*, *47*(1), 413–418.

Erickson, P. G., & Cheung, Y. W. (1999). Harm reduction among cocaine users: Reflections on individual intervention and community social capital1. *International Journal of Drug Policy*, *10*(3), 235–246.

First, M. B., Spitzer, R. L., Gibbon, M., & Williams, J. B. W. (1995). The structured clinical interview for DSM-III-R personality disorders (SCID-II), part I: Description. *Journal of Personality Disorders*, *9*(2), 83–91.

First, M. B., Spitzer, R. L., Gibbon, M., & Williams, J. B. W. (1996). *Structured clinical interview for DSM-IV Axis I disorders, clinician version (SCID-CV)*. Washington, DC: American Psychiatric Press, Inc.

First, M. B., Spitzer, R. L., Gibbon, M., Williams, J. B. W., & Benjamin, I. (1994). *Structured clinical interview for DSM-IV Axis II personality disorders (SCID-II). Version 2.0*. Retrieved from: New York.

Frisher, M., Crome, I., Macleod, J., Bloor, R., & Hickman, M. (2007). *Predictive factors for illicit drug use among young people: A literature review*. Retrieved from United Kingdom.

Gupta, A. (2018). Smoking addiction among young women working at night at international call centres in India. *Tobacco Induced Diseases*, *16*(s1), 342. https://doi.org/10.18332/tid/84388.

Hashem, N. Z. (2002). *Substance use disorders among female secondary school students in Cairo*. Cairo: Ain Shams University.

Hatata, H., Khalil, A., Asaad, T., Abo Zeid, M., & Okasha, T. (2004). *Dual diagnosis in substance use disorders*. Cairo: Ain Shams University.

Haug, S., Nunez, C. L., Becker, J., Gmel, G., & Schaub, M. P. (2014). Predictors of onset of cannabis and other drug use in male young adults: Results from a longitudinal study. *BMC Public Health*, *14*, 1202. https://doi.org/10.1186/1471-2458-14-1202.

Hser, Y. I., Hoffman, V., Grella, C. E., & Anglin, M. D. (2001). A 33-year follow-up of narcotics addicts. *Archives of General Psychiatry*, *58*(5), 503–508. https://doi.org/yoa20042.

Hser, Y. I., Huang, D., Chou, C. P., & Anglin, M. D. (2007). Trajectories of heroin addiction: Growth mixture modeling results based on a 33-year follow-up study. *Evaluation Review*, *31*(6), 548–563. https://doi.org/10.1177/0193841X07307315.

Jadidi, N., & Nakhaee, N. (2014). Etiology of drug abuse: A narrative analysis. *Journal of Addiction, 2014*, 352835. https://doi.org/10.1155/2014/352835.

Kendler, K. S., Prescott, C. A., Myers, J., & Neale, M. C. (2003). The structure of genetic and environmental risk factors for common psychiatric and substance use disorders in men and women. *Archives of General Psychiatry, 60*(9), 929–937. https://doi.org/10.1001/archpsyc.60.9.929.

Khantzian, E. J. (1985). The self-medication hypothesis of addictive disorders: Focus on heroin and cocaine dependence. *The American Journal of Psychiatry, 142*(11), 1259–1264.

Lin, C.-C., Tamí-Maury, I., Ma, W.-F., Lam, C., Tsai, M.-H., Lin, M.-T., … Gritz, E. R. (2017). Social and cultural context of betel quid consumption in Taiwan and implications for prevention and cessation interventions. *Substance Use & Misuse, 52*(5), 646–655. https://doi.org/10.1080/10826084.2016.1246572.

Loffredo, C. A., Boulos, D. N. K., Saleh, D. A. A., Jillson, I. A., Garas, M., Loza, N., … Amr, S. (2015). Substance use by Egyptian youth: Current patterns and potential avenues for prevention. *Substance Use & Misuse, 50*(5), 609–618. https://doi.org/10.3109/10826084.2014.997391.

Maruf, M. M., Khan, M. Z. R., & Jahan, N. (2016). Pattern of substance use: Study in a de-addiction clinic (original article) (report). *Oman Medical Journal, 31*(5), 327. https://doi.org/10.5001/omj.2016.66.

Masson, C., Shopshire, M., Sen, S., Hoffman, K., Hengl, N., Bartolome, J., … Iguchi, M. (2013). Possible barriers to enrollment in substance abuse treatment among a diverse sample of Asian Americans and Pacific Islanders: Opinions of treatment clients. *Journal of Substance Abuse Treatment, 44*(3), 309–315.

Merikangas, K. R., Stolar, M., Stevens, D. E., Goulet, J., Preisig, M. A., Fenton, B., … Rounsaville, B. J. (1998). Familial transmission of substance use disorders. *Archives of General Psychiatry, 55*(11), 973–979.

Mosayeb Yarmohamadi, V., Mahmoud, T., Fatemeh, G., Ali, F., & Mohamad Hasan, F. (2015). Factors associated with the transition from drug abuse to initiation of injection drug use. *Iranian Rehabilitation Journal, 13*(2), 16–22.

Niškanović, J. (2013). Opiate addiction in Republic of Srpska: Characteristics and etiology. *Specijalna Edukacija i Rehabilitacija, 12*(4), 539–557. https://doi.org/10.5937/specedreh12-4498.

Nithin, K., Tanuj, K., Bhaskaran, U., Rekha, T., Prasanna, M., Vaman, K., … Saran, S. (2013). Profile of substance use among patients attending de-addiction centres in a coastal city of southern India. *PLoS ONE, 8*(2), e57824. https://doi.org/10.1371/journal.pone.0057824.

Prabhu, P., Srinivas, R., Vishwanathan, K., & Raavi, A. (2014). Factors influencing alcohol and tobacco addiction among patients attending a de-addiction Centre, South India (original article) (report). *Journal of International Society of Preventive and Community Dentistry, 4*(2), 103. https://doi.org/10.4103/2231-0762.137645.

Rahman, M. M., Rahaman, M. M., Hamadani, J. D., Mustafa, K., & Shariful Islam, S. M. (2016). Psycho-social factors associated with relapse to drug addiction in Bangladesh. *Journal of Substance Use, 21*(6), 627–630. https://doi.org/10.3109/14659891.2015.1122099.

Rhee, S. H., Hewitt, J. K., Young, S. E., Corley, R. P., Crowley, T. J., & Stallings, M. C. (2003). Genetic and environmental influences on substance initiation, use, and problem use in adolescents. *Archives of General Psychiatry, 60*(12), 1256–1264. https://doi.org/10.1001/archpsyc.60.12.1256.

Rumpold, G., Klingseis, M., Dornauer, K., Kopp, M., Doering, S., Hfer, S., … Schler, G. (2006). Psychotropic substance abuse 1 among adolescents: A structural equation model on risk and protective factors. *Substance Use & Misuse, 41*(8), 1155–1169. https://doi.org/10.1080/10826080600752136.

Sau, M., Mukherjee, A., Manna, N., & Sanyal, S. (2013). Sociodemographic and substance use correlates of repeated relapse among patients presenting for relapse treatment at an addiction treatment center in Kolkata, India. *African Health Sciences*, *13*(3), 791. https://doi.org/10.4314/ahs.v13i3.39.

Simons-Morton, B., Haynie, D. L., Crump, A. D., Eitel, P., & Saylor, K. E. (2001). Peer and parent influences on smoking and drinking among early adolescents. *Health Education & Behavior*, *28*(1), 95–107. https://doi.org/10.1177/109019810102800109.

Sinha, R. (2001). How does stress increase risk of drug abuse and relapse? *Psychopharmacology*, *158*(4), 343–359. https://doi.org/10.1007/s002130100917.

Studer, J., Baggio, S., Deline, S., N'Goran, A. A., Henchoz, Y., Mohler-Kuo, M., ... Gmel, G. (2014). Drinking locations and alcohol-related harm: Cross-sectional and longitudinal associations in a sample of young Swiss men. *International Journal of Drug Policy*, https://doi.org/10.1016/j.drugpo.2014.11.004.

Tafesh, M. R. H. (2013). *Assessment of tramadol abuse among clients who are attending private psychiatric clinics—KAP study*. Islamic University of Gaza.

Ventura, J., Liberman, R. P., Green, M. F., Shaner, A., & Mintz, J. (1998). Training and quality assurance with the structured clinical interview for DSM-IV (SCID-I/P). *Psychiatry Research*, *79*(2), 163–173. pii: S0165-1781(98)00038-9.

Youssef, I., Fahmy, M., & Haggag, W. (2005). Psychological profile and prevalence of psychoactive substance use among Suez Canal university students in Ismailia. *Current Psychiatry*, *12*, 269–285.

Chapter 12

Accessing addiction recovery capital via online and offline channels: The role of peer-support and shared experiences of addiction

Ana-Maria Bliuc[a], David Best[b,c], Ahmed A. Moustafa[d]

[a]*School of Social Sciences and Psychology, Western Sydney University, Sydney, NSW, Australia,* [b]*Sheffield Hallam University, Sheffield, United Kingdom,* [c]*The Australian National University, Canberra, ACT, Australia,* [d]*Marcs Institute for Brain, Behaviour, and Development and School of Psychology, Western Sydney University, Penrith, NSW, Australia*

Introduction

Drug addiction is often associated with a range of everyday life problems such as impaired social functioning, poor physical and mental health and unemployment (Meulenbeek, 2000). One of the reasons underlying drug addiction is the lack of social support. Individuals with drug use disorders commonly lack support from family members and friends, who may perceive them as irresponsible and beyond help, leading to experiences of isolation (see discussion below on this point). Havassy, Hall, and Wasserman (1991) argue that a lack of social support can result in social exclusion, which in their study led to negative psychological states (e.g., depression) and incapacity to imagine future health-promoting behaviors such as abstinence.

The lack of social support may start in childhood. For example, weak social parent-child attachment, family conflict at home, and reduced supervision were identified as factors that encourage drug use. In adolescents, individuals that live with families who experience elevated levels of conflict are more likely to engage in using drugs.

One form of lack of social support is social isolation. Some studies have found that socially isolated rats (in adolescence) were more likely to consume drugs than a control group (Whitaker, Degoulet, & Morikawa, 2013). Social isolation, abandonment, and neglect (especially during early stages of life) increase risk of developing drug addiction (Lu, Shepard, Hall, & Shaham, 2003; Meyer-Lindenberg & Tost, 2012; Sinha, 2008).

Cognitive, Clinical, and Neural Aspects of Drug Addiction. https://doi.org/10.1016/B978-0-12-816979-7.00012-1

Besides lack of social support, other social factors underlying drug use is peer pressure. Peer pressure was found to contribute to heavy marijuana and alcohol use. Adolescents and adults often consume drugs to 'fit in' within their circle of friends (Kandel, 1985). On university campuses, students may consume alcohol to be socially accepted by their peers. The effects of peer pressure on drug use were also studied on animals. In a series of experiments, Professor Mark Smith and colleagues found that rats would self-administer cocaine if another rat in the same cage was self-administering cocaine as well. Socially housed rats would not self-administer cocaine if their partner was not using cocaine (Peitz et al., 2013; Smith, 2012). Along these same lines, studies in humans found that participating in a drug associated activity as a means of social interaction was highly correlated with continued drug use (Galea, Rudenstine, & Vlahov, 2005). Some studies also found that reasons for mixing drugs is social gathering and peer pressure (Yedinak et al., 2016).

Further, social support also plays an important role for treatment outcomes. Social support can be provided by clinicians, friends, family, and even other drug using individuals during group therapies. A trust-based relationship with professionals can help people in recovery have more confidence in their own ability to care for themselves and recover. By allowing drug using individuals to commit themselves to their treatment (such as seeing a counselor regularly and keeping appointments), positive feelings about accessing the support available are reinforced, leading to a sense of control over their lives, confidence and self-esteem.

Some group therapy methods integrate regular group activities, such as spending time in groups playing games, watching movies, and attending social activities. These kind of activities help people in recovery form new social circles and transition into a different lifestyle with friends—a lifestyle not centered on drug use.

Family therapy is a type of socially-based therapy in which families of an addict are also involved in the treatment process. Family therapies open lines of communication between the caregivers and drug using individual. The outcome of this treatment is often that adolescents, for example, experience a greater sense of warmth and support from parents, resulting in a decrease in substance use (Bertrand et al., 2013). By fostering this support, adolescents can improve their ability to self-disclose and confide in their parents, which in turn strengthens the parent-child relationship (Bertrand et al., 2013). This is important to develop in order to maintain a supportive environment once an adolescent and their family start to move away from a life centered around the adolescent's substance abuse and towards a stable family routine. Further, other studies found that once treatment was completed, individuals who experience family problems by the end of their treatment were shown to report higher levels of drug use compared to those without family conflict (Fish, Maier, & Priest, 2015; Skeer et al., 2011).

Recovery support through belonging to offline pro-recovery groups

The process of recovery is defined as a "voluntary maintained lifestyle characterised by sobriety, personal health and citizenship" (Betty Ford Institute, 2007, p. 22) and "personal transformation evident in observable outcomes across multiple domains of functioning and supported by abstinence or increased control over substance use" (Best et al., 2016). In the domain of recovery from drug and alcohol addiction, the role of face-to-face (offline) social networks is widely recognized (White & Kelly, 2010). This association is particularly relevant during young adulthood when substance use fulfills a social facilitation function (Chung, 2013a, 2013b). For instance, in a recent study on the role of social networks on recovery outcomes (i.e., post-treatment days abstinent) in young adults undergoing residential treatment, Eddie and Kelly (2017) found that both the time spent with and number of high-risk friends were negatively associated with the percentage of abstinent days post-recovery. On the contrary, the number of and time spent with low-risk friends were associated with positive recovery outcomes (see also Mawson, Best, Beckwith, Dingle, & Lubman, 2015).

Formal, professionally-directed post-treatment care (after-care continuous programs) was shown to be highly effective in relapse prevention. Programs based on assertive approaches, as for example, the Assertive Continuing Care approach to adolescent post-treatment (ACC, see Godley, Godley, Dennis, Funk, & Passetti, 2007)—where support is provided even if the clients do not successfully complete their primary treatment and clinicians meet with clients in their homes and other community settings (thus minimizing missed appointments)—were shown to yield positive recovery outcomes. In the case of this program, the recovery outcomes were reflected in longer term abstinence and better adherence to the criteria of continuous post-treatment care (Godley et al., 2007). However, these formal post-treatment programs are often highly expensive and difficult to implement; thus, there is a pressing need to develop and trial more sustainable solutions (including those capitalizing on peer-support). The issue of sustainability is compounded by emerging evidence that for a primary treatment to be effective, post-treatment services such as monitoring and support should be provided for between 1 and 5 years (DuPont, Compton, & McLellan, 2015; White, 2009).

Thus, as the importance of developing networks of pro-recovery peers is increasingly being recognized, evidence on the positive effects of post-primary treatment recovery peer-support is emerging. For example, in a recent randomized controlled trial of adolescents in the US following discharge from residential treatment, Godley and colleagues (Godley, Passetti, Hunter, Greene, & White, 2019) found that being allocated to a voluntary peer-support program (as opposed to only being provided standard continuing after-treatment care) had significant effects of recovery outcomes such as increased involvement with low-risk, pro-recovery peers and increased participation in recovery management activities, which in turn were associated with reduced substance use.

In understanding the psychological processes which help the recovery process in the context of peer-support, addiction theorists propose that a key mechanism is building *recovery capital*, a concept defined as "the breadth and depth of internal and external resources that can be drawn upon to initiate and sustain recovery from AOD [alcohol and other drug] problems" (Granfield & Cloud, 1999). There is evidence that greater recovery capital indicates recovery progress at the same time acting as a predictor of sustained recovery (Groshkova, Best, & White, 2013). The concept of recovery capital provides a functional framework to assess the diverse resources of individuals to recover and interpret individual recovery experiences with both formal treatment and mutual help meetings as forms of social recovery capital (Bliuc, Doan, & Best, 2018). There are four types of recovery capital (Cloud & Granfield, 2009):

1. *Social capital:* all the resources that a person has as a result of their social relationships (i.e., both support from and obligations to groups to which they belong);
2. *Physical capital:* concrete assets such as property and money that may provide people with more recovery options (e.g., through accessing expensive rehabilitation services);
3. *Human capital:* skills, positive health, aspirations, and personal resources that will enable the individual to thrive (e.g., education and high intelligence have been regarded as key aspects of human capital);
4. *Cultural capital:* mainstream values, beliefs and attitudes that are linked to the ability to fit into dominant (socially accepted) social behaviors (see Best & Laudet, 2010).

Two primary ways to build recovery capital are proposed in the literature: (a) via links with both group members (peers or other people in recovery) and people from other social networks (for example when connecting to people who are not going themselves to the process of recovery, but instead they can be classified as advocates or supporters from the broader community); and (b) via community involvement, including engagement in meaningful activities such as, not only employment and training, but also childcare and volunteering (Best & Laudet, 2010; Groshkova et al., 2013); this broader definition of what meaningful activities may represent highlights some of the key mediators of recovery such as the family, local community and workplaces.

A rapidly expanding body of research provides increased support for the argument that belonging to *many supportive social networks* represents a key resource in the process of addiction recovery and more generally for maintaining good health (Cruwys et al., 2013; Cruwys, Haslam, Dingle, Haslam, & Jetten, 2014: Haslam, Cruwys, & Haslam, 2014; Jetten, Haslam, & Alexander, 2012; Litt, Kadden, Kabela-Cormier, & Petry, 2009). In particular, according to the *Social Identity Model of Recovery* (SIMOR, Best et al., 2016), during the process of recovery, there is a transitioning phase from groups associated with "using" to groups that are supportive of recovery; this transition is conducive to a positive change in values, beliefs, attitudes, and behaviors. This change process is gradual and characterized by a weakening of the bonds that the individual in recovery has with using groups while engagement with recovery groups develops. As Fig. 1 illustrates, SIMOR proposes that the internalization of recovery values and norms from supportive groups shapes future attitudes and behaviors related to substance use, in particular involving reduced exposure to conditioned cues that increase the risk of relapse and avoidance of substance use (Best, Irving, Collinson, Andersson, & Edwards, 2017). This means that individuals have the opportunity to develop and establish positive identities regardless of continuing symptoms, and SIMOR affords a plausible explanation for high relapse risk in the early stages of the recovery process (which may diminish as the individual becomes embedded in multiple groups supportive of recovery).

The maintenance of a recovery-oriented identity (i.e., the formation of new identity following and during recovery) requires ongoing and active engagement with recovery-oriented groups (Best et al., 2016) due to the fact that changing beliefs and attitudes around substance use represents a complex process (Bathish et al., 2017). These processes are often observed in studies of Alcoholics Anonymous highlighting the role of supportive social networks in promoting positive change and long-term well-being (Kaskutas, Bond, & Humphreys, 2002; Kelly, Hoeppner, Stout, & Pagano, 2012). The interpersonal relationships which naturally develop in these networks were found to be associated to positive health outcomes (Berkman, Kawachi, & Glymour, 2014; Ozbay et al., 2007; Umberson & Karas Montez, 2010).

Importantly, in recent years, the Internet has changed the way in which people access and discuss health information as well as how they seek peer-support in coping with a range of health conditions (Amante, Hogan, Pagoto, English, & Lapane, 2015; Chaffey, 2017; Chung, 2013a, 2013b; Hasler, Ruthven, & Buchanan, 2014; Thomas, McLeod, Jones, & Abbott, 2015). Increasingly, the Internet been used as a source of health information which is accessed via both health information websites and online networks in the form of online forums and message boards. The latter enables individuals to interact in a safe virtual space to share knowledge, experiences and skills related to health concerns, including addiction (Moorhead et al., 2013; Westerman, Spence, & Van Der Heide, 2014). In the next section, we discuss the role of online channels to access addiction recovery support.

Online channels of addiction recovery support

There are obvious practical advantages of using online communication to access health information and support and this is particularly true when the conditions of concern are highly stigmatized (as it is the case with addiction as, according

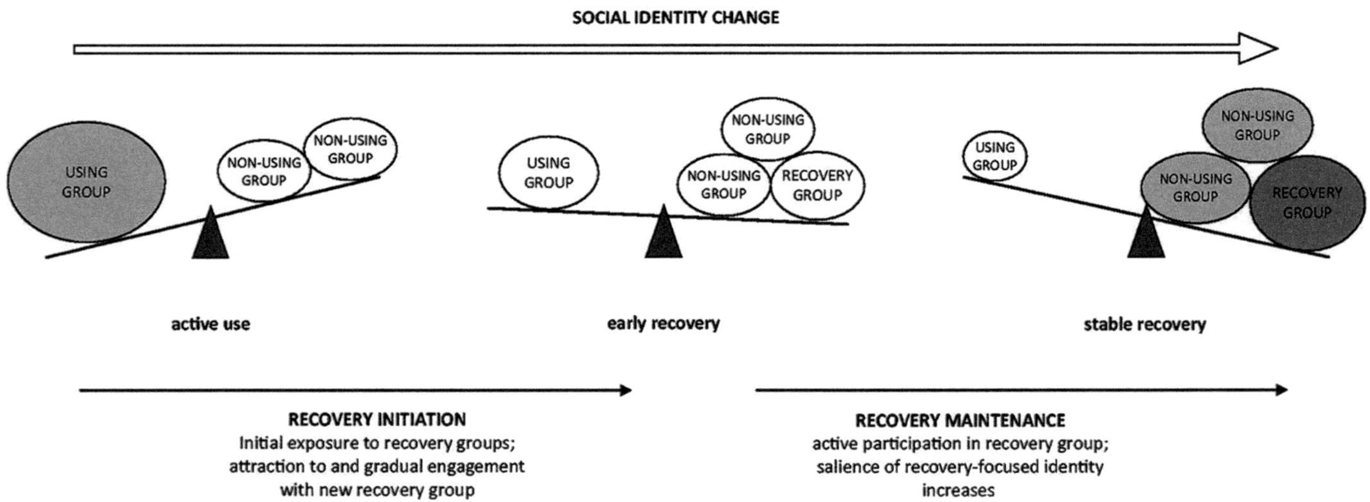

FIG. 1 Representation of the social identity change processes in SIMOR. *(Adapted from Best, D., Beckwith, M., Haslam, C., Alexander Haslam, S., Jetten, J., Mawson, E., & Lubman, D. I. (2016). Overcoming alcohol and other drug addiction as a process of social identity transition: The social identity model of recovery (SIMOR). Addiction Research & Theory, 24(2), 111–123. https://doi.org/10.3109/16066359.2015.1075980.)*

to the World Health Organization/WHO, drug use is considered one of the most stigmatized health conditions in the world and alcohol addiction the fourth most stigmatized, WHO, 2001). Due to the highly stigmatized nature of these conditions, people with substance-use disorders are often blamed for their condition (Glass, Mowbray, Link, Kristjansson, & Bucholz, 2013; Kulesza, Ramsey, Brown, & Larimer, 2014; Schomerus et al., 2011). For this reason, people attempting to recover from addiction may avoid attending face-to-face group meetings or other forms of offline peer-support. Thus, in these cases, in addition to anonymity, online communication can provide opportunities to access information and support for people who otherwise will not access standard, face-to-face services due to psychological social, geographical, and mobility-related isolation (Rodham, McCabe, & Blake, 2009; Savic, Best, Rodda, & Lubman, 2013).

Going beyond these practical benefits, online communication in social networks can facilitate meaningful social interaction, which in turn can provide people with a sense of belonging, purpose and support (Oh, Ozkaya, & LaRose, 2014). There is increasing evidence of the significant health benefits potentially brought by online engagement; for example, a large recent US study shows that people who are well-integrated in online social networks are likely to have lower mortality rates (Hobbs, Burke, Christakis, & Fowler, 2016). Our recent research investigating the role of participation in online social networks in addiction recovery shows that these networks can assist recovery by helping people in recovery build *recovery capital* which supports the maintenance of a drug-free lifestyle (see Best, Bliuc, Iqbal, Upton, & Hodgkins, 2018; Bliuc, Best, Iqbal, & Upton, 2017). Along these lines, in a study of an online community which represents an extension of a standard (face-to-face) addiction recovery program in the UK, we examined the structure of an online social network supportive of recovery as well as markers of social identity change in participants, as potential predictors of retention in the recovery program (Bliuc, Best, Iqbal, & Upton, 2017). In this research, we found evidence that positive online interactions between members of an online recovery community lead to positive recovery outcomes through a process of supporting participants to develop social recovery capital (both bonding and bridging capital) which bound them to a group which was supportive of positive change. More specifically, in this study, we found that retention in the program was determined by indicators of an emerging, strong recovery identity as captured by:

- the number of "likes" received on comments and posts on the recovery group's Facebook page (a form of online validation from the recovery community);
- the position in the social network of the participants (how central someone was in the social network indicating how connected someone was within the online community); and
- the linguistic content reflecting a positive group identity (the frequency in the use of the pronoun "we" and achievement words as captured by computerized linguistic analysis).

Moreover, in the Bliuc, Best, Beckwith, and Iqbal (2017) study, our observation of the evolution of the social network over time (8 months) allowed us to identify those participants in the online community who have undergone the most change in their position in the social network; in other words, those participants who moved from the periphery (from being largely un-connected) to the center of the social network (to becoming 'central nodes' or the most connected members of the online community). These participants were invited to participate in in-depth interviews which generated rich data on the lived experiences of social identity change (transitioning to a functional and well-connected recovery identity). This data illustrates particularly well how recovery capital is built in the context of members' interactions in the online space, and how this process is experienced by some of the participants in the online community. For instance, the interview data shows how different sub-categories of recovery capital were built in this case: bonding capital (reflected in resources that were made available through linkages between group members) and bridging capital (reflected in resources built through linkages with outgroups or members of the broader community). Bonding recovery capital was valued by participants in the form of "life social connectivity" or the ability to instantly access peer support as well as "potentially life-saving" information. Bridging recovery capital was manifesting as access to the wider community which was used as an opportunity to challenge stereotypes and negative perceptions about people going through addiction recovery. Finally, a key resource built through participation in the online community was the ability to make positive identity change visible as a way of reaching to and helping others in recovery (the accumulated, widely accessibly content of communication in the online community represents a recovery resource for others who may not be part of the recovery program per se).

Mapping the structure of the social network and its evolution over 8 months gave us some insight into changes in the patterns of interaction and cohesiveness of the overall online community. For example, as can be seen in Fig. 2, at the start, the members of the online community were mostly disconnected and the program staff and a small number of program participants (clients) were the ones driving the online activity. However, this had gradually changed, so that after 8 months, the clients were the most connected (central) with the highest number of connections in the network. Fig. 2 also provides a powerful illustration of how the online community as a *collective* is becoming more cohesive (with most members communicating with each other).

Other research that we conducted in the context of an exclusively virtual community of support (Bliuc et al., 2018)—with no one physical recovery community in its background as in the case of our previous study—is endorsing the earlier argument that positive engagement with supportive recovery networks is key to an effective and sustainable addiction recovery journey. Specifically, in this study we analyzed naturally occurring linguistic data from SoberRecovery (https://www.soberrecovery.com/forums/), an online community which describes itself as a community of "recovering alcoholics, addicts, co-dependents,

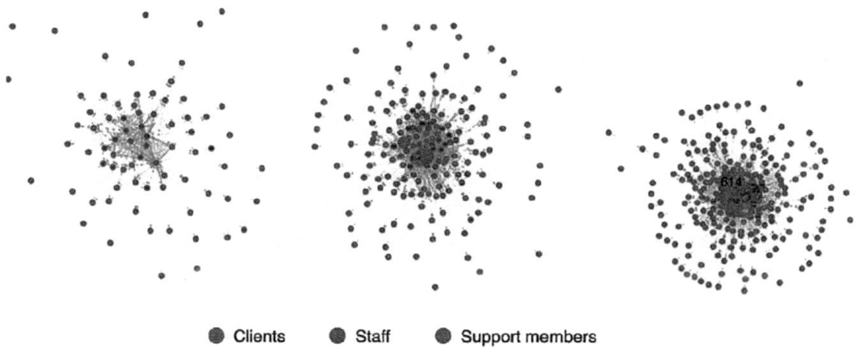

FIG. 2 The evolution of the structure of the online community at Time 1 (first month), Time 2 (fourth month), and Time 3 (eighth month). *(Adapted from Bliuc, A.-M., Best, D., Beckwith, M., & Iqbal, M. (2017). Online support communities in addiction recovery: Capturing social interaction and identity change through analyses of online communication. In S. A. Buckingham & D. Best (Eds.), Addiction, behavioural change and social identity: The path to resilience and recovery. Abingdon, Oxon/New York, NY: Routledge.)*

their friends and family". We examined both positive (recovery identity and self-efficacy) and negative (self-stigma) aspects of recovery capital as they relate to levels of well-being and found that self-stigma negatively predicted self-efficacy and well-being, while social identification with a recovery identity mediated the adverse effects of these associations.

Beyond providing evidence on the importance of connection and belonging to supportive networks, these studies suggest that in an online environment, the meaning of a "recovery community" includes not only peers (who are going through similar stages in the processes of recovery), but also pro-recovery advocates (including staff members in the case of Best et al. (2018) and Bliuc, Best, Beckwith, and Iqbal (2017) studies) and supporters from the broader community (family, friends, colleagues, etc.). Moreover, these online channels of addiction recovery also allow for 'lurkers' or people who do not directly engage with other online participants to access relevant content and identify nevertheless with these online communities—thus deriving a sense of belonging and connection (which comes with associated positive outcomes described in SIMOR). However, the potential benefits of 'lurking' recovery communities remain largely unexplored.

Conclusion

The power of online communication in the context of addiction recovery primarily rests on the possibility to readily interact with people who share the same experiences and concerns regardless of proximity, geographical barriers and even cultural contexts. In this way, borderless "online communities of support" including not only those people recovering from addiction, but also their supporters and advocates are established and easily accessed. In similar ways to the socio-political domain where social movements can emerge and flourish online, but produce social change "in real life" (Alberici & Milesi, 2016; Odağ, Uluğ, & Solak, 2016), addiction recovery online communities have the potential to support more than individual change. In other words, they can enable and energize social change about how people position themselves in regards to addiction recovery—replacing a model where the person in recovery is a relatively passive recipient of treatment and therapy (with the health professionals being central in shaping the process of recovery) with a model where, through ready access to a wealth of information and support, people in recovery are empowered to decide how their recovery journey would proceed.

Research on both the roles of offline and online social networks on recovery from addiction provides increasingly convincing evidence on the importance of belonging and identification as mediators of positive effects in addiction recovery, and more generally in health. With the increasingly wide use of the Internet to access health information and peer support around health issues in general and addiction recovery in particular, we note the new opportunities that online health engagement is opening for research in this domain. Shneiderman (2008) noted more than a decade ago that, as researchers, we need to adapt our approach to doing science in response to changes in technology and behaviors around technology; in his words: "traditional scientific methods need to be expanded to deal with complex issues that arise as social systems meet technological innovation" (Shneiderman, 2008, p. 1349). In agreement with this view, we suggest that by studying how people participate in online communities of recovery (by harvesting online data), we can gain insights into spontaneous individual social interactions as well as emerging socio-psychological processes—such as recovery capital development, social identity formation, norm development and validation, etc. Most importantly, by observing changes over time in the structure of online networks (as both structure/connections between members and content of communication between members), we can systematically capture the dynamical evolution of these communities with its potential effects on members. These new opportunities to use non-intrusive and inexpensive methods to access large amounts of data have obvious benefits, but at the same time, we need to be aware and cautious as they also create ethical gray areas which may require further and careful consideration by researchers pursuing these avenues.

As with broader mobile-based interventions for the prevention and treatment of addiction (e.g., Fowler, Holt, & Joshi, 2016; Kazemi et al., 2017; Quanbeck, Chih, Isham, Johnson, & Gustafson, 2014), emerging research suggests that access to online communities of recovery from addiction has a significant positive effect on recovery outcomes for members of those communities. However, this research is mostly exploratory at this stage and strong evidence of these effects is yet to emerge—for example, there are no studies directly investigating the effects of interventions based around building recovery capital in online communities as such interventions have not yet been developed. In similar ways to the growing use of mobile health apps and interventions, it is likely that the online communities of recovery will become more prevalent and will shape how addiction support may be delivered in the future. Therefore, one key area that practitioners and researchers in the field could focus on developing in the future is finding effective and sustainable ways to integrate offline, standard (treatment and post-treatment) programs with online channels that will foster positive identity change and long-term maintenance. Most such programs have, already some online resources available to participants; however, we suggest that for such resources to reach maximum effectiveness, they need to be carefully considered in the design of the program and integrated with the other activities and its aims. Finally, professional advice on how to access and sort reliable health information on the Internet needs to be more systematically implemented as part of post-treatment care programs.

References

Alberici, A. I., & Milesi, P. (2016). Online discussion, politicized identity, and collective action. *Group Processes and Intergroup Relations, 19*, 43–59.

Amante, D. J., Hogan, T. P., Pagoto, S. L., English, T. M., & Lapane, K. L. (2015). Access to care and use of the Internet to search for health information: Results from the US National Health Interview Survey. *Journal of Medical Internet Research, 17*(4).

Bathish, R., Best, D., Savic, M., Beckwith, M., Mackenzie, J., & Lubman, D. I. (2017). "Is it me or should my friends take the credit?" the role of social networks and social identity in recovery from addiction. *Journal of Applied Social Psychology, 47*(1), 35–46. https://doi.org/10.1111/jasp.12420.

Berkman, L. F., Kawachi, I., & Glymour, M. M. (Eds.), (2014). *Social epidemiology*. New York: Oxford University Press.

Bertrand, K., Richer, I., Brunelle, N., Beaudoin, B., Lemieux, A., & Ménard, J. M. (2013). Substance abuse treatment for adolescents: How are family factors related to substance use change? *Journal of Psychoactive Drugs, 45*(1), 28–38.

Best, D., Beckwith, M., Haslam, C., Alexander Haslam, S., Jetten, J., Mawson, E., et al. (2016). Overcoming alcohol and other drug addiction as a process of social identity transition: The social identity model of recovery (SIMOR). *Addiction Research & Theory, 24*(2), 111–123. https://doi.org/10.3109/16066359.2015.1075980.

Best, D., Bliuc, A.-M., Iqbal, M., Upton, K., & Hodgkins, S. (2018). Mapping social identity change in online networks of addiction recovery. *Addiction Research & Theory*, 1–11. https://doi.org/10.1080/16066359.2017.1347258.

Best, D., Irving, J., Collinson, B., Andersson, C., & Edwards, M. (2017). Recovery networks and community connections: Identifying connection needs and community linkage opportunities in early recovery populations. *Alcoholism Treatment Quarterly, 35*(1), 2–15.

Best, D., & Laudet, A. (2010). *The potential of recovery capital*. London: RSA.

Betty Ford Institute Consensus Panel. (2007). What is recovery? A working definition from the Betty ford institute. *Journal of Substance Abuse & Treatment, 33*, 221–228.

Bliuc, A.-M., Best, D., Beckwith, M., & Iqbal, M. (2017). Online support communities in addiction recovery: Capturing social interaction and identity change through analyses of online communication. In S. A. Buckingham & D. Best (Eds.), *Addiction, behavioural change and social identity: The path to resilience and recovery*. Abingdon, Oxon/New York, NY: Routledge.

Bliuc, A. M., Best, D., Iqbal, M., & Upton, K. (2017). Building addiction recovery capital through online participation in a recovery community. *Social Science & Medicine, 193*, 110–117. https://doi.org/10.1016/j.socscimed.2017.09.050.

Bliuc, A. M., Doan, T. N., & Best, D. (2018). Sober social networks: The role of online support groups in recovery from alcohol addiction. *Journal of Community & Applied Social Psychology*.

Chaffey, D. (2017). *Global social media research summary 2017*. Retrieved from: http://www.smartinsights.com/social-media-marketing/social-media-strategy/new-global-social-media-research/.

Chung, T. (2013a). *Adolescent substance use: Symptoms and course*. In *Principles of addiction* (pp. 97–106).

Chung, J. E. (2013b). Social interaction in online support groups: Preference for online social interaction over offline social interaction. *Computers in Human Behavior, 29*(4), 1408–1414. https://doi.org/10.1016/j.chb.2013.01.01.

Cloud, W., & Granfield, R. (2009). Conceptualizing recovery capital: Expansion of a theoretical construct. *Substance Use & Misuse, 43*(12–13), 1971–1986. https://doi.org/10.1080/10826080802289762.

Cruwys, T., Dingle, G. A., Haslam, C., Haslam, S. A., Jetten, J., & Morton, T. A. (2013). Social group memberships protect against future depression, alleviate depression symptoms and prevent depression relapse. *Social Science and Medicine, 98*, 179–186.

Cruwys, T., Haslam, S. A., Dingle, G. A., Haslam, C., & Jetten, J. (2014). Depression and social identity an integrative review. *Personality and Social Psychology Review, 18*, 215–238.

DuPont, R. L., Compton, W. M., & McLellan, A. T. (2015). Five-year recovery: A new standard for assessing effectiveness of substance use disorder treatment. *Journal of Substance Abuse Treatment, 58*, 1–5.

Eddie, D., & Kelly, J. F. (2017). How many or how much? Testing the relative influence of the number of social network risks versus the amount of time exposed to social network risks on post-treatment substance use. *Drug and Alcohol Dependence, 175*, 246–253.

Fish, J. N., Maier, C. A., & Priest, J. B. (2015). Substance abuse treatment response in a Latino sample: The influence of family conflict. *Journal of Substance Abuse Treatment, 49*, 27–34. https://doi.org/10.1016/j.jsat.2014.07.011 [Epub 2014 Aug 4].

Fowler, L. A., Holt, S. L., & Joshi, D. (2016). Mobile technology-based interventions for adult users of alcohol: A systematic review of the literature. *Addictive Behaviors, 62*, 25–34.

Galea, S., Rudenstine, S., & Vlahov, D. (2005). Drug use, misuse, and the urban environment. *Drug and Alcohol Review, 24*(2), 127–136.

Glass, J. E., Mowbray, O. P., Link, B. G., Kristjansson, S. D., & Bucholz, K. K. (2013). Alcohol stigma and persistence of alcohol and other psychiatric disorders: A modified labeling theory approach. *Drug and Alcohol Dependence, 133*(2), 685–692.

Godley, M. D., Godley, S. H., Dennis, M. L., Funk, R. R., & Passetti, L. L. (2007). The effect of assertive continuing care on continuing care linkage, adherence and abstinence following residential treatment for adolescents with substance use disorders. *Addiction, 102*(1), 81–93.

Godley, M. D., Passetti, L. L., Hunter, B. D., Greene, A. R., & White, W. L. (2019). A randomized trial of volunteer recovery support for adolescents (VRSA) following residential treatment discharge. *Journal of Substance Abuse Treatment*.

Granfield, R., & Cloud, W. (1999). *Coming clean: Overcoming addiction without treatment*. NYU Press.

Groshkova, T., Best, D., & White, W. (2013). The assessment of recovery capital: Properties and psychometrics of a measure of addiction recovery strengths. *Drug and Alcohol Review, 32*(2), 187–194. https://doi.org/10.1111/j.1465-3362.2012.00489.x.

Haslam, C., Cruwys, T., & Haslam, S. A. (2014). "The we's have it": Evidence for the distinctive benefits of group engagement in enhancing cognitive health in aging. *Social Science and Medicine*, *120*, 57–66.

Hasler, L., Ruthven, I., & Buchanan, S. (2014). Using internet groups in situations of information poverty: Topics and information needs. *Journal of the Association for Information Science & Technology*, *65*(1), 25–36. https://doi.org/10.1002/asi.22962.

Havassy, B. E., Hall, S., & Wasserman, D. A. (1991). Social support and relapse: Commonalities among alcoholics, opiate users, and cigarette smokers. *Addictive Behaviors*, *16*(5), 235–246.

Hobbs, W. R., Burke, M., Christakis, N. A., & Fowler, J. H. (2016). Online social integration is associated with reduced mortality risk. *Proceedings of the National Academy of Sciences*, *113*(46), 12980–12984. https://doi.org/10.1073/pnas.1605554113.

Jetten, J., Haslam, C., & Alexander, S. H. (Eds.), (2012). *The social cure: Identity, health and well-being*. Hove, England: Psychology Press.

Kandel, D. B. (1985). On processes of peer influence in adolescent drug use: A developmental perspective. *Advances in Alcohol and Substance Abuse*, *4*, 139–163.

Kaskutas, L. A., Bond, J., & Humphreys, K. (2002). Social networks as mediators of the effect of alcoholics anonymous. *Addiction*, *97*, 891–900.

Kazemi, D. M., Borsari, B., Levine, M. J., Li, S., Lamberson, K. A., & Matta, L. A. (2017). A systematic review of the mHealth interventions to prevent alcohol and substance abuse. *Journal of Health Communication*, *22*(5), 413–432.

Kelly, J. F., Hoeppner, B., Stout, R. L., & Pagano, M. (2012). Determining the relative importance of the mechanisms of behavior change within alcoholics anonymous: A multiple mediator analysis. *Addiction*, *107*, 289–299.

Kulesza, M., Ramsey, S., Brown, R., & Larimer, M. (2014). Stigma among individuals with substance use disorders: Does it predict substance use, and does it diminish with treatment. *Journal of Addictive Behaviors*, *3*(1), 1000115. https://doi.org/10.4172/2324-9005.1000115. Therapy & Rehabilitation.

Litt, M. D., Kadden, R. M., Kabela-Cormier, E., & Petry, N. M. (2009). Changing network support for drinking: Network support project 2-year follow-up. *Journal of Consulting and Clinical Psychology*, *77*(2), 229–242.

Lu, L., Shepard, J. D., Hall, F. S., & Shaham, Y. (2003). Effect of environmental stressors on opiate and psychostimulant reinforcement, reinstatement and discrimination in rats: A review. *Neuroscience & Biobehavioral Reviews*, *27*(5), 457–491.

Mawson, E., Best, D., Beckwith, M., Dingle, G., & Lubman, D. (2015). Social identity, social networks and recovery capital in emerging adulthood: A pilot study. *Substance Abuse Treatment, Prevention, and Policy*, *10*(1), 45. https://doi.org/10.1186/s13011-015-0041-2.

Meulenbeek, P. A. (2000). Addiction problems and methadone treatment. *Journal of Substance Abuse Treatment*, *19*(2), 171–174.

Meyer-Lindenberg, A., & Tost, H. (2012). Neural mechanisms of social risk for psychiatric disorders. *Nature Neuroscience*, *15*(5), 663–668. https://doi.org/10.1038/nn.3083.

Moorhead, S. A., Hazlett, D. E., Harrison, L., Carroll, J. K., Irwin, A., & Hoving, C. (2013). A new dimension of health care: Systematic review of the uses, benefits, and limitations of social media for health communication. *Journal of Medical Internet Research*, *15*, e85.

Odağ, Ö., Uluğ, Ö.M., & Solak, N. (2016). "Everyday I'm çapuling": Identity and collective action through social network sites in the Gezi Park protests in Turkey. *Journal of Media Psychology: Theories, Methods, and Applications*, *28*, 148–156.

Oh, H. J., Ozkaya, E., & LaRose, R. (2014). How does online social networking enhance life satisfaction? The relationships among online supportive interaction, affect, perceived social support, sense of community, and life satisfaction. *Computers in Human Behavior*, *30*, 69–78.

Ozbay, F., Johnson, D. C., Dimoulas, E., Morgan, C. A., III, Charney, D., & Southwick, S. (2007). Social support and resilience to stress: From neurobiology to clinical practice. *Psychiatry (Edgmont)*, *4*(5), 35.

Peitz, G. W., Strickland, J. C., Pitts, E. G., Foley, M., Tonidandel, S., & Smith, M. A. (2013). Peer influences on drug self-administration: An econometric analysis in socially housed rats. *Behavioural Pharmacology*, *24*(2), 114–123. https://doi.org/10.1097/FBP.0b013e32835f1719.

Quanbeck, A., Chih, M. Y., Isham, A., Johnson, R., & Gustafson, D. (2014). Mobile delivery of treatment for alcohol use disorders: A review of the literature. *Alcohol Research: Current Reviews*, *36*(1), 111.

Rodham, K., McCabe, C., & Blake, D. (2009). Seeking support: An interpretative phenomenological analysis of an Internet message board for people with complex regional pain syndrome. *Psychology and Health*, *24*, 619–634.

Savic, M., Best, D., Rodda, S., & Lubman, D. I. (2013). Exploring the focus and experiences of smartphone applications for addiction recovery. *Journal of Addictive Diseases*, *32*(3), 310–319.

Schomerus, G., Corrigan, P. W., Klauer, T., Kuwert, P., Freyberger, H. J., & Lucht, M. (2011). Self-stigma in alcohol dependence: Consequences for drinking-refusal self-efficacy. *Drug and Alcohol Dependence*, *114*(1), 12–17. https://doi.org/10.1016/j.drugalcdep.2010.08.013.

Shneiderman, B. (2008). Copernican challenges face those who suggest that collaboration, not computation are the driving energy for socio-technical systems that characterize Web 2.0. *Science*, *319*(5868), 1349–1350.

Sinha, R. (2008). Chronic stress, drug use, and vulnerability to addiction. *Annals of the New York Academy of Sciences*, *1141*, 105–130. https://doi.org/10.1196/annals.1441.030.

Skeer, M. R., McCormick, M. C., Normand, S. T., Mimiaga, M. J., Buka, S. L., & Gilman, S. E. (2011). Gender differences in the association between family conflict and adolescent substance use disorders. *Journal of Adolescent Health*, *49*(2), 187–192.

Smith, M. A. (2012). Peer influences on drug self-administration: Social facilitation and social inhibition of cocaine intake in male rats. *Psychopharmacology*, *224*(1), 81–90.

Thomas, N., McLeod, B., Jones, N., & Abbott, J.-A. (2015). Developing internet interventions to target the individual impact of stigma in health conditions. *Internet Interventions*, *2*(3), 351–358. https://doi.org/10.1016/j.invent.2015.01.003.

Umberson, D., & Karas Montez, J. (2010). Social relationships and health: A flashpoint for health policy. *Journal of Health and Social Behavior*, *51*(Suppl. 1), S54–S66.

Westerman, D., Spence, P. R., & Van Der Heide, B. (2014). Social media as information source: Recency of updates and credibility of information. *Journal of Computer-Mediated Communication*, *19*(2), 171–183. https://doi.org/10.1111/jcc4.12041.

Whitaker, L. R., Degoulet, M., & Morikawa, H. (2013). Social deprivation enhances VTA synaptic plasticity and drug-induced contextual learning. *Neuron*, *77*(2), 335–345. https://doi.org/10.1016/j.neuron.2012.11.022.

White, W. L. (2009). The mobilization of community resources to support long-term addiction recovery. *Journal of Substance Abuse & Treatment*, *36*, 146–158.

White, W. L., & Kelly, J. F. (2010). Recovery management: What if we really believed that addiction was a chronic disorder? In J. F. Kelly & W. White (Eds.), *Addiction recovery management* (pp. 67–84). New York: Humana Press.

World Health Organization. (2001). *The world health report. Mental health: New understandings, new hope*. Geneva: World Health Organization.

Yedinak, J. L., Kinnard, E. N., Hadland, S. E., Green, T. C., Clark, M. A., & Marshall, B. D. (2016). Social context and perspectives of non-medical prescription opioid use among young adults in Rhode Island: A qualitative study. *The American Journal on Addictions*, *25*(8), 659–665. https://doi.org/10.1111/ajad.12466. [Epub 2016 Nov 8].

Further reading

Godley, M. D., Godley, S. H., Dennis, M. L., Funk, R. R., Passetti, L. L., & Petry, N. M. (2014). A randomized trial of assertive continuing care and contingency management for adolescents with substance use disorders. *Journal of Consulting and Clinical Psychology*, *82*(1), 40.

Jetten, J., Haslam, C., Haslam, S. A., Dingle, G., & Jones, J. M. (2014). How groups affect our health and well-being: The path from theory to policy. *Social Issues and Policy Review*, *8*(1), 103–130. https://doi.org/10.1111/sipr.12003.

Kelly, J. F. (2017). Is alcoholics anonymous religious, spiritual, neither? Findings from 25 years of mechanisms of behavior change research. *Addiction*, *112*(6), 929–936. https://doi.org/10.1111/add.13590.

Kelly, J. F., Magill, M., & Stout, R. L. (2009). How do people recover from alcohol dependence? A systematic review of the research on mechanisms of behavior change in alcoholics anonymous. *Addiction Research & Theory*, *17*(3), 236–259.

Rains, S. A. (2014). The implications of stigma and anonymity for self-disclosure in health blogs. *Health Communication*, *29*(1), 23–31. https://doi.org/10.1080/10410236.2012.714861.

Shahab, L., & McEwen, A. (2009). Online support for smoking cessation: A systematic review of the literature. *Addiction*, *104*, 1792–1804.

Simpson, D. D., & Sells, S. B. (1990). *Opioid addiction and treatment: A 12-year follow-up*. Malabar, Florida: Krieger Publishing Company.

Tarrant, M., & Butler, K. (2011). Effects of self-categorization on orientation towards health. *British Journal of Social Psychology*, *50*(1), 121–139.

Vauth, R., Kleim, B., Wirtz, M., & Corrigan, P. W. (2007). Self-efficacy and empowerment as outcomes of self-stigmatizing and coping in schizophrenia. *Psychiatry Research*, *150*(1), 71–80.

Wilson, G. B., Kaner, E. F., Crosland, A., Ling, J., McCabe, K., & Haighton, C. A. (2013). A qualitative study of alcohol, health and identities among UK adults in later life. *PLoS ONE*, *8*(8), e71792.

Zhang, Z., Friedmann, P. D., & Gerstein, D. R. (2003). Does retention matter? Treatment duration and improvement in drug use. *Addiction*, *98*, 673–684.

Chapter 13

The benefits and limitations of methadone: A comparison to other opioid replacement treatments

Yousif S. Mahdi[a], Ahmed A. Moustafa[a,b]

[a]*Qatar University, Doha, Qatar,* [b]*Marcs Institute for Brain, Behaviour, and Development and School of Psychology, Western Sydney University, Penrith, NSW, Australia*

Introduction

Addition is a chronic, relapsing disorder impacting the central nervous system (Lutz & Kieffer, 2013). Relapse frequently occurs in opioid addicted individuals, as they are particularly vulnerable to certain stimuli, known as "triggers", such as the exposure to the substance itself or paraphernalia related to the use of the opiate. Even after successful treatment, the exposure to such triggers may lead to craving of the drug in substance abusers. Importantly, it is these stimulants that have been reported to challenge a person's ability to remain abstinent from opiates, such as heroin. The most widely accepted explanation for opiate dependency is traced to a neurobiological basis, which describes endorphins as having a central role in the development of opiate addiction (Gold et al., 1982; Gold & Rea, 1983).

Endorphins are endogenous opioid peptides that function as inhibitory neurotransmitters, produced in the pituitary gland and hypothalamus when the body experiences pain or stress (Amir, Brown, & Amir, 1979). They assist the human body to experience an analgesic effect and therefore acts as a mood stabilizer, however an opiate addicted individuals endorphin system has been damaged due to excessive use of drugs (Goldstein, 1976). In view of the fact that opiates cause a significant imbalance in the endorphin-receptor system, the brain is no longer effective in producing the suitable response in times of stress (Avoy, 2006). Accordingly, the drug-using patient will use drugs to manage pain symptoms.

The use of synthetic opioids such as methadone continue to interact with the opiate receptors throughout treatment, yet the endorphin system's balance

Cognitive, Clinical, and Neural Aspects of Drug Addiction. https://doi.org/10.1016/B978-0-12-816979-7.00013-3

cannot be restored while on therapeutic doses of methadone (Gold et al., 1981). Methadone is an anti-addictive analgesic used to manage addiction symptoms in patients with opioid dependency. It was developed in 1964 as a medical response to the post-World War II heroin epidemic in New York City. The American Medical Association needed to produce a pharmaceutical drug that could combat the deleterious effects of heroin addiction, which did not only lead to increased mortality, but also crime, hepatitis B/C, and the spread of HIV/AIDS (Joseph, Stancliff, & Langrod, 2000). Methadone was put forward as the solution through systematic investigation by Dole and Nyswander (1965).

Methadone is an acyclic of opiate drugs, as it acts on the same opioid receptors, producing similar euphoric effects; this reduces withdrawal symptoms and diminishes the feelings of addiction. Methadone provides a means for individuals with opiate use disorder to relinquish themselves from the highly harmful side effects and addictive properties of illegal opiates with a state-controlled replacement drug therapy. Formerly, methadone was seen as a progressive treatment for chronic opiate addiction. Several empirical studies supported the reliability of methadone in reducing death and crime rates (Russolillo, Moniruzzaman, & Somers, 2018). However, as longitudinal studies began to emerge, methadone appeared to produce side effects in users and did not fully eradicate their addiction symptoms, but rather maintained their dependency, allowing for the potential of an impending relapse to take place. Arguably, methadone is not a comprehensive treatment to adequately cure opiate dependence, as its purpose is not to overcome addiction symptoms, but rather subdue the use of other illegal opiate based drugs. This claim is strengthened by the fact that research has shown that methadone treatment has been used for as long as 22 years in some patients, without providing them with a complete successful recovery (Goldstein & Herrera, 1995). In addition, methadone is linked to negative side effects, including a potential risk of lethal overdose, continuing addiction, harmful effects during pregnancy and high treatment discontinuation as patients report unbeneficial therapeutic effects.

Methadone is currently costing the National Health Service (NHS) in the UK approximately £17,174 per quality-adjusted life year based on a UK health technology assessment model (Kenworthy et al., 2017). The cost is quite similar to that in other countries, including the USA, Australia, and Canada (Zaric, Brennan, Varenbut, & Daiter, 2012). In order to advance the treatment of opiate addiction, a new versatile treatment is needed or the use of other available treatment options. Currently, other substitute interventions for opiate addiction have shown promising results for the treatment of opiate dependence; this includes levo-alpha-acetylmethadol (LAAM), buprenorphine, and naltrexone. However, they all have side effects. It is for this reason, that even alternative treatments to methadone pose their own unique disadvantages in treating opiate addicted individuals.

The aim of this chapter is to evaluate the advantages and disadvantages of methadone for the treatment of drug addiction. Furthermore, this chapter aims to discuss the use of other opiate treatments that may have greater advantages in treating opiate addiction. Through this evaluation, it will become clear whether issues concerning treatment safety, retention rates, and treatment satisfaction could be resolved. These concerns will be addressed using recent empirical investigations.

Opiate addiction

Illicit opioid users continue abusing drugs, despite their associated significant social and health problems, such as potential imprisonment, negative social implications related to career and family, exposure to blood-borne viruses that increase the potential for contracting infectious diseases, and the risk of a lethal opioid overdose (Brown, Stimmel, Taub, Kochwa, & Rosenfield, 1974). Moreover, opiate addiction has implications on the society in terms of health care costs. The abuse of opioids contribute to mortality in many parts of the world, and has continued to become a growing concern. An investigation conducted by Jones, Campopiano, Baldwin, and McCance-Katz (2015) estimated the national and state trends in opioid agonist medication-assisted treatment needs and capacity. Using data from the US National Survey on Drug Use and Health between 2009 and 2012, Jones et al. found that at the national level, opioid misuse had increased from an estimated 507,130 people aged 12 years and older in 2003 to 2,319,213 individuals in 2012. As a result, the high number of opiate addicted individuals exceeds the national rates of maximum potential buprenorphine and methadone treatment capacity. Therefore, in the USA, there exists significant gaps between treatment needs and the capacity available to provide treatment at the state and national levels.

Several individuals and societies have marked a stigma on 'chronic addiction', as they believe that it is a character flaw that can be easily overcome. However, research has shown that addiction has physiological and psychological aspects that cause drug dependence. The presence of stigmas may explain why treatments for opiate addiction are limited.

Several studies have shown that despite abstinence from illegal opiates, neurobiological changes inflicted by long-term substance abuse can lead to relapse upon exposure to certain environmental stimuli (Chaudhri, Sahuque, & Janak, 2009; Perry, Zbukvic, Kim, & Lawrence, 2014). For instance, exposure to paraphernalia, returning to a place associated with the addiction or being around other individuals when drug use took place, can cause a strong craving for opiates and tests the persons resolve to remain abstinent (Crombag, Bossert, Koya, & Shaham, 2008; Dole, 1988). The satisfaction of taking opiates relieves feelings of distress and depression that one may hold, temporarily resolving their personal issues and uplifting their mood (Ahmed & Koob, 1997). The associative learning theory known as 'instrumental conditioning' has provided a fundamental explanation for drug addiction (Everitt & Robbins, 2005; Robinson & Berridge, 1993). According to this theory, when a certain behavioral act is followed by a rewarding outcome, the individual will tend to repeat this behavior, this is known as the "law of effect" (Mackintosh, 1983). In relation to opiate use, the pleasant euphoric stimulation and the temporary removal of stress due to the positive reinforcement of opiate use, make an addicted individual further seek opiates. Accordingly, an addict's attention becomes fixated on repeating the occurrence of pleasurable outcomes, that is, to get the drug. Behavioral control and the ability to abstain from the substance becomes increasingly more difficult with repeated use.

The biological basis of chronic opiate addiction is rooted in the deterioration of the endorphin-receptor system (Gold et al., 1982; Gold & Rea, 1983). Endorphins are endogenous opioids and are among the neurotransmitters, which transmit electrical signals within the brain and are released to reduce our perception of pain and stress by creating intense feelings of euphoria (Loh & Ross, 1979). When exogenous opioids are introduced into the brain, they induce membrane hyperpolarization and neuronal inhibition in multiple structures. The illicit opiates bind to the μ-opioid receptor (MOR) in the nucleus accumbens. The blockade of MORs creates an excess of endorphins, as well as disinhibit the dopamine pathways, causing dopamine to be released due to input from the ventral tegmental area (VTA), which modulates the activity of neurons in the nucleus accumbens. These structures comprise the cortical-pallidal-striatal circuit, which contain pathways that are implicated in the mediation of goal-directed behavior. In order to reduce the level of endorphins in the brain to enable a state of homeostasis, a neural adaptation forces the body to reduce its own production of endogenous opioids, resulting in a shortage of naturally occurring endorphins and can lead to an insufficient responses in the endorphin-receptor system in times of stress; this process is known as aberrant synaptic plasticity (Koneru, Satyanarayana, & Rizwan, 2009). Due to endorphins operating as mood stabilizers (as well as being essential for bodily functions such as regulating blood pleasure, body temperature, respiration, sleep and digestion), addicted individuals are more prone to stress, and as a result, a dependence on opiates becomes the solution in order to normalize the level of endorphins in the altered brain. The lack of endorphins in the body creates a greater susceptibility to environmental "triggers" that can increase drug craving (Gold et al., 1982).

The methadone intervention

Introduced in the United States in 1947, methadone is the most widely used clinical opiate replacement therapy. The synthetic opioid is prescribed as an analgesic for patients with opioid dependency who suffer from withdrawal symptoms. To date, methadone is the most systematically studied, most researched and most politically polarizing drug, than any other pharmacotherapy for the treatment of opiate dependency (Moore, Ramcharitar, Levine, & Fowler, 2010). Methadone's main therapeutic purpose is to prevent or at least substantially reduce the use of illicit opioids (e.g. heroin, morphine, and codeine), with the goal is to improve the health status, psychological and social wellbeing of opioid-dependent individuals. Methadone works by binding to opiate receptors in several brain areas (He et al., 2009). This effect is thought to contribute to the prevention of relapse as it suppresses the need for opiate narcotics by precipitating withdrawal symptoms, allowing individuals to function normally. Methadone's main route of administration is oral and is metabolized by enzymes (CYP3A4, CYP2B6, and CYP2D6) found in the liver. Pharmacological studies suggest that methadone acts as an antagonist at the MOR receptor. Methadone modulates synaptic transmission and releases neurotransmitters such as dopamine, substance P, acetylcholine, and norepinephrine (Anderson & Kearney, 2000). It is this process that allows methadone to work similarly to other opioids; however, the effects of methadone on the brain are milder, that is, they are not strong as the effect of heroin. This is perhaps why some heroin addicts on methadone treatment may relapse.

An individual who is admitted into a Methadone Maintenance Treatment (MMT) goes through preliminary steps in order to achieve a balance between adequate relief of withdrawal symptoms and avoidance of toxicity and the potential risk of death during the induction phase of MMT. The induction phase is important as deaths during the initial dosage has been associated with a number of reasons, including inadequate assessment of tolerance by practitioners, commencement on doses that are too high for the level tolerance, and inadequate observation and supervision of dosing by both the clinic and patient (Baxter Sr et al., 2013). The first dose is usually less than 30 mg; however, this is based on an individual history of use, the route of administration of illicit opioids, and body mass index (Baxter et al., 2013). The first 2 weeks of treatment is known as the stabilization phase as the focus is to prevent the patient from oscillating between intoxication and withdrawal. Clinicians will review and support patients daily during this stage. After the appropriate level of methadone dosage is achieved, compliance is then measured by daily visits. The final phase is monitoring drug use; in this phase staff will decrease or increase the level of methadone, and this continues until both the patient and practitioner are satisfied with the level of improvement in withdrawal symptoms and addiction (Day, 2012). The idea is that reducing the dose slowly will be effective in eventually completely discontinuing MMT. Methadone is therefore not a cure per se, but is rather a tool that assists in improving the quality-of-life of heroin addicts (Joseph et al., 2000).

Further, there is the possibility of sex differences that apply to the successfulness of methadone treatment, as shown by Bawor et al. (2015). In this study, 503 participants (54% males and 46% females) received MMT at the time of the experiment. Results show that men were significantly more likely than women to abuse cannabis and smoke cigarettes. The researchers concluded that there is variable patterns in substance misuse, social functioning, and health among both men and women receiving MMT. However, these findings may be explained by gender differences that are apparent among individuals with opioid use disorder despite receiving treatment. Further, Campbell et al. (2018) found that out of 570 participants seeking treatment for opioid use disorder, females were significantly younger, financially dependent, live with a sexual partner, and unemployed in comparison to male counterparts. These findings stress the importance of the different levels of economic and mental health status prior to opioid use disorder treatment.

The benefits of methadone

MMT has been shown to be an effective and beneficial therapy in certain respects. Mattick, Breen, Kimber, and Davoli (2009) evaluated the effects of methadone and compared it with treatments that did not involve synthetic opioid replacement, such as placebo medication and drug-free detoxification. The researchers conducted a meta-analysis of randomized clinical trials of methadone compared with non-opioid treatments, which included 11 studies with a total of 1969 subjects. Methadone was significantly more effective than non-pharmacological approaches in reducing the use of heroin; on average 30.4% of the patients did not experience relapse while on methadone. However, the results also show that there was no significant difference between methadone and non-opioid alternatives in criminal activity and preventing mortality. This review highlights that methadone treatment has significant effects in preventing individuals from using illicit opiates if patients fully follow the program. However, a limitation of the current meta-analysis is that the studies sampled used self-report as opposed to urine or blood tests to collect information regarding patient's use of heroin. Therefore, the results may have been affected by social desirability response bias, as individuals may conceal their use of opiates. Nonetheless, this investigation more importantly shows that methadone is perhaps more effective then no treatment at all.

In terms of mortality linked to the use of MMT, this topic was addressed in a longitudinal investigation in Tel Aviv, Israel, which followed up 890 patients from a previous study from between 15 and 24 years (Peles, Schreiber, Sason, & Adelson, 2018). It was found that 237 of the subjects passed away (116 of which were on MMT); the rates of mortality were the same for other patients not in MMT. Overall, the retention and mortality rates after 24 years were non-significantly different after 15 years and cocaine abuse was a predictor of whether recovering opiate users would relapse or not. Controlling for the effects of cocaine, methadone demonstrates a peak level of effectiveness.

Another notable strength of MMT is that it reduces the risk of contracting the human immunodeficiency virus (HIV). Millson et al. (2007) assessed injection-related HIV risk behavior among opioid addicts who were prescribed methadone. The participants were 183 opiate dependent patients from Canada, who were enrolled in a government scheme to provide methadone treatment in needle-exchange clinics. In this prospective observational cohort study, self-reports were used to assess HIV risk behavior during methadone treatment.

Six months after treatment, it was found that the number of subjects injecting opiates was reduced by 16.9%, sharing needles was decreased by 6.6% and sharing drug equipment was significantly reduced by 14.5%. The researchers concluded that methadone programs have the potential to reduce the risk of contracting HIV. This demonstrates that methadone can prevent the risk of contracting HIV, which is a common risk in illicit opioid practice.

Furthermore, besides reducing withdrawal symptoms, methadone does not reverse the dire physiological and psychological effects of illegal opiate drugs, as individuals can develop a life-long addiction to methadone itself. In the next section, we discuss the negative outcomes and limitations of methadone.

The limitations of methadone

Opioid addiction places a woman and her infant's development at increased risk of negative outcomes. For instance stillbirth, fetal growth reduction, prematurity and neonatal mortality are common side effects in women who use heroin while pregnant (Little et al., 1990). Similar studies were reported in heroin using mothers in Taiwan showing that heroin use 2–3 months before delivery can negatively affect the infant (Su et al., 2012). In addition to the effects of drugs, there are also severe risks associated with methadone during pregnancy (Blandthorn, Forster, & Love, 2011). Methadone-exposed infants have the potential to develop symptoms of methadone addiction after birth, which include the development of tremors, sleep disturbance (Allocco et al., 2016), and hypertonicity (stiff muscles; Velez, Jansson, Schroeder, & Williams, 2009). This constellation of symptoms in an opioid exposed infant is known as the Neonatal Abstinence Syndrome (NAS) and in a large cohort of more than 10,000 infants suffering from NAS, methadone was used by 31% of the mothers (Tolia et al., 2015). This demonstrates that methadone has the potential to produce aversive effects on the infants. Importantly, the current treatment for NAS is morphine, methadone, buprenorphine, tincture of opium and paregoric of opium solutions that are administered orally during feeding (Jansson & Velez, 2012). Lim, Prasad, Samuels, Gardner, and Cordero (2009) examined the effects of high-dose methadone on the health status of pregnant women and its effects on the susceptibility of NAS. The investigators conducted a retrospective investigation of 68 neonates and their mothers who were given methadone during pregnancy. The last dosage of methadone and the duration of treatment for NAS were assessed as outcome measures. The data showed that that higher doses of methadone are related to an increase in NAS as well as that 77% of the infants were treated for NAS. Therefore, it is evident that methadone has detrimental effects on infants. However, for some individuals, discontinuing from methadone may lead to many other side effects, which we describe below.

Another fundamental flaw of methadone is that it has a high discontinuation rate among its users. Fathollahi, Torkashvand, Najmeddin, and Rezaeian (2016) measured the retention rate of 1396 patients admitted in MMT clinics over a

12-month period in Iran. The results revealed that the mean retention duration was 193.22 days, with retention rates at three-month (66.0%) being greater than at one-year (34.4%). The researchers found that predictors of one-year retention were high methadone dosage, treatment with state (and not private) clinics, and polysubstance abuse. Due to low retention rates, those that discontinue methadone treatment have a significantly increased risk of an opiate relapse and excessive use of heroin. This illustrates the difficulty of the addiction recovery process and the fact that individuals may need multiple sessions of treatment over time. Moreover, this limitation of methadone also highlights the dissatisfaction among its users. However, these results are limited by gender bias, as 93.8% of the subjects were male.

One other limitation of methadone is that patients are required to visit the clinic daily, which limits their movement and independence. The inconvenience of daily dosing and clinic visits may be unattractive to certain patients, and the restrictions imposed by the frequent dosing schedule, may weaken clients prospects to sustain employment and other social commitments. For example, some studies have suggested that methadone dosage should be titrated as needed depending on patients' use of heroin (Trafton, Minkel, & Humphreys, 2006). By doing so, the patients may not be required to visit the clinic daily to get their methadone treatment, and allows for some freedom and independence. Soyka, Zingg, Koller, and Kuefner (2008) examined the retention rate of methadone after a 6-month treatment program. Subjects were 140 opioid-dependent patients who were given daily-doses of methadone (50 mg). The predictors of outcome were withdrawal symptoms, assessed with the Opiate Withdrawal Scale, clinical staff reports of patient side effects, and weekly urine tests to confirm or disconfirm the presence of drugs. The findings revealed that approximately 52.1% of the patients remained in the treatment program. It was concluded that methadone has a low retention rate. This study demonstrates that methadone is not very effective and the issue of treatment response requires further attention. Even if the majority of patients remained in the methadone program, it should be clear that patients have views of the treatments they are offered and these views affect their willingness to enroll and stay in treatment. In order to advance the treatment of chronic opiate addiction, there needs to be a resolution between information concerning acceptability of the treatment and the effects achieved. This will allow a fully informed decision by policy makers and health care workers to produce an appropriate intervention for the management of opioid dependence. Furthermore, it is important to note that when higher-doses of methadone are given to some patients, higher retention can be achieved. Fareed et al. (2009) found that giving patients 100 mg of methadone lead to a better outcome for illicit opiate use, as the detection for other opiates and cocaine was 14% in the first urine drug screening and 8% in the last screening. However, a higher dose of methadone may subsequently lead to the development of health problems and greater difficulty in discontinuing MMT.

It is also quite common for methadone users to use other illegal substance while on MMT. Roux et al. (2016) investigated the frequency of cocaine use among patients receiving MMT and how this affects their treatment response. Their longitudinal investigation used case files from 183 self-reported occasional and regular cocaine clients who were concurrently using methadone. It was found that time using methadone significantly reduced occasional cocaine use, although most patients did not stop using cocaine. However, clinical predictors were also identified and participants who regularly used cocaine also presented symptoms of depression and ADHD and were more likely to relapse and resume using drugs. Methadone therefore, can be seen as insufficient in accommodating those who also use cocaine regularly. Accordingly, this can perhaps be resolved by means of tailoring pharmacological and behavioral interventions by screaming for psychological disorders prior to starting MMT.

Another negative effect of methadone is that it affects many physiological functions, including the immune system response. Chan et al. (2015) examined the immune system functioning of heroin addicted patients undergoing MMT ($n=34$) and compared them to a healthy control sample ($n=20$). By testing the tumor necrosis factor (TNF)- α and cytokine production of IL-1β, IL-6,IL-8,IL-10, it was found that the production of IL-1β, IL-6 and IL-8 and Plasma TNF-α was higher in methadone-maintained patients. The researchers concluded that methadone has the potential to induce long-term systemic inflammation as the immune systems response is influenced by the duration and dosage of methadone use.

Arguably, the most detrimental impact of methadone is its prevention of normal daily functioning for many users. Despite the fact that the main purpose of methadone is to combat opiate addiction, methadone has been a significant source of opiate dependence itself. Methadone is a full opioid agonist, therefore has the potential to produce dependence. The duration of methadone use is individually and clinically determined; therefore, treatment continues as long as the physician/clinician and the patient agree it is appropriate. The implications for allowing patients to have a direct involvement in the continuation of treatment may reinforce addiction, as patients have an essentially endless supply of methadone to continue their addiction. Methadone is highly addictive due to the fact that it interacts with the central nervous system similarly to narcotic opiate drugs. An individual's brain can become well adjusted to the pain-reliving effects of methadone, leading to dependence and addiction. Like heroin use, the endorphins produced endogenously in the brain can become deadened or degenerated as a result of methadone use; this will cause addicted individual's to seek relief through the continued use of methadone. Overtime, methadone users can become physically dependent, as they build a tolerance and begin to experience withdrawal symptoms if they stop using methadone. With this in mind, the methadone intervention appears to mainly replace illegal opioids and acts as a state administrated legal synthetic opioid treatment. On this basis, it can be argued that methadone can successfully prevent the use of illegal opioids and the physiological damage they cause (with varying degrees of success), but does not act upon the main ailing effect of chronic addiction. Arguably, there are challenges regarding methadone, suggesting a modification or abandonment of the treatment is needed. Supporting the claim that methadone is highly addictive, Rhodin, Grönbladh, Nilsson, and Gordh (2006) recruited patients with problematic opioid use in Sweden who have been long-term prescribes of opioid treatment (mean = 7.1 years). Patients were given a structured interview by clinical psychologists to assess their continued use of methadone and to understand factors underlying their dependence. Patients reported aspects such as withdrawal symptoms, increased feelings of chronic pain without methadone, and an inability to function adequately at work. Substance abuse disorder was reported in 32% of the patients, and 68% required methadone for musculoskeletal pain. The researchers concluded that methadone has the ability to facilitate addiction. The current findings show that methadone is an addictive drug. Furthermore, some health care services are not questioning whether it is suitable to be prescribing only methadone and not offering alternative treatments if the MMT fails. As a result, interest in the development of alternatives to broaden the range of therapies has been the focus of increasing research in recent years.

Alternative treatments for drug addiction

As discussed above, methadone has been shown to have many limitations and thus, is not a comprehensive treatment for opiate addiction; this is based on the findings that methadone does not improve the quality-of-life of the patients. Medication has been considered to be the best possible approach, as physiological symptoms require immediate attention in the short-term to prevent relapse. With this in mind, a drug that can function in a similar fashion to methadone and provide greater independence to patients may be a good step towards ultimately "curing" opiate addiction. There are a number of alternatives to methadone as a maintenance medicine for the management of opioid dependence.

An alternative opioid analgesic is levo-alpha-acetylmethadol (LAAM). LAAM is also a synthetic opioid that has similar properties to methadone in that it suppresses withdrawal symptoms and blocks the intoxicating effects of illicit opiates. However, LAAM holds a major advantage compared to methadone as it has a longer half-life and patients can be given LAAM every 48 h, rather than every 24 h as required with methadone. In some cases, three-day dosing has been achieved satisfactorily (McCance-Katz et al., 2006). The benefit of greater suppression of withdrawal symptoms may have an impact on the patient's period of recovery and allows less interference with their day-to-day activities, unlike methadone, which requires daily attention and management. This would suggest that addicted individuals might better accept LAAM, and this may lead to greater treatment satisfaction and retention rates. Randomized control investigations have shown that LAAM has superior effects over methadone. Conner, Anglin, Annon, and Longshore (2009) identified the longitudinal effects of

LAAM and methadone maintenance on heroin addiction behavior. By recruiting 315 heroin addicted patients, the participants were either given LAAM or methadone for the duration of 12 months, with the option of continuing treatment for a further 6 months. The study involved the assessments of drug use, criminal behavior, HIV risk behaviors, employment, and residential status using structured interviews, self-report questionnaires, and urine tests to examine opiate use were obtained at the treatment intake period and during treatment. The results showed that LAAM demonstrated greater treatment retention and in-treatment suppression of heroin than the methadone group. This demonstrates that LAAM is a more effective treatment than methadone in terms of controlling the patient's addiction as well as withdrawal symptoms. Furthermore, subjects placed in the LAAM treatment group were also less likely to contract HIV and to get involved in criminal behavior; these subjects were also more likely to be employed and have secure residential status, in comparison to methadone patients. The results allowed the researchers to conclude that LAAM has better effects on opiate addicted patients in relation to their use of illicit opioids, treatment satisfaction and quality of life.

However, LAAM is considered the second-line treatment for the management of opioid dependence due to its association with cardia arrhythmias, and is therefore unstable for treatment (Kinlock, Battjes, & Schwartz, 2002). However, refuting evidence has shown that when comparing LAAM and methadone over a 52-week period, it was found that participants in the LAAM group were more likely to adhere to the intervention throughout the investigation (57.4%) than the methadone participants (46.2%). Further, no cardiological or other adverse events were observed in the LAAM group. The researchers concluded that LAAM is more effective than methadone in promoting retention and reducing opiate use during treatment (Anglin, Conner, Annon, & Longshore, 2007). Along these lines, Clark et al. (2002) conducted a review that involved 15 studies, reporting that LAAM is more effective than methadone at reducing heroin use. Furthermore, Longshore, Annon, Anglin, and Rawson (2005) also reported that in addition to the reduction of heroin use, LAAM is more effective than methadone due to its alternate-day dosing schedule.

However, Wieneke et al. (2009) conducted a clinical trial using Electrocardiography (ECG) recordings to evaluate the tolerability and efficacy of maintenance treatment using LAAM in comparison with racemic methadone. These recordings were completed as baseline and after a 24-week program of randomly assigned groups. The findings revealed that in the 53 patients who completed the trial, no clinical cardiac complications were observed in either group. However, patients receiving LAAM treatment had prolonged QTc-interval that was significantly higher than the methadone group and no statistical change was observed in QTc-dispersion. It was concluded that LAAM has the potential to induce heart failure to a higher degree in comparison to methadone and it is recommended that ECG-monitoring be used under LAAM medication. This shows that LAAM has a higher risk of cardiac related death. Therefore, the longer half-life advantage of LAAM does not necessarily justify its clinical use as an opiate addition intervention and in fact carriers a higher potential of mortality. Also, one study found that the discontinuation of using LAAM leads to anxiety symptoms (Hamilton, Harris, & Gewirtz, 2013). Ultimately, LAAM was withdrawn from the European market entirely in 2001 due to prolonged QT intervals and torsade de pointes (Krantz et al., 2002). It is important to also point out that recent investigations also show that high dose methadone carriers the risk of QT prolongation and torsade de pointes (Shashoua, 2018; Traficante, Feibish, & Kashani, 2017).

Another medication that has been the subject of interest and that has been shown to have desirable analgesic properties is buprenorphine. While methadone is still the most commonly used pharmaceutical opiate deferent, in recent years, increased rates and preference to buprenorphine-based treatments in Australia have been recorded (Australian Institute of Health & Welfare, 2012). Pharmacologically, buprenorphine has morphine-like subjective effects and produces cross-tolerance to other opioids. Buprenorphine is registered as the treatment of choice for opioid dependency in France (Kahan, Srivastava, Ordean, & Cirone, 2011). When compared to methadone, buprenorphine holds several benefits, and may also address certain constraints of methadone, which has been empirically demonstrated. Kakko, Heilig, and Sarman (2008) compared the effects of buprenorphine and methadone in treating pregnant opiate addicts. The procedure used a population-based comparison to retrospectively analyze buprenorphine or methadone-exposed pregnancies in Stockholm. Participants were 39 women in a buprenorphine maintenance treatment between 2001 and 2006 and 35 women in the methadone interventions between 1982 and 2006, all of which, during these periods had given birth at least once. The outcome measures were assessed on the basis of birth outcome, neonatal adaptation, withdrawal syndrome and infant mortality. The result showed that buprenorphine-exposed pregnancies resulted in 19 (40.4%) cases of NAS symptoms, with only 7 (14.9%) of these infants requiring withdrawal treatment. Compared to 35 (77.8%) infants exhibiting NAS and 19 (52.8%) requiring withdrawal treatment in the methadone group. The researchers concluded that this non-randomized comparison suggested that buprenorphine offers advantages over methadone in terms of infant's health during and after pregnancy, as the treatment is less likely to lead to NAS and produces milder symptoms and reduces the need for withdrawal treatment.

Moreover, research has demonstrated that buprenorphine has far-reaching effects and may hold potential in fully reliving patients from their chronic addiction. Possibly the most beneficial feature of buprenorphine is its ability to allow patients to discontinue the treatment without major withdrawal symptoms. This would demonstrate that opiate dependent patients have to some degree recovered and feel unattached to opiate drugs after being prescribed buprenorphine. Researchers suggest that this is due to the receptor kinetics that causes a long duration of action and subsequently allows for alternate-day dosing

(McQuay, Moore, & Bullingham, 1986). Studies have shown that abrupt buprenorphine discontinuation produces limited signs and symptoms of opiate withdrawal. Tompkins, Smith, Mintzer, Campbell, and Strain (2014) systematically examined spontaneous withdrawal symptoms after the discontinuation of buprenorphine in comparison to morphine. Seven subjects were either given buprenorphine or morphine for 9 days, after participants underwent an 18-day period of abrupt treatment, during which, four double-blind placebo injections were administered daily. The investigators assessed the subjects' spontaneous opioid withdrawal using researcher observer-ratings of psychomotor and cognitive performance to measure the antagonist effects of opioids. Opiate withdrawal was also assessed using the Clinical Opiate Withdrawal Scale (COWS). The results showed that morphine withdrawal symptoms were significantly greater than those of buprenorphine withdrawal, as morphine patients relapsed after 7 days, whereas the buprenorphine treatment group remained abstinent from opioids during the abrupt period of the study. It was concluded that spontaneous withdrawal from high dose buprenorphine appears to be milder compared to morphine after drug cessation. Arguably, this finding demonstrates that buprenorphine provides a potentially easier withdrawal phase that causes a long duration of action allowing for alternate-day dosing and complete discontinuation. However, the Tompkins et al. (2014) study has some methodological flaws that the researchers have not addressed. The investigators did not employ a longitudinal study design; accordingly, it is unclear whether subjects relapsed after the intervention, and whether buprenorphine would provide long-term effects to prevent illicit opioid use.

One study examined the characteristics of first-time methadone and buprenorphine users in New South Wales, Australia (Burns et al., 2015). Using a retrospective investigation based on information collected from governmental databases from 15,600 patients between 2001 and 2010, it was found that patients who were prescribed buprenorphine switched medications more frequently and required more treatment episodes, and overall had poorer retention over methadone prescribed clients in the short-term. This shows that individuals starting methadone treatment show higher retention rates than their buprenorphine counterparts. This investigation found no interaction between demographic variables and medication received. Interestingly, a similar investigation conducted by Hansen, Siegel, Wanderling, and DiRocco (2016) in New York City found that buprenorphine was adopted more commonly in all social areas as well as in moderate-income and mixed-ethnicity areas. Further, buprenorphine has been shown to reduce drug-related mortality in a retrospective cohort of opioid dependent patients in New South Wales (Australia) who started either treatment between 2001 and 2010. However, after the 4-week period of starting either treatment, mortality risk did not differ for all-cause mortality but non-drug related mortality was lower for methadone. Patients who switched from buprenorphine to methadone had lower mortality in the first 4-weeks when compared to controls who received methadone treatment. Yarborough et al. (2016)

investigated patients' preferences and how their previous experience influenced their choice of treatment between methadone and buprenorphine. By recruiting 283 opioid dependent patients in semi-structured interviews, Yarborough et al. (2016) have focused on preferences for opiate dependent treatment, prior experience with treatment, knowledge of treatment options and their experience with the treatment of chronic pain in the context of problems with opioids. From the interviews, key themes emerged, including that patients highlighted in their choice between treatments; this included awareness of treatment alternatives with participants stating that they were not provided with information about the treatments options available to them. Results also show that there are differences in treatment expectations between clinicians and patients; specifically patients reported negative experience with both methadone and buprenorphine, which influenced their choice. For example, while there is stigma regarding the efficacy of methadone clinics, buprenorphine was perceived to be less addictive than methadone. The researchers concluded from these findings that medication options need to be made more clear to bridge the gap in communication between the patient and clinician and that patients need to have greater involvement in the decision-making process which may in turn enhance treatment outcomes and adherence.

Finally, naltrexone has also been used in the treatment of opiate addiction. Naltrexone is an opioid antagonist that works by blocking opiate receptors in the brain. However, unlike the previously mentioned medications, its attachment to opiate receptors does not lead to addiction, which is supported by several studies that have shown that it does not produce physical dependence. Naltrexone is used for the prevention of opiate relapse following opioid detoxification. Therefore, this requires that patients to be opiate free at the time of initial naltrexone administration. Research demonstrates that naltrexone has a high potential in helping to prevent relapse in opiate dependent patients. Kunøe et al. (2009) evaluated the safety and effectiveness of naltrexone. The procedure involved a 6-month naltrexone administration in reducing opioid use after in-patient treatment of 56 abstinence-oriented patients while measuring illicit opioid use and other outcomes after the experiment. The results show that patients receiving naltrexone had on average 60 days less heroin use than controls in the 180-day period. These findings conclude that naltrexone treatment is a safe and significantly effective treatment in reducing opioid use. Therefore, it is reasonable to assume that naltrexone can provide substantially long abstinence of illicit opioids. Arguably, the researchers have failed to provide an assessment period that is adequate in reviewing patients' opiate use, due to only examining the patients after 6-months. The way in which opiates are metabolized in the liver, additional opiates become less observable in the body after 1-week (Boerner, Abbott, & Roe, 1975). Due to infrequent blood tests, it is difficult to assume that the findings are reflective of patient's actual opiate use based on blood tests and self-report. Moreover, retention rates have also been flagged as a recurring issue when using naltrexone medication, as compliance among patients is low. it is not known why this is the case, but it is perhaps due to the non-addictive properties of the medications.

One may assume that there are more broad explanations as to why naltrexone is disliked among patients. For instance, naltrexone is administered orally and usually opioid addicts inject opiates directly into their bloodstream. A long-action, injectable version of naltrexone (XR-naltrexone) that can be administered once per month has been developed. Krupitsky et al. (2011) assessed the efficacy, safety, and patient-reported outcomes of an injectable form of naltrexone after patients went through detoxification. Using a double-blind, placebo-controlled, randomized, 24-week trial on 250 patients with opioid dependence disorder in Russia, patients either received naltrexone (380 mg) or placebo. All patients also received biweekly counseling sessions. Abstinence from opiates was measured using urine drug tests and self-report; the study also included measures of opioid craving and number of days of retention. The results showed that the proportion of weeks of confirmed abstinence was 90.0% in the naltrexone condition, compared to 35.0% in the placebo condition. There was no discontinuation of treatment in the naltrexone condition. The researchers concluded that naltrexone represents an effective treatment option and that an injectable version of the drug is far more appealing to patients, as it mimics injection heroin and thus acts as a viable alternative to illicit opiate use. The fact that injecting the drug can increase retention highlights the importance of how treatment administration can have a significant impact on patient retention. These findings are supported by a comparative randomized investigation comparing injection naltrexone and buprenorphine (Sullivan et al., 2017). Results showed that based on 150 opioid-dependent adults, patients were more likely to choose XR-naltrexone in comparison to the buprenorphine. It was concluded that XR-naltrexone is a safe, effective, and well tolerated in low doses. Alternatives to methadone offer promising results in terms of attrition and relapse rates currently seen opiate agonist tapers in both outpatient and inpatient settings.

Conclusion

In conclusion, in response to the question as to whether methadone is a comprehensive therapy to adequately assist individuals that are addicted to opioids, it can be argued that methadone is not a sufficient treatment in controlling opiate dependency, and is linked to many negative effects. Other drug interventions, such as Levo-alpha-acetylmethadol, buprenorphine, and naltrexone, demonstrate advantages for the opiate dependent patients, although they also have some limitations and can cause side effects. A key aim of this chapter was to assess the efficacy of methadone in health care services by reviewing and evaluating the recent literature regarding its safety and ability to manage addiction symptoms.

The most substantially valuable contribution methadone is providing is its ability to reduce the use of illegal opiate narcotics. It was shown that methadone is much more beneficial than simply using opiate-free treatments (Mattick et al., 2009). Therefore, methadone is useful for some individuals in preventing the use of illicit opioids. However, for methadone to prevent relapse depends on whether the drug using patients can be committed to the intervention, as the therapy is often most effective after 1-year of use. Therefore, methadone is not well tolerated by patients who are dissatisfied with the treatment. One may suggest that further research is needed to assess what features of the methadone program can be changed to better suit patients. Peles et al. (2018) has showed that if the use of other illicit substances such as cocaine is controlled for, methadone is highly effective in reducing opiate addiction. Another positive aspect of methadone that has proven useful to society and health care services is the prevention of the spread of HIV from using heroin. Research found that the use of methadone has a significant impact in the level of HIV risk behavior that an addict may face, as methadone eliminates the use and sharing of needles and other illicit opiate related equipment (Millson et al., 2007).

However, despite its positive contributions, methadone has been found to pose many risks. For instances, in relation to the issue of women's opiate addiction during pregnancy, methadone does not adequately address this problem. Research showed that low-doses of methadone (5.5 mg) could cause NAS for the majority of infants that are born to mothers who are prescribed methadone (Lim et al., 2009). This highlights that methadone is a restricted treatment and cannot be widely applied to at risk populations. It is advised that if methadone is prescribed to pregnant women, doses below 5.5 mg should be used, as this may reduce the risk of infants being born with NAS, but is significantly lower than the usual dosage of 30 mg. Another limitation of methadone is that is has low retention rates. Research has shown that methadone has retained up to 52.1% of patients; the others who discontinued treatment reported dissatisfaction with the therapy (Soyka et al., 2008). This issue is likely to lead to greater number of patients relapsing and being readmitted to the methadone program, and therefore exhausting health care recourses. Finally, the greatest disadvantage in using methadone is its addictive property. Methadone binds to opiate receptor much in the same way as illicit opiates do thus producing a less intense euphoric effect, and reinforcing addiction. Research has demonstrated that there are individuals who admit to continuing methadone treatment, due to substance abuse disorder (32%; Rhodin et al., 2006) and therefore methadone is highly addictive and can be difficult to cease treatment once started. It is also common for patients in the MMT program to abuse other drugs while using methadone (Roux et al., 2016). Methadone can also induce long-term systemic inflammation in the immune system (Chan et al., 2015).

In addressing the other key aim of this investigation, several studies investigated whether other alternative medicines have greater benefits over methadone. LAAM holds the advantage of having a longer half-life, meaning that patients will not need daily dosing; further, research showed that LAAM indeed had much more beneficial effects over methadone in treatment retention, quality-of-life and treatment satisfaction (Conner et al., 2009). However, LAAM is linked to increased risk of heart failure problems and anxiety symptoms (Hamilton et al., 2013). It is advisable that LAAM remains to be withdrawn from the market (Krantz et al., 2002) even if careful cardiac assessments are carried out (Wieneke et al., 2009). However, this risk of prolonged QT intervals and torsade de pointes by methadone (Shashoua, 2018; Traficante et al., 2017).

Another alternative is buprenorphine; this drug has been successfully shown to reduce the effect of NAS symptoms in infants, an issue methadone has not sufficiently addressed (Kakko et al., 2008). Moreover, buprenorphine has also been shown to be successful in allowing patients to completely discontinue the treatment without showing withdrawal symptoms (Tompkins et al., 2014). A refuting investigation showed that in comparison to methadone, buprenorphine was less effective in retaining patients in the short-term in a longitudinal study (Burns et al., 2015) but is better able to reduce drug-related mortality (Hansen et al., 2016). However, patient awareness and involvement in opiate treatment needs to be improved in order to possibly enhance treatment outcome and adherence (Yarborough et al., 2016).

Finally, naltrexone has positive long-term effects if injected, rather then taken orally, allowing patients to only take one-dose per month; therefore subjects are not required to be seen on a daily bases, unlike methadone which requires visits to a clinic every 24 h (Krupitsky et al., 2011). Even more importantly, the success of the injectable version of naltrexone raises insights into the importance of how opiate treatment is administered and how this affects patient satisfaction. However, naltrexone carries an increased risk of overdose that is greater than methadone (Kelty & Hulse, 2012). Sullivan et al. (2017) demonstrated that patients are more likely to still choose XR-naltrexone if given the choice of buprenorphine.

Overall, opiate addiction is a challenging disease that has multiple features that need to be resolved; it is difficult for any treatment to completely treat opiate dependence. However, methadone is unsatisfactory for treating addiction and health care services need to utilize the use of more diverse therapies that are currently available. One may advice that alternative treatment is used in conjunction with psychosocial treatments to better improve the effectiveness of the treatment and insure that patients are not tempted to relapse. Health care services need to question in greater detail whether methadone is suitable and why alternatives are not explored more by clinical practitioners.

References

Ahmed, S. A., & Koob, G. F. (1997). Cocaine- but not food-seeking behavior is reinstated by stress after extinction. *Psychopharmacology*, *132*, 289–295.

Allocco, E., Melker, M., Rojas-Miguez, F., Bradley, C., Hahn, K. A., & Wachman, E. M. (2016). Comparison of neonatal abstinence syndrome manifestations in preterm versus term opioid-exposed infants. *Advances in Neonatal Care: Official Journal of the National Association of Neonatal Nurses*, *16*(5), 329.

Amir, S., Brown, Z. W., & Amir, Z. (1979). The role of endorphins in stress: Evidence and speculations. *Neuroscience & Biobehavioral Reviews*, *4*, 77–86.

Anderson, I. B., & Kearney, T. E. (2000). Medicine cabinet: Use of methadone. *Western Journal of Medicine*, *172*(1), 43.

Anglin, M. D., Conner, B. T., Annon, J., & Longshore, D. (2007). Levo-alpha- acetylmethadol (LAAM) versus methadone maintenance: 1-year treatment retention, outcomes and status. *Addiction*, *102*(9), 1432–1442.

Australian Institute of Health & Welfare. (2012). *Health expenditure Australia 2010–11*. AIHW. (No. 47).

Avoy, D. (2006). Methadone treatment-a broad perspective. *Professional Perspectives on Addiction Medicine*, *13*(2), 222–224.

Bawor, M., Dennis, B. B., Varenbut, M., Daiter, J., Marsh, D. C., Plater, C., et al. (2015). Sex differences in substance use, health, and social functioning among opioid users receiving methadone treatment: A multicenter cohort study. *Biology of Sex Differences*, *6*(1), 21.

Baxter, L. E., Sr., Campbell, A., Deshields, M., Levounis, P., Martin, J. A., McNicholas, L., et al. (2013). Safe methadone induction and stabilization: Report of an expert panel. *Journal of Addiction Medicine*, *7*(6), 377–386. https://doi.org/10.1097/01.ADM.0000435321.39251.d7.

Blandthorn, J., Forster, D. A., & Love, V. (2011). Neonatal and maternal outcomes following maternal use of buprenorphine or methadone during pregnancy: Findings of a retrospective audit. *Women and Birth*, *24*(1), 32–39. https://doi.org/10.1016/j.wombi.2010.07.001.

Boerner, U., Abbott, S., & Roe, R. L. (1975). The metabolism of morphine and heroin in man. *Drug Metabolism Reviews*, *4*(1), 39–73.

Brown, S. M., Stimmel, B., Taub, R. N., Kochwa, S., & Rosenfield, R. E. (1974). Immunologic dysfunction in heroin addicts. *Archives of Internal Medicine*, *134*(6), 1001–1006.

Burns, L., Gisev, N., Larney, S., Dobbins, T., Gibson, A., Kimber, J., et al. (2015). A longitudinal comparison of retention in buprenorphine and methadone treatment for opioid dependence in New South Wales, Australia. *Addiction*, *110*(4), 646–655.

Campbell, A. N., Barbosa-Leiker, C., Hatch-Maillette, M., Mennenga, S. E., Pavlicova, M., Scodes, J., et al. (2018). Gender differences in demographic and clinical characteristics of patients with opioid use disorder entering a comparative effectiveness medication trial. *The American Journal on Addictions*, *27*(6), 465–470.

Chan, Y. Y., Yang, S. N., Lin, J. C., Chang, J. L., Lin, J. G., & Lo, W. Y. (2015). Inflammatory response in heroin addicts undergoing methadone maintenance treatment. *Psychiatry Research*, *226*(1), 230–234.

Chaudhri, N., Sahuque, L. L., & Janak, P. H. (2009). Context-induced relapse of conditioned behavioral responding to ethanol cues in rats. *Biological Psychiatry*, *64*(3), 203–210.

Clark, N., Lintzeris, N., Gijsbers, A., Whelan, G., Dunlop, A., Ritter, A., et al. (2002). LAAM maintenance vs methadone maintenance for heroin dependence. *Cochrane Database of Systematic Reviews*, *2*, CD002210.

Conner, B. T., Anglin, M. D., Annon, J., & Longshore, D. (2009). Effect of religiosity and spirituality on drug treatment outcomes. *Journal of Behavioral Health Services & Research*, *36*(2), 189–198.

Crombag, H. S., Bossert, J. M., Koya, E., & Shaham, Y. (2008). Context-induced relapse to drug seeking: A review. *Philosophical Transactions of the Royal Society of London. Series B, Biological Sciences*, *363*(1507), 3233–3243.

Day, E. (2012). Methadone treatment as palliation or an individualized process of recovery management? *Addiction*, *107*(5), 953–954.

Dole, V. P. (1988). Implications of methadone maintenance for theories of narcotic addiction. *JAMA*, *260*(20), 3025–3029.

Dole, V. P., & Nyswander, M. (1965). A medical treatment for diacetylmorphine (heroin) addiction: A clinical trial with methadone hydrochloride. *JAMA*, *193*(8), 646–650.

Everitt, B. J., & Robbins, T. W. (2005). Neural systems of reinforcement for drug addiction: From actions to habits to compulsion. *Nature Neuroscience*, *8*, 1481–1489.

Fareed, A., Casarella, J., Roberts, M., Sleboda, M., Amar, R., Vayalapalli, S., et al. (2009). High dose versus moderate dose methadone maintenance: Is there a better outcome? *Journal of Addictive Diseases*, *28*(4), 399–405.

Fathollahi, M. S., Torkashvand, F., Najmeddin, H., & Rezaeian, M. (2016). Predictors of one-year retention in methadone maintenance treatment (MMT) in Iran, Rafsanjan. *International Journal of High Risk Behaviors & Addiction*, *5*(3).

Gold, M. S., Dackis, C. A., Pottash, A. L., Sternbach, H. H., Annitto, W. J., Martin, D., et al. (1982). Naltrexone, opiate addiction, and endorphins. *Medicinal Research Reviews*, *2*(3), 211–246.

Gold, M. S., Pottash, A. L. C., Extein, I. R. L., Martin, D. A., Finn, L. B., Sweeney, D. R., et al. (1981). Evidence for an endorphin dysfunction in methadone addicts: Lack of ACTH response to naloxone. *Drug & Alcohol Dependence*, *8*(3), 257–262.

Gold, M. S., & Rea, W. S. (1983). The role of endorphins in opiate addiction, opiate withdrawal, and recovery. *Psychiatric Clinics of North America*, *6*(3), 489–520.

Goldstein, A. (1976). Opioid peptides endorphins in pituitary and brain. *Science*, *193*(4258), 1081–1086.

Goldstein, A., & Herrera, J. (1995). Heroin addicts and methadone treatment in Albuquerque: A 22-year follow-up. *Drug and Alcohol Dependence*, *40*(2), 139–150.

Hamilton, K. L., Harris, A. C., & Gewirtz, J. C. (2013). Affective and neuroendocrine effects of withdrawal from chronic, long-acting opiate administration. *Brain Research*, *1538*, 73–82. https://doi.org/10.1016/j.brainres.2013.09.026.

Hansen, H., Siegel, C., Wanderling, J., & DiRocco, D. (2016). Buprenorphine and methadone treatment for opioid dependence by income, ethnicity and race of neighborhoods in New York City. *Drug and Alcohol Dependence*, *164*, 14–21.

He, L., Kim, J., Ou, C., McFadden, W., van Rijn, R. M., & Whistler, J. L. (2009). Methadone antinociception is dependent on peripheral opioid receptors. *Journal of Pain*, *10*(4), 369–379.

Jansson, L. M., & Velez, M. (2012). Neonatal abstinence syndrome. *Current Opinion in Pediatrics*, *24*(2), 252–258.

Jones, C. M., Campopiano, M., Baldwin, G., & McCance-Katz, E. (2015). National and state treatment need and capacity for opioid agonist medication-assisted treatment. *American Journal of Public Health*, *105*(8), e55–e63.

Joseph, H., Stancliff, S., & Langrod, J. (2000). Methadone maintenance treatment (MMT). *The Mount Sinai Journal of Medicine*, *67*(5), 6.

Kahan, M., Srivastava, A., Ordean, A., & Cirone, S. (2011). Buprenorphine new treatment of opioid addiction in primary care. *Canadian Family Physician*, *57*(3), 281–289.

Kakko, J., Heilig, M., & Sarman, I. (2008). Buprenorphine and methadone treatment of opiate dependence during pregnancy: Comparison of fetal growth and neonatal outcomes in two consecutive case series. *Drug and Alcohol Dependence*, *96*(1), 69–78.

Kelty, E., & Hulse, G. (2012). Examination of mortality rates in a retrospective cohort of patients treated with oral or implant naltrexone for problematic opiate use. *Addiction*, *107*(10), 1817–1824.

Kenworthy, J., Yi, Y., Wright, A., Brown, J., Maria Madrigal, A., & Dunlop, W. C. N. (2017). Use of opioid substitution therapies in the treatment of opioid use disorder: Results of a UK cost-effectiveness modelling study. *Journal of Medical Economics*, *20*(7), 740–748. https://doi.org/10.1080/13696998.2017.1325744.

Kinlock, T. W., Battjes, R. J., & Schwartz, R. P. (2002). A novel opioid maintenance program for prisoners: Preliminary findings. *Journal of Substance Abuse Treatment*, *22*(3), 141–147.

Koneru, A., Satyanarayana, S., & Rizwan, S. (2009). Endogenous opioids: Their physiological role and receptors. *Global Journal of Pharmacology*, *3*(3), 149–153.

Krantz, M. J., Lewkowiez, L., Hays, H., Woodroffe, M. A., Robertson, A. D., & Mehler, P. S. (2002). Torsade de pointes associated with very-high-dose methadone. *Annals of Internal Medicine*, *137*(6), 501–504.

Krupitsky, E., Nunes, E. V., Ling, W., Illeperuma, A., Gastfriend, D. R., & Silverman, B. L. (2011). Injectable extended-release naltrexone for opioid dependence: A double-blind, placebo-controlled, multicentre randomised trial. *The Lancet*, *377*(9776), 1506–1513.

Kunøe, N., Lobmaier, P., Vederhus, J. K., Hjerkinn, B., Hegstad, S., Gossop, M., et al. (2009). Naltrexone implants after in-patient treatment for opioid dependence: Randomised controlled trial. *The British Journal of Psychiatry*, *194*(6), 541–546.

Lim, S., Prasad, M. R., Samuels, P., Gardner, D. K., & Cordero, L. (2009). High-dose methadone in pregnant women and its effect on duration of neonatal abstinence syndrome. *American Journal of Obstetrics and Gynecology*, *200*(1), 70–71.

Little, B. B., Snell, L. M., Klein, V. R., Gilstrap, L. C., Knoll, K. A., & Breckenridge, J. D. (1990). Maternal and fetal effects of heroin addiction during pregnancy. *Journal of Reproductive Medicine*, *35*(2), 159–162.

Loh, H. H. & Ross, D. H. (Eds.), (1979). *Neurochemical mechanisms of opiates and endorphins*: Raven Press.

Longshore, D., Annon, J., Anglin, M. D., & Rawson, R. A. (2005). Levo-alpha-acetylmethadol (LAAM) versus methadone: Treatment retention and opiate use. *Addiction*, *100*(8), 1131–1139.

Lutz, P. E., & Kieffer, B. L. (2013). Opioid receptors: Distinct roles in mood disorders. *Trends in Neurosciences*, *36*(3), 195–206.

Mackintosh, N. J. (1983). *Conditioning and associative learning*. Oxford: Clarendon Press.

Mattick, R. P., Breen, C., Kimber, J., & Davoli, M. (2009). Methadone maintenance therapy versus no opioid replacement therapy for opioid dependence. *Cochrane Database*, *22*(3), 199–201.

McCance-Katz, E. F., Rainey, P. M., Smith, P., Morse, G. D., Friedland, G., Boyarsky, B., et al. (2006). Drug interactions between opioids and antiretroviral medications: Interaction between methadone, LAAM, and delavirdine. *American Journal on Addictions*, *15*(1), 23–34.

McQuay, H. J., Moore, R. A., & Bullingham, R. E. S. (1986). *Buprenorphine kinetics*. In *Vol. 8. Advances in pain research and therapy* (pp. 271–278).

Millson, P., Challacombe, L., Villeneuve, P. J., Strike, C. J., Fischer, B., Myers, T., et al. (2007). Reduction in injection-related HIV risk after 6 months in a low-threshold methadone treatment program. *AIDS Education & Prevention*, *19*(2), 124–136.

Moore, K. A., Ramcharitar, V., Levine, B., & Fowler, D. (2010). Advocates for prescription opioid reform. *Journal of Managed Care Pharmacy*, *16*(1), 26–27.

Peles, E., Schreiber, S., Sason, A., & Adelson, M. (2018). Similarities and changes between 15-and 24-year survival and retention rates of patients in a large medical-affiliated methadone maintenance treatment (MMT) center. *Drug and Alcohol Dependence*, *185*, 112–119.

Perry, C., Zbukvic, I., Kim, J. H., & Lawrence, A. J. (2014). Role of cues and contexts on drug-seeking behavior. *British Journal of Pharmacology*, *171*(20), 4636–4672.

Rhodin, A., Grönbladh, L., Nilsson, L. H., & Gordh, T. (2006). Methadone treatment of chronic non-malignant pain and opioid dependence–a long-term follow-up. *European Journal of Pain*, *10*(3), 271–278.

Robinson, T. E., & Berridge, K. C. (1993). The neural basis of drug craving: An incentive-sensitization theory of addiction. *Brain Research Reviews*, 247–291.

Roux, P., Lions, C., Vilotitch, A., Michel, L., Mora, M., Maradan, G., et al. (2016). Correlates of cocaine use during methadone treatment: Implications for screening and clinical management (ANRS Methaville study). *Harm Reduction Journal, 13*(1), 12.

Russolillo, A., Moniruzzaman, A., & Somers, J. M. (2018). Methadone maintenance treatment and mortality in people with criminal convictions: A population-based retrospective cohort study from Canada. *PLoS ONE.*

Shashoua, A. (2018). *Torsades de pointes after methadone treatment.* In *Challenging cases and complication management in pain medicine* . (pp. 27–30) Cham: Springer.

Soyka, M., Zingg, C., Koller, G., & Kuefner, H. (2008). Retention rate and substance use in methadone and buprenorphine maintenance therapy and predictors of outcome: Results from a randomized study. *International Journal of Neuropsychopharmacology, 11*(05), 641–653.

Su, P. H., Chang, Y. Z., Yang, C., Ng, Y. Y., Chen, J. Y., & Chen, S. C. (2012). Perinatal effects of combined use of heroin, methadone, and amphetamine during pregnancy and quantitative measurement of metabolites in hair. *Pediatrics and Neonatology, 53*(2), 112–117. https://doi.org/10.1016/j.pedneo.2012.01.008 [Epub 2012 Mar 2].

Sullivan, M., Bisaga, A., Pavlicova, M., Choi, C. J., Mishlen, K., Carpenter, K. M., et al. (2017). Long-acting injectable naltrexone induction: A randomized trial of outpatient opioid detoxification with naltrexone versus buprenorphine. *American Journal of Psychiatry, 174*(5), 459–467.

Tolia, V. N., Patrick, S. W., Bennett, M. M., Murthy, K., Sousa, J., Smith, P. B., et al. (2015). Increasing incidence of the neonatal abstinence syndrome in US neonatal ICUs. *New England Journal of Medicine, 372*(22), 2118–2126.

Tompkins, D. A., Smith, M. T., Mintzer, M. Z., Campbell, C. M., & Strain, E. C. (2014). A double blind, within subject comparison of spontaneous opioid withdrawal from buprenorphine versus morphine. *Journal of Pharmacology and Experimental Therapeutics, 348*(2), 217–226.

Traficante, D. C., Feibish, G., & Kashani, J. (2017). Methadone-induced Torsades de pointes masquerading as seizures. *Clinical Practice and Cases in Emergency Medicine, 1*(1), 40.

Trafton, J. A., Minkel, J., & Humphreys, K. (2006). Determining effective methadone doses for individual opioid-dependent patients. *PLoS Medicine, 3*(3), e80.

Velez, M. L., Jansson, L. M., Schroeder, J., & Williams, E. (2009). Prenatal methadone exposure and neonatal neurobehavioral functioning. *Pediatric Research, 66*(6), 704.

Wieneke, H., Conrads, H., Wolstein, J., Breuckmann, F., Gastpar, M., Erbel, R., et al. (2009). Levo-α-acetylmethadol (LAAM) induced QTc-prolongation-results from a controlled clinical trial. *European Journal of Medical Research, 14*(1), 7.

Yarborough, B. J. H., Stumbo, S. P., McCarty, D., Mertens, J., Weisner, C., & Green, C. A. (2016). Methadone, buprenorphine and preferences for opioid agonist treatment: A qualitative analysis. *Drug and Alcohol Dependence, 160*, 112–118.

Zaric, G. S., Brennan, A. W., Varenbut, M., & Daiter, J. M. (2012). The cost of providing methadone maintenance treatment in Ontario, Canada. *American Journal of Drug and Alcohol Abuse, 38*(6), 559–566.

Further reading

Kimber, J., Larney, S., Hickman, M., Randall, D., & Degenhardt, L. (2015). Mortality risk of opioid substitution therapy with methadone versus buprenorphine: A retrospective cohort study. *The Lancet Psychiatry, 2*(10), 901–908.

Chapter 14

Motivational interviewing for the treatment of addiction

Eid Abohamza[a], Ahmed A. Moustafa[b]
[a]*Department of Social Sciences, College of Arts and Sciences, Qatar University, Doha, Qatar,*
[b]*Marcs Institute for Brain, Behaviour, and Development and School of Psychology, Western Sydney University, Penrith, NSW, Australia*

Introduction

Below, we will discuss the following topics: group treatment of substance abuse, the motivational enhancement therapy (MET), five principles of motivational interviewing, and practical stages for conducting MET.

Group treatment of substance abuse

Several paradigms exist for the treatment of adolescents who struggle with substance abuse, including individual, group, and family counseling. The kind of treatment is likely to vary depending on the type of facility. For example, Spirito et al. (2017) also found that combined motivational enhancement therapy and family check-up are more effective at treating marijuana abuse than other treatments. Further, Washton (2005) reported that most drug users and abusers need a combination of treatment modalities in order to obtain the maximum benefits from therapy. Still, regardless of the type of facility or treatment program, individuals enroll in treatment with different levels of interest in eliminating substances from their lives. Of particular importance is the individual's stage of readiness for change (Prochaska & DiClemente, 1992), a variable that can directly impact the type of treatment modality, intervention, and outcomes. Aldemir, Berk, and Coşkunol (2018) found that motivational interviewing increases the readiness of patients to quit drugs as well as decreases drug craving.

Group counseling and substance abuse treatment are natural allies. Flores and Georgi (2005) mentioned that "people who abuse substances are often more likely to remain abstinent and committed to recovery when treatment is provided in groups, apparently because of rewarding and therapeutic forces such as affiliation, confrontation, support, gratification, and identification" (p. 2). The ability for group

Cognitive, Clinical, and Neural Aspects of Drug Addiction. https://doi.org/10.1016/B978-0-12-816979-7.00014-5

counseling to connect clients to treatment is an important asset because the greater the amount, quality, and duration of treatment, the better the client's prognosis.

The use of group counseling continues to grow, with time and money savings being the key reasons (Finn, 2002; Morgenstern & Longabaugh, 2000; Rotgers & Nguyen, 2006; Schroer, Fuhrmann, & Jong-Meyer, 2004). Stewart, Christner, and Freeman (2007) also noted that group counseling can facilitate a sense of common purpose, particularly when this view is modeled by the group leader. Group leaders help achieve this goal by encouraging members to participate in defining goals, setting the session's agenda, utilizing different skills in group exercises, offering and receiving feedback and support to others, and developing homework to continue practicing new skills outside of session and the treatment milieu.

The use of group counseling as an intervention for individuals struggling with depression, anxiety, anger management, and addiction (Bieling, Mccabe, & Antony, 2006; Stewart et al., 2007) as well as with youth with substance abuse problems (Flores & Georgi, 2005) has increased over the years. Moreover, Dies (1991) reports other factors such as strong/positive peer and adult role models and the availability of a supportive social system that can help keep them from relapsing. Although there are many theoretical and structural approaches to group counseling and therapy, one method considered particularly useful for group work with adolescent substance abuse clients is the motivational enhancement therapy (MET) (Miller, 1995).

The motivational enhancement therapy (MET)

The motivational enhancement therapy (MET) is a systematic intervention approach for evoking change (Miller, 2008). MET began with a natural recovery approach introduced by Prochaska and colleagues (Prochaska & DiClemente, 1982, 1984; Prochaska, DiClemente, & Norcross, 1998). Prochaska and DiClemente (1992) illustrated a theoretical system that demonstrates how people can change their addictive behaviors whether they are in treatment or not. MET is closely related to motivational psychology and the existential-humanistic tradition (Ingersoll, Wanger, & Gharib, 2002; Miller & Rollnick, 2002). This includes biomedical treatments, with MET helping to facilitate client compliance with medications or abstinence from substance use. The use of motivational techniques also helps clients engage in 12-step programs accept new identities, remain abstinent, and by helping them continue to work on their recovery.

Operationally, group MET offers a systemic format where members move toward understanding and addressing their problems. Several strategies are utilized throughout the process, including confrontation and peer pressure. Cognitive behavioral and social learning models are also employed with motivational techniques to help clients improve and modify their behaviors (Ingersoll et al., 2002; Tomlin & Richardson, 2004). Within each of these models, it is assumed that human beings possess an intrinsic motivation that can lead them to make positive life changes (Feld, Woodside, Kaplan, Olmsted, & Carter, 2001; Miller & Rollnick, 2002).

The systemic framework of MET is increasingly recognized as an effective strategy for substance abuse treatment, particularly in the last decade (Brown et al., 2009; Magill, Apodaca, Barnett, & Monti, 2010; Stotts, DiClemente, & Dolan-Mullen, 2001). Examples of MET's efficacy have been noted with the treatment of hospitalized adult smokers, outpatient and inpatient adolescent smokers, opiate users, and other individuals with addictive behaviors (Miller & Rollnick, 2002; Stotts et al., 2001). One aspect of these positive responses is viewed as subsequent to the personalized and objective feedback delivered in the supportive, nonconfrontational context characterized by the MET approach. As such, these strategies promote change by allowing clients to process and resolve their own ambivalence and move to a point of decision and commitment to change (Miller & Rollnick, 2002). Ultimately, MET is geared toward helping clients reach their own decision to change their addictive behavior by increasing their intrinsic motivation. In effect, "The client, rather than the therapist, presents the argument for change" (Feld et al., 2001, p. 394). In this approach, the clinician makes "no direct attempt to dismantle denial, confront irrational or maladaptive beliefs, or convince or persuade" (Resnicow et al., 2002, p. 444; Resnicow, Baskin, Rahotep, Periasamy, & Rollnick, 2004).

Over the last decade, the effectiveness of MET has been validated and supported as a treatment option for substance users and abusers (Brown & Miller, 1993; Burke, Arkowtiz, & Dunn, 2002; Burke, Arkowtiz, & Menchola, 2003; Miller & Wilbourne, 2002; Noonan & Moyers, 1997; Project MATCH, 1997; Saunders, Wilkinson, & Phillips, 1995; Stephen, Roffman, & Curtrain, 2000). Several studies have addressed the point concerning whether MET or motivational interviewing (MI) could promote better results and treatment in substance abuse patients. For example, Brown and Miller (1993) conducted a study to compare MI with the standard treatment population in a residential alcoholism facility. The study found that patients who received MI participated actively in treatment and showed significantly lower alcohol consumption at a 3-month follow-up interview. The study also found the beneficial effects of MI were mediated by increased participation in treatment strategies. In another study, Saunders et al. (1995) used MI with heroin patients in a methadone maintenance program. They found that those who received MI had lower attrition rates (30%) in comparison to those who received educational session about drug abuse/use (49%). Another important result of their study was that the participants who received MI had longer periods of sobriety prior to relapse than those in the control treatment. Furthermore, the study found that MI is more effective with clients who start treatment with lower initial motivation than with those who were already motivated to recover. In a sample of Iranian patients with methamphetamine addiction, motivational interviewing and MET was found to increase treatment attendance (Alammehrjerdi, Ezard, & Dolan, 2018). In addition to alcohol, methamphetamine, and heroin addiction, motivational interviewing was also found that to be effective treatments for patients with ketamine drug abuse (Siu, Ko, & Mak, 2018). Importantly, in a recent review, it was reported that in comparison to other treatments, motivational interviewing is effective at treating several disorders including marijuana addiction as well as gambling (DiClemente, Corno, Graydon, Wiprovnick, & Knoblach, 2017).

In a recent study, Rasekh et al. (2018) show that drug addicts who completed motivational interviewing training before admission to the hospital were able to complete drug addiction treatment. Satre, Delucchi, Lichtmacher, Sterling, and Weisner (2013) also found that in comparison to a placebo treatment that involve reading about drug and alcohol risk, three sessions of motivational interviewing can reduce excessive drug and alcohol use in patients with depression. Rani, Maheshwari, and Arora (2014) studied the effects of motivational enhancement therapy in drug addiction in India. They divided drug addicts into two groups: one that receives motivational enhancement therapy and another group that receives control treatment that does not include motivational processes. It was found that motivational enhancement therapy enhances recognition of the drug addiction problems, taking steps forward as well as decreases ambivalence. In Project MATCH (1997), a group of clinical researchers conducted a large multi-site study to compare different treatment modalities. These included three groups, each receiving a different treatment: (1) 4 sessions of motivational intervention, (2) 12 sessions of cognitive behavioral substance treatment, and (3) 12 sessions of the 12-step treatment program. Results indicated no significant difference between the distinct types of treatment on the outcome variables, suggesting that the 4 sessions of motivational intervention is as effective as the 12 sessions in either cognitive behavioral treatment or the 12-step treatment. Further, Burke et al. (2002) found that in the treatment of alcohol problems, even very brief motivational intervention (one to four sessions) has a positive and lasting effect on drinking outcomes. Subsequently, in a meta-analysis study of controlled trials of motivational intervention, Burke et al. (2003) described MI as a "promising" treatment paradigm in treating different health problems including drug substance treatment. Specifically, they reported that MI yields better follow-up results than other treatments or no treatment. Specifically, 52% of the patients receiving MI treatments have shown significant improvements at follow-up compared to 37% of clients who had reduced their drinking behaviors by 56%. These results provided further evidence that motivational interventions are effective longer-term treatment paradigms (Burke et al., 2003; Stephen et al., 2000). Further, Witkiewitz, Hartzler, and Donovan (2010) acknowledged that clients with low baseline motivation in outpatient setting and males with low dependence or females in aftercare setting may benefit more from MET techniques than from cognitive behavioral techniques. However, recent studies suggest otherwise. For example, in another study, Riper et al. (2014) have provided a meta-analysis study showing that combined cognitive behavioral therapy and motivational interviewing is better than either alone at treating comorbid depression and drug addiction. A more recent meta-analysis also lends support to the effectiveness of motivational interviewing (Magill et al., 2018). It is also important to note that motivational interviewing has been done in many different ways, including using either relational and directional aspects of motivational interviewing (Morgenstern et al., 2017), which can perhaps lead to different results (Hallgren et al., 2018).

Overall, the MET approach is based on several assumptions regarding the nature of motivation, such as motivation as a key to change, motivation is multidimensional, motivation is a dynamic and fluctuating state, motivation is interactive, motivation can be modified, and no motivation equals to no treatment (Miller, 2008; Tomlin & Richardson, 2004). In doing so, it provides an effective means for helping clients move through the unique phases of change including process of understanding, addressing, and maintaining abstinence from abusive patterns of substance use. However, implementation of MET requires an understanding of its five main principles.

Five principles of motivational interviewing

According to Miller and Rollnick (2002) and Miller (2008), there are five principles underlying MET. These five principals are expressing empathy, developing discrepancies, avoiding argument, agreeing with the client's resistance, and supporting self efficacy.

The first principle involves expressing empathy through reflective listening. According to this view, empathy is a learnable skill (Martino & Moyers, 2008; Miller & Rollnick, 2002), and one in which counselors need to be able to relate to the subtleties of client experiences (Miller, 2008; Tomlin & Richardson, 2004; Walters, Ogle, & Martin, 2002). Empathy has also been shown to have a positive impact on reducing negative drinking patterns. Miller and Rollnick (2002), for instance, demonstrated that counselors who utilize an empathetic counseling techniques show reduced alcohol drinking, in comparison to another group who did not receive empathetic counseling (77% vs. 55% respectively).

Miller (1995) noted that counselors should utilize an empathic style to respect client's feeling and encourage them to speak their minds in a non-judgmental way. Martino and Moyers (2008), Miller (2008), and Westra and Dozois (2008) noted that MET is most productive when counselors establish rapport, and a safe, open atmosphere. Establishing this type of environment is a fundamental component of motivational interviewing. Counselors recognize that every client is unique in his or her feelings and values. Therefore, counselors should unconditionally accept the clients' views and listen without imposing any direction or judgment on their words because this can create barriers between counselors and clients that impair the therapeutic relationship. It has been reported that there are 12 examples of non-empathetic responses: ordering direction and giving the voice of authority, warning or threatening, giving advice or making suggestions, persuading with logic or lecturing, moralizing or telling the clients their duties, judging, criticizing or blaming, agreeing or praising (if the reinforcement implies agreement on whatever the client said), labeling, shaming, imposing counselor" interpretation on the client's statements to find something hidden, reassuring or consoling, questioning, withdrawing or using humor, or changing the subject. Tracking the client's story and counselor's self-awareness can limit these mistakes.

The second principle is working to develop discrepancies between client goals or values and their current behaviors. Walters et al. (2002), Martino and Moyers (2008), and Westra and Dozois (2008) pointed out that helping clients become aware of their discrepancies is a sensitive, strategic, and important stage in the treatment process. The motivation to change can be enhanced when clients recognize the discrepancies between what they are doing and what the ideal behavior is. Discrepancies can be raised by developing the client's awareness of their problems and the consequences that can happen to them, their family, or their communities. Martino and Moyers (2008), Miller (2008), and Westra and Dozois (2008) stated that counselors should work on disparities between the person and the behavior and help the client look at how important personal goals are in relationship to personal well-being or a healthy family life, and to see how these goals are being undermined by their current addiction problems. This requires counselors to carefully track the client's story while remaining cognizant of their own values and support systems. Once clients understand the potential consequences of their current problem, counselors should support and help them until they recognize their problems.

The third principle is to avoid arguments and direct confrontation. When counselors try to convince clients that they have a problem, this can result in resistance. Thus, the client-counselor relationship should be collaborative rather than argumentative. Martino and Moyers (2008) and Wagner and Sanchez (2002) noted that exploring client's values are very important. This exploration can be enhanced by a collaborative relationship between the client and counselor. Counselors use these values to help clients change and define their "ideal self". Focusing on the ideal self can help counselors detect what current behavior clients have that is inconsistent with their ideal self (Ingersoll et al., 2002; Martino & Moyers, 2008; Miller, 2008; Resnicow et al., 2004; Walters et al., 2002).

The fourth principle is to adjust to rather than oppose a client's resistance. Several researchers (Ingersoll et al., 2002; Miller, 2008; Miller & Rollnick, 2002; Resnicow et al., 2004; Westra & Dozois, 2008) mentioned that opposing client resistance does impede treatment. Therefore, when the level of resistance intensifies, counselors should adapt to the resistance and work to rebuild empathy with clients to encourage them to stay involved in treatment. Miller (2008) acknowledged that resistance is a signal to the counselors to change their MET techniques and to listen carefully to what the clients are saying. Miller and Rollnick (2002) reported that there are four types of client's resistance: arguing with counselors, interrupting the counselors as a defensive mechanism, denying the problem, and ignoring or not following the counselors.

The fifth and final principle is to support self-efficacy and optimism. Counselors and clinicians should encourage client self-efficacy. Encouraging clients to have higher self-efficacy will allow reduce their worriness about their problems and their abilities to overcome their addictions. If the counselors do not support client self-efficacy, their defense mechanisms, such as denial or rationalization, may intensify in order to protect themselves from emotional pain. Developing self-efficacy requires encouragement, optimism, and feasibility to accomplish change (Westra & Dozois, 2008). There are techniques that counselors can utilize to prompt client self-efficacy, such as sharing personal experiences if the counselor has experienced similar situations and were able to overcome them (Westra & Dozois, 2008). Additionally, educating clients about self-efficacy and addiction can increase their self-efficacy (Gauggel & Hoop, 2004; Ingersoll et al., 2002; Miller, 2008; Walters et al., 2002).

In addition to knowing and understanding the principles of motivational interviewing, counselors must understand the three primary phases in MET. These phases include building motivation for change, strengthening the commitment to change, and follow-through strategies. Moreover, each stage requires an understanding of the proper therapeutic use of MET and counselor self-awareness in response to clients at different stages of the change (treatment) process. To that end, an understanding of the change process is fundamental to the counselor utilizing the MET format.

Practical stages for conducting MET

Phase one: Building motivation for change

Miller and Rollnick (2002) stated that the overall goal for this phase is building the client's motivation to change. Therefore, this stage will rely on the starting point as provided by clients. In other words, counselors need to "*meet clients where they are*". Miller and Rollnick (2002) focused on the initial screenings to clarify where clients are in the change process. In doing so, counselors must utilize MET in an open format; in other words, they should work to eliminate closed-ended questions that would minimize the potential for clients to elaborate on their story. Additionally, this type of question-answer trap will make the client passive and the counselor more active and will not give the counselor the opportunity to explore the client's motivation. Therefore, Miller and Rollnick (2002) recommended counselors to avoid this trap by using a questionnaire administered ahead of time, by asking open-ended questions, and by using reflective listening rather than responding with questions. It is also important to remember that clients often enter treatment in a state of ambivalence and equivocation between wanting treatment and not wanting treatment (Ingersoll et al., 2002; Miller, 2008; Miller & Rollnick, 2002; Tomlin & Richardson, 2004; Treasure & Schmidt, 2008; Velasquez, Maurer, Crouch, & DiClemente, 2001). As a result, Miller and Rollnick (2002) stated that counselors should be more motivational than confrontational.

Counselors should also avoid beginning their treatment with telling the clients that they have a serious problem with alcoholism or drug abuse. Such responses are likely to elicit client responses such as "*The problem is not really that bad*," and "*I don't really need to change that much.*" (Miller & Rollnick, 2002, p. 56). Counselors making these erroneous opening statements may find clients to be particularly reluctant to engage in treatment. Counselors should also avoid the "expert" trap, suggested by Miller and Rollnick (2002) as a means of giving clients answers or solutions for their problems. Here, counselors turn into problem-solvers more than motivational counselors. In cases where counselors give clients solutions for their problems, clients may become passive, an inconsistent role for the client within the motivational therapy paradigm. According to motivational therapy, clients ultimately know themselves and their own problems, values, goals, abilities, and competencies better than the counselor. Therefore, MET encourages counselors to be the facilitator more than a problem solver.

Labeling trap

Miller and Rollnick (2002) and Zuckoff, Swartz, and Grote (2008) viewed a labeling trap as occurring when counselors try to convince clients that they are an "addict" or "alcoholic". They go on to say that "because such labels often carry a certain stigma in the public mind, it is not surprising that people with reasonable self-esteem resist them" (p. 60–61). Miller and Rollnick (2002) and Zuckoff et al. (2008) recommended not using labeling and avoiding bringing up issues for clients the counselor feels needs to be addressed. When clients do verbalize items related to substance abuse, the best path is to use reflection and reframing techniques. If clients express concern about being stigmatized due to substance use or abuse, counselors should frame a response in terms of identifying and working on the client's concerns and placing the client's energy where it can be used more effectively. However, there are some counselors who believe labeling can be an important step to helping clients recover (Ingersoll et al., 2002).

Premature focus trap

Several studies (Miller & Rollnick, 2002; Tomlin & Richardson, 2004; Westra & Dozois, 2008) noted that premature focus is one of the traps that many counselors encounter. Often, counselors who lack self-awareness or are utilizing more directive approaches may focus on a topic they see as a priority. Still, clients often have broader concerns than counselors, particularly those who over- or under-identify with a client's issues or concerns. Motivational counselors are more interested in listening to situations that the clients claim to be important to them. To that end, Miller and Rollnick (2002) recommended counselors to avoid struggling with clients about the initial topic for discussion. In short, counselors should start with the client's concerns. Within their story, the counselor will be better able to discern their level of motivation to change.

Blaming trap

Individuals with substance abuse or dependence often blame other issues or persons for their problem. Miller and Rollnick (2002), Tomlin and Richardson (2004), and Westra and Dozois (2008) all noted the importance of recognizing and avoiding patterns of discussion that are incongruent with recovery or show an external locus of control. In other words, engaging in conversations where clients are seeking to discover which external factors are to blame for their substance abuse is a not productive given that clients must accept their own responsibility to change their substance abuse or dependence patterns. One method of refocusing the client is by using a "no-fault" policy where counselors use a comment such as, "I'm not interested in looking for who's responsible, but rather what's troubling you, and what you might be able to do about it" (p. 63). Another intervention technique commonly used by MET counselors is defined by the acronym "FRAMES", which stands for Feedback, Rolling with resistance, Advice about change, Menu of options for change, Empathy, and Supporting self-efficacy (Miller & Rollnick, 2002).

Feedback

Counselors in most settings will have objective assessments such as court records, urine screens, blood tests, or lab results showing elevated liver enzymes to assist in determining a substance related diagnosis. Using these types of assessments in addition to the counselor's own assessments can prove more powerful than using psychological assessments or opinions alone. Physician reports, however, can be very useful in helping compare a client's scores or ratings with established diagnostic standards or empirically based instruments used for identifying substance use (Arkowitz & Miller, 2008; Conrad, Wong, Jones, & Stitzer, 2004; Miller, 2008; Tomlin & Richardson, 2004; Treasure & Schmidt, 2008; Walters et al., 2002; Westra & Dozois, 2008; Zuckoff et al., 2008). These physiological assessments may include tests related to substance abuse, physical health, risk factors for HIV, family history, how drug abuse may impact the familial homeostasis, and various medical exams (Ingersoll et al., 2002; Miller, 2008; Miller & Rollnick, 2002).

Responsibility

Miller (2008) mentioned that MET gives clients the right to be active, not passive. It is the client's responsibility to choose the direction of treatment. Clients motivated by the perceived personal control. Therefore, MET commonly emphasizes the patient's responsibility and choice for stop using drugs or alcohol. NIAAA (National Institute on Alcohol Abuse and Alcoholism, 1999) gave an example, in which patients should know from the doctor that "No one can make you change or make you decide to change. What you do about your drinking is up to you."

Advice

Miller and Rollnick (2002) and Westra and Dozois (2008) confirmed the importance of giving clients gentle and simple advice. However, advising people and making resolute, declarative statements to clients are different techniques, with the latter being avoided in most instances. Miller (2008) noted that advice within MET can be directive or educational, such as making suggestions or by explaining relevant information. Overall, any advice given should be simple, match the client's mental and developmental level, and be relevant to the situation.

Menu of options for change

Another key element of the FRAMES model is offering clients who are ready to change a menu of choices. Some research noted that these menus should contain several treatment options (Arkowitz & Miller, 2008; Ingersoll et al., 2002). Examples may include self-help groups, numerous professional references, low intensity treatment, or intensive treatment services such as residential or inpatient treatment facilities. Counselors should not impose any choice on clients. Rather, counselors should work collaboratively with clients to select the types of service they may find beneficial.

Empathy

Arkowitz and Miller (2008), Ingersoll et al. (2002), and Westra and Dozois (2008) noted that empathy means the counselors are able to see the world through the client's eyes. Counselors will be able to think about client's problems in similar ways as their clients. Expressing empathy to clients helps facilitate change, which in turn often encourages clients to become more engaged and willing to share their experiences. This will allow the counselor to assess when and where clients need support, and what difficulties may need to be focused on as a part of the change planning process.

Supporting self-efficacy

Perceived self-efficacy involves "beliefs in one's capabilities to organize and execute a course of action required producing a given attainment" (Bandura, 1977, p. 3). Self-efficacy can be impacted by past experiences of success and beliefs about particular tasks. Individuals move toward change only when they perceive the chance of success (Arkowitz & Miller, 2008; Ingersoll et al., 2002). Further, these abilities may be physiological, behavioral, emotional, or intellectual (Bandura, 1977). For example, individuals who perceive they have control or power over their own affect and cognitions may feel more motivated to act. In doing so, they may feel more efficacious in managing or executing certain tasks. In short, it has been emphasized that clients will not be motivated or ready to change unless certain processes have occurred which are precontemplation, contemplation, preparation, action, and maintenance stage.

Phase two: Strengthening commitment to change

According to the MET approach, once the client shows motivation and desire to change, counselors should work to strengthen the client's commitment to change. Miller (2008) highlighted several indicators that clients are becoming more motivated to change. These indicators include less client resistance or ambivalence, fewer questions and inquires, more verbal and/or non-verbal expressions of peace and maturation, and words and behaviors demonstrating a willingness to change. Several techniques are valuable and beneficial when these indicators are present.

Asking key questions

Miller and Rollnick (2002) and Zuckoff et al. (2008) noted the importance for counselors to provide clear summaries to clients in order to facilitate their understanding of all the reasons they have given for changing their behavior. At the end of the summary, counselors should acknowledge points of ambivalence and ask key question including the following: What plan do you have? Or what do you believe you will do?

Discussing a plan

Miller (2008), Gauggel and Hoop (2004) and Zuckoff et al. (2008) noted that the most important point in the treatment plan is when counselors switch from focusing on the reasons for change to the planning for change. This transition can happen when clients show a desire to change and ask about the next step in the process. In such cases, the counselor's goal is to help provide options that can help them establish a solid plan for change. Another part of the counselor's role is to provide useful insights regarding interventions that could help clients temper and eventually abstain entirely from problematic substance use patterns. Importantly, it is not the counselor's task at this point to advise a plan for change and gives it to the client; rather, the ultimate goal is for clients to generate a plan that will help them change their drug use behaviors (Miller, 2008). Miller (2008) suggests that counselors should continue to use reflective listening, paraphrasing, summarizing techniques, and persist in promoting and generating self-motivational statements and plans for change that can be integrated into each client's individualized treatment plan.

Communicating free choice

Change is the client's responsibility, and freedom of choice is a vital principal in MET. Counselors should stress the principle of freedom to choose in their conversations with clients (Miller, 2008; Miller & Rollnick, 2002).

Consequences of action and inaction

Ingersoll et al. (2002) noted that making clients weigh the pros and cons for making changes in their lives is crucial. One strategy is to ask clients to write down the negative and positive consequences of their decisions. Miller and Rollnick (2002) and Arkowitz and Miller (2008) also promoted the idea of having discussions with clients about their fears related to change. This discussion will help clients and counselors develop a more complete picture of the client's current position and both current and future concerns related to changing their behaviors. Techniques considered especially useful in this regard are reflective listening, reframing, paraphrasing, and summarizing.

Information and advice

Clients often direct questions and concerns about the change process at counselors (Miller, 2008). Many times, these questions may include inquiries as to the state of their psychological wellness or substance abuse issues. Counselors should not shy away from these discussions and provide information they know to be accurate. However, it is often more helpful if the counselor reflects the questions back to the client, as this allows the client to think about and process their own questions and concerns (Miller, 2008).

Rolling with resistance

Ingersoll et al. (2002), Resnicow et al. (2004) and Tomlin and Richardson (2004) noted that client resistance can arise under numerous circumstances. Counselors can manage resistance by expressing empathy and being less argumentative. Examples of resistance may include, but are not limited to: interrupting, ignoring, arguing, denying, and or derailing the conversation. From the counselor's standpoint, it is important to remember that resistance manifesting within a therapeutic setting also provides valuable information about issues such as the strength of the alliance, level of comfort or safety, or willingness to change. In other words, resistance is important information and should be conceptualized within the larger notion of the change process.

The change plan worksheet

This worksheet plan will help the group members determine their progress through the stages of change (Arkowitz & Miller, 2008; Ingersoll et al., 2002; Miller, 2008; Miller & Rollnick, 2002; Tomlin & Richardson, 2004). Worksheets can also help group members design more concrete plans for change, including changes the client wants to make, key reasons for making changes, procedures for making changes, connecting with support systems, expected benefits of change, and potential obstacles to change (Arkowitz & Miller, 2008; Miller, 2008; Miller & Rollnick, 2002; Tomlin & Richardson, 2004). Another element of the worksheet, called recapitulating, emphasizes the use of the note section to summarize for clients what has occurred during treatment (Miller, 2008; Tomlin & Richardson, 2004). Counselors should focus on reminding clients of self-motivational words and the positive and negative consequences of changing.

Asking for commitment

Arkowitz and Miller (2008) and Miller (2008) acknowledged the likelihood that clients might have concerns regarding the progress they are experiencing as they move toward true change. Counselors should be ready to discuss these changes with clients, as well as how their worksheet aids this process. One way to reduce this issue is for counselors to discuss all aspects of the worksheet, including potential barriers to progress and change (Miller & Rollnick, 2002; Tomlin & Richardson, 2004). Further, they should discuss the additional help that clients might need or want in the future from other treatment agencies. It is suggested that when counselors feel clients are ready to sign the commitment to change plan, counselors should ask clients to sign the original plan and place a copy in their files.

Phase three: Follow-through strategies

In the first phase, clients establish a strong base of motivation for change. In the second phase, clients have a serious commitment to change. The third phase involves three elements: reviewing progress, renewing motivation, and redoing commitment.

Reviewing progress

Miller (2008) and Tomlin and Richardson (2004) suggested that the role of counselors in this phase includes review and follow-up with clients. It is also important to discuss the commitment and plans that were previously made. Moreover, counselors should look at what improvement clients have made toward future change. Throughout this process, counselors should continue to work on renewing the client's motivation and their commitment to change. Moreover, when and if a client requests additional treatment, counselors should work collaboratively with them to ensure all efforts are made to attain positive treatment outcomes.

As mentioned previously, MET can be used in individual, group, and systemic (i.e., family counseling) frameworks. However, counselors are encouraged to recognize the utility of MET in all types of therapeutic settings as well as the unique curriculum or treatment plan considerations subsequent to the implementation of MET.

Stages of change

Why do some people change their behaviors and others do not? Miller and Rollnick (2002) argue that the answer lies with the individual and note that some people quit smoking or recover from addiction problems without help from professionals. People go through certain stages of change, regardless of whether they employ the services of a professional counselor. However, the use of empirically tested interventions, such as MET, has been shown to be effective in facilitating the motivation to change negative behaviors. As a result, the use of MET in a group settings may be an important element for enhancing the motivation to change substance use behaviors among adolescents. Changes occur in five stages (Prochaska & DiClemente, 1984), with each involving different time frames, levels of motivation, and treatment strategies. According to DiClemente (2006), "*These stages depict the motivational and dynamic fluctuations of the processes of change over time*" (p. 26).

Precontemplation stage

The first stage, precontemplation involves individuals who are not aware of their problem and, as such, are unwilling to change (Ingersoll et al., 2002; Miller, 1995; Miller & Rollnick, 2002; Schroer et al., 2004; Tomlin & Richardson, 2004; Treasure et al., 1999; Velasquez et al., 2001). For people in this stage, the benefits from their behavior outweigh any benefit from change. From a therapeutic standpoint, then, "*the key change process for people in this stage include consciousness raising, dramatic relief, self-re-evaluation, environmental re-evaluation, and tipping the decisional balance*" (Velasquez et al., 2001, p. 11). Overall, the goal of the precontemplation stage is to help people engage in the counseling process and begin to consider changing their behavior as well as considering the patterns and potential effects of their substance use and abuse (DiClemente, 2006; Ingersoll et al., 2002; Tomlin & Richardson, 2004).

Contemplation stage

In the contemplation stage, patients are able to admit they have a problem and start to think seriously about solving it (DiClemente, 2006; Ingersoll et al., 2002; Schroer et al., 2004; Tomlin & Richardson, 2004). Further, people continue to struggle with ambivalence about change and often have trouble identifying clear-cut reasons for their substance abuse or potential resolutions to these patterns. Despite gathering information about potential solutions, clients may spend years in this stage struggling with their mixed feelings (Miller, 1995; Miller & Rollnick, 2002; Tomlin & Richardson, 2004; Velasquez et al., 2001). Overall, the goal of the contemplation stage is to help people resolve their ambivalence and, in doing so, begin the true process of making decisions that will positively impact their lifestyle and prepare them for making permanent changes to their substance use behaviors (Ingersoll et al., 2002).

Preparation stage

The preparation stage has been labeled the "decision-making" stage (DiClemente, 2006) and is marked by the development of a plan of action and commitment to follow through with the behavior change (DiClemente, 2006; Schroer et al., 2004; Tomlin & Richardson, 2004). In essence, the task of the preparation stage is for an individual to develop competence and self-efficacy to move to action. Typically, individuals in this stage have multiple times attempted to change but eventually failed. Therefore, they are aware of their successes and failures from previous attempts (Prochaska et al., 1998). At this point, clients recognize that the benefits of change outweigh the cost of the behavior. Miller (1995) and Miller and Rollnick (2002) suggested that many counselors can easily make the mistake of thinking that an individual who is contemplating change is ready to make a commitment to change. The best example of this is smokers who express the desire to quit smoking. When counselors confront them to cease their smoking, the client can find several reasons to justify that this is not an appropriate time to quit smoking. Therefore, it is important for counselors to remain supportive, yet maintain an understanding of the difficulties inherent in making the full commitment to change. One way to facilitate this commitment is for clients to design a specific plan of action that will help them accomplish their desire to change their behavior (Ingersoll et al., 2002; Miller, 1995; Miller & Rollnick, 2002; Tomlin & Richardson, 2004; Velasquez et al., 2001).

Action stage

The action stage involves the implementation of the plan. It is the busiest stage, and it requires the highest level of commitment, time, energy, and action. The ultimate goal in this stage is to help clients make major and efficient changes in their lives. Therefore, counselors should work on the motivational strategies that will be most effective for each individual client or group of clients (Miller, 2008; Miller & Rollnick, 2002; Schroer et al., 2004; Tomlin & Richardson, 2004). Velasquez et al. (2001) suggested that in this stage "*there are several key processes that help individuals to move forward and implement their treatment plan such as self-efficacy, self-liberation, stimulus control, counter-conditioning, reinforcement management, and helping relationship*" (p. 12). The time period for an individual to be in the action stage is from 1 day to 6 months of changing and sustaining the behavior change (Prochaska & DiClemente, 1992), with the goal being the beginning of true change, the ability to cope with difficulties in the change process, and the presence of social support for new ways of being (Ingersoll et al., 2002; Tomlin & Richardson, 2004).

Maintenance stage

Maintenance is the last step in the process of change. In this stage, individuals are working on preventing relapse and integrating the behavior change into different life areas and can last from as little as 6 months to as long as a lifetime (DiClemente, 2006; Prochaska & DiClemente, 1992). It should be noted, however, that relapse is common in this stage because clients are often exposed to the triggers and temptations that might move them back to the first stage.

In sum, understanding the stages of change can help clients and counselors work together and select the appropriate motivational strategies for clients (Miller, 1995, 2008; Miller & Rollnick, 2002). Individuals may go through the stages several times or become stuck in one stage for a long period of time (Prochaska & DiClemente, 1992). Some individuals may experience problems that send them back to an earlier stage; identified as "relapse." Prochaska, Norcross, and DiClemente (2006) described the process of going back through an earlier stage as 'recycling' rather than relapsing. Overall, it is imperative for the counselor to understand the stages of change and the individual nature of the change process for each client. In the group setting, counselors need to be especially cautious to consider the likelihood that clients are at different stages and use the group setting as a means for providing support and motivation to all clients regardless of their current level of change.

The group counseling format

Group counseling is recognized as an effective treatment modality for working with people struggling with substance abuse issues (Pita, 1992; Royce & Scratchley, 1996). Among the benefits of group counseling are time and cost efficiency as well as a sense of belonging and acceptance for clients. Several studies have explored the group MET format. Foote et al. (1999) examined the effectiveness of four sessions of group motivational intervention utilizing the FRAMES technique. Their results suggest that key motivational processes are affected by group motivational treatment, that participants exposed to the group motivational intervention perceived significantly greater support for personal autonomy from the group leader. Participants also indicated a more realistic assessment of the benefits of abstinence. Importantly, clients who receive group motivational intervention have better attendance than treatment as usual group without motivational intervention.

Easton, Swan, and Sinha (2000) examined the effectiveness of group MET for clients who struggle with anger management and substance abuse problems. Results showed that clients exposed to motivational sessions are more likely to consider changing their behaviors than those who receive standard treatments. Lincourt, Kuettel, and Bombardier (2002) found that substance abuse clients who participate in an MET group as a preceding session before starting the regular treatment showed better attendance and greater rates of treatment completion than those who received treatment without MET.

The group approach is also an effective means for providing therapy to adolescents. Several types of groups exist that can easily integrate the MET format, including psycho-educational and pure motivational interview psychotherapy groups. The main function for the psycho-educational group is to enhance the motivation of the members. This goal can be reached through the educational content or the interpersonal process. In some cases, peer-driven motivational groups can be developed. In this structure, a group member can lead others in exploring topical issues. However, the pure MET psychotherapy group should be provided by professional counselors who adhere closely with the philosophy of the motivational interview (Ingersoll et al., 2002; Schroer et al., 2004; Velasquez et al., 2001; Walters et al., 2002).

Suggested format for motivational groups

Ingersoll et al. (2002) and Webb, Scudder, Kaminer, and Kadden (2003) suggested a design for the structure group MET. First, group MET requires that counselors be aware of and balance the needs of several group members simultaneously as opposed to a single individual's needs. Counselors should pay attention to the group process during the session, such as the stage of the group development, group interactions, and whether one person dominates the session (Ingersoll et al., 2002; Schroer et al., 2004; Tomlin & Richardson, 2004; Walters et al., 2002). Several studies (Ingersoll et al., 2002; Miller, 2008; Walters et al., 2002) further noted that counselors should invite all members into the discussion by reframing what has been shared and relate it to the experiences of all group members to the extent possible.

Groups can be categorized into open and closed groups, with each having their own advantages and disadvantages. Open groups can be accommodating to individuals and bring a continual influx of new ideas and energies from the constant rotation of new members. However, the open group can also make members more sensitive to confidentiality problems, therefore making them less willing to self-disclose. Conversely, closed groups utilize a framework where members remain stable from the beginning to the end. When managed well, closed groups allow members develop strong alliances. However, inconsistent attendance among group members in this format could also have negative implications for the alliance and, ultimately, group outcome. Thus, it is important to assess the commitment of the group members before joining the group.

Another important element is group size. Corey (2008) suggested the optimal group should be between 6 and 12 members, with the group leader having to consider the ideal group size so all members have the opportunity to be engaged (Ingersoll et al., 2002; Walters et al., 2002). When co-leaders are employed, they are strongly encouraged to collaborate on the structure, goals, and leadership they expect of the group. This can help offset incongruent goals and therapeutic styles that could lead to environments where clients feel unsafe and, therefore, do not engage in the group process.

Another important group factor is the duration of the group, generally viewed as either time-limited group or open-ended groups. The time-limited group involves meetings that last for a defined number of sessions, while open-ended groups are ongoing and do not have any set termination date. In cases where mandated clients are among those in the group, a time-limited group may be suitable in order to help them complete designated programming or therapeutic goals. An open-ended group can be appropriate for clients who are waiting for the next phase of treatment, such as residential placement.

Group motivational enhancement therapy

Group MET should be provided by MET trained counselors only. It is important to have a consent form for every member before the group starts. Counselors should come to sessions well prepared and have a discussion topic in mind. For open groups, it is important to track when members entered the group to avoid duplicating the activities.

Providing structure for group MET

When planning to use structured exercises, counselors should provide materials for every member. Chairs should be arranged in a circle to facilitate group interaction. Ingersoll et al. (2002), Webb et al. (2003), and Fields (2004) suggested that group MET leader(s) should invite and escort the group members into the group room. Once every leader and member has been introduced (Fields, 2004; Ingersoll et al., 2002; Webb et al., 2003), the group leader(s) should explain the process, procedures, and goals and expectations of the group in tandem with group members. Additionally, Ingersoll et al. (2002) suggested that a topic should be selected for each session. Examples of goals common to substance abuse MET groups with substance abusing clients include confidentiality, safety, no verbal abuse or threatening, being prompt and sober for each session, and have respect for the abilities and individual differences of group members.

After the rules have been discussed and agreed upon, the group leader can show them the dates, times and topics of remaining groups. The leader (s) should then provide each member with their name, contact number, and the emergency contact number. Ingersoll et al. (2002) and Webb et al. (2003) illustrated that after the group leader ensures the safety of the group, the counselor should check with the group members about issues of concern. It is important to get a sense of the issues that are important to clients before the group facilitator delves into the group agenda. The group facilitator should carefully observe the verbal and nonverbal language of the clients. This will help the group leader determine if the planned topics fit well with the group, or whether it is necessary to move to another topic area or session that better meet the group's needs.

Ingersoll et al. (2002) and Fields (2004) suggested that once the group has settled, the group leader(s) should introduce content. For example, if the group leader(s) was planning on doing an "awareness window" it would be essential to use client words in addition to the group leader in presenting the significance of the theme. Once the content is introduced, group leaders should work to ensure that all group members are included in the exercises. Counselors should bear in mind that some members are sensitive and resistant to the tasks, so patience should be shown by the counselor. Once the group members complete the exercise, counselors should facilitate group discussions, emphasizing that there is no right or wrong answer. This sequence provides a good opportunity for counselors to understand the client's perspectives and observe the interpersonal skills and styles of group members.

Conflict and confrontation

Ingersoll et al. (2002) reminded counselors to distinguish between disruptive and challenging behaviors. Tomlin and Richardson (2004) further noted the potential for group members to confront a peer in a negative manner. Such interchanges may have positive or negative implications, so counselors should remain present and intercede when conflict overwhelms the ability of the group members to remain present and productive with each other in terms of communication or behavior. Once the conflict is tempered, leaders can use the opportunity to talk with group members about their reactions and how to handle the conflict and/or other difficult experiences in the future.

MET group curriculum

The MET curriculum is well suited for the group setting for several reasons. First, the group format is widely used within substance abuse and dependence treatment programs. To that end, several studies have shown MET as a particularly effective group intervention (Easton et al., 2000; Foote et al., 1999; Lincourt et al., 2002; Pita, 1992; Royce & Scratchley, 1996). This is particularly important given that group therapy has been found to be more effective than individual psychotherapy for adolescents (Tillitski, 1990). Further, group learning creates a climate in which adolescents, through conversation with peer group members, acquire insight into their own thoughts, feelings and behaviors. As in Tillitski (1990), Glodich and Allen (1998) found that adolescents benefit from shared experiences and insights of other adolescents, that youths are more receptive to feedback from peers than from adults, and that group interaction provides healthy modeling even when adolescents choose only to listen (Glodich & Allen, 1998). Moreover, Muuss (1996) made the claim that social feedback can be used to provide adolescents with first-hand information on how others perceive them in real time interactions. As knowledge of self, reflection (insight into thoughts, feelings, choice of behaviors) and self-motivation are essential features of a successful MET process, it follows logically that the use of MET with adolescent substance abusing clients is especially sound.

Treatment as usual versus MET

In understanding treatment for adolescents at in-patient and residential treatment programs, it is important to first assess which treatment options are effective. O'Brien and Perfas (2005) noted that the main goals for residential programs include using effective methods to shape and manage behaviors, maintaining psychological change and spiritual healing through specific program and activities, and maintaining self-competence and self-reliance through daily activities on the unit. However, the most common approach at residential substance abuse facilities are modalities characterized by the use of a multidisciplinary team format. This framework generally involves programming that offers clients a clear structure, with clearly defined member roles within the unit, while administrators set the unit norms, codes of conduct, and the value systems that guide the facility's philosophy. In many cases, treatment curriculums teach clients the elements, processes, and methods of the unit as well as the maintenance of sobriety, social skills, health, and safe lifestyle (O'Brien & Perfas, 2005).Whatever model of treatment is used in the residential facilities, there are several common treatment programs used in the addiction treatment. These include substance abuse counseling, relapse prevention, cognitive behavioral therapy, 12-step program and the matrix model (Rawson & McCann, 2011). The National Survey of Substance Abuse treatment Services (N-SSATS) identified the matrix model as a framework used by 63% of substance abuse facilities (N-SSATS, 2009, 2010). Specific interventions are also noted in the N-SSATS (2010) including several clinical approaches utilized by substance abuse facilities. More specifically, in the N-SSATS (2010), the majority of substance abuse treatment facilities used substance abuse counseling (96%) and relapse prevention (87%). Other common interventions in treatment are the 12-step based treatment paradigms such as Alcoholics Anonymous (Rawson & McCann, 2011; Sharma & Branscum, 2010) and Narcotic Anonymous (Galanter, Dermatis, Mansky, McIntyre, & Perez-Fuentes, 2007; Rawson & McCann, 2011). According to N-SSATS (2010), 56% of these programs utilize motivational interviewing, over 39% use anger management, 35% use brief intervention techniques; over 25% employ contingency management/motivational incentives, and 21% use trauma-related counseling. In addition, 85% never used or rarely used community reinforcement vouchers and rational emotive behavioral therapy is used by 51% of substance abuse facilities. Still other approaches used in treatment facilities include yoga therapy, group counseling (Finn, 2002; Flores & Georgi, 2005), individual counseling (Rawson & McCann, 2011; Rounsaville, Carroll, & Back, 2005), family counseling, psychoeducation, creative art counseling, spirituality, nutritional and exercise therapy (Rawson & McCann, 2011).

References

Alammehrjerdi, Z., Ezard, N., & Dolan, K. (2018). Methamphetamine dependence in methadone treatment services in Iran: The first literature review of a new health concern. *Asian Journal of Psychiatry*, *31*, 49–55. https://doi.org/10.1016/j.ajp.2018.01.001.

Aldemir, E., Berk, G., & Coşkunol, H. (2018). The effects of the addiction programme of probation on treatment motivation, abstinence and quality of life: A comparative study with motivational interviewing and individual intervention. *Noro Psikiyatri Arsivi*, *55*(3), 261–270. https://doi.org/10.5152/npa.2017.19440.

Arkowitz, H., & Miller, W. R. (2008). Learning, applying, and extending motivational interviewing. In H. Arkowitz, H. A. Westra, W. R. Miller, & S. Rollnick (Eds.), *Motivational interviewing in the treatment of psychological problems* (pp. 1–25). New York, NY: The Guilford Press.

Bandura, A. (1977). *Social learning theory*. Oxford, England: Prentice-Hall.

Bieling, P. J., Mccabe, R. E., & Antony, M. M. (2006). *Cognitive-behavioral therapy in groups*. New York, NY: Guilford Press.

Brown, J. M., & Miller, W. R. (1993). Impact of motivational interviewing on participation and outcome in residential alcoholism treatment. *Psychology of Addictive Behaviors*, *7*(4), 211–218. https://doi.org/10.1037/0893-164X.7.4.211.

Brown, R. A., Strong, D. R., Abrantes, D. R., Myers, M. G., Ramsey, S. E., Christopher, W., et al. (2009). Effects on substance use outcomes in adolescents receiving motivational interviewing for smoking cessation during psychiatric hospitalization. *Addictive Behaviors*, *34*(10), 887–889. https://doi.org/10.1016/j.addbeh.2009.03.003.

Burke, L. B., Arkowtiz, H., & Dunn, C. (2002). The efficacy of motivational interviewing and its adaptations: What we know so far. In W. R. Miller & S. Rollnick (Eds.), *Motivational interviewing: Preparing people for change*. (2nd ed.)(pp. 217–250). New York, NY: Guilford.

Burke, L. B., Arkowtiz, H., & Menchola, M. (2003). The efficacy of motivational interviewing: A meta-analysis of controlled clinical trials. *Journal of Consulting and Clinical Psychology*, *71*(5), 843–861. https://doi.org/10.1037/0022-006X.71.5.843.

Conrad, J., Wong, C. J., Jones, H. E., & Stitzer, M. L. (2004). Community reinforcement approach and contingency management interventions for substance abuse. In W. M. Cox & E. Klinger (Eds.), *Handbook of motivational counseling: Concepts, approaches, and assessment* (pp. 421–438). New Jersey, NJ: John Wiley & Sons Inc.

Corey, G. (2008). *Theory and practice of group counseling* (7th ed.). Belmont, CA: Thomson, Brooks, Cole.

DiClemente, C. (2006). *Addiction and change: How addictions develop and addicted people recover*. New York, NY: The Guilford Press.

DiClemente, C. C., Corno, C. M., Graydon, M. M., Wiprovnick, A. E., & Knoblach, D. J. (2017). Motivational interviewing, enhancement, and brief interventions over the last decade: A review of reviews of efficacy and effectiveness. *Psychology of Addictive Behaviors*, *31*(8), 862–887. https://doi.org/10.1037/adb0000318.

Dies, K. R. (1991). A model of adolescent group psychotherapy. *Journal of Child & Adolescent Group Therapy*, *1*(1), 59–70.

Easton, C., Swan, S., & Sinha, R. (2000). Motivation to change substance use among offenders of domestic violence. *Journal of Substance Abuse Treatment*, *19*(1), 1–5.

Feld, R., Woodside, D. B., Kaplan, A. S., Olmsted, M. P., & Carter, J. C. (2001). Pretreatment motivational enhancement therapy for eating disorders: A pilot study. *International Journal of Eating Disorders*, *29*(4), 393–400.

Fields, A. (2004). *Motivational enhancement therapy for problem & pathological gamblers: A five session curriculum-based group intervention*. Washington, DC: Hollifield Associates.

Finn, A. (2002). Group counseling for people with addictions. In D. Capuzzi & D. R. Gross (Eds.), *Introduction to group counseling*. (3rd ed.)(pp. 351–376). Denver, CO: AFU.

Flores, P. J., & Georgi, J. M. (2005). *Substance abuse treatment: Group therapy, a treatment improvement protocol TIP 41*. Rockville, MD: U.S Department of Health and Human Services.

Foote, J., DeLuca, A., Magura, S., Warner, A., Grand, A., Rosenblum, A., et al. (1999). A group motivational treatment for chemical dependency. *Journal of Substance Abuse Treatment*, *17*(3), 181–192. https://doi.org/10.1016/S0740-5472(99)00003-3.

Galanter, M., Dermatis, H., Mansky, P., McIntyre, J., & Perez-Fuentes, G. (2007). Substance-abusing physicians: Mentoring and twelve-step-based treatment. *The American Journal on Addictions*, *16*, 117–123. https://doi.org/10.1080/10550490601184456.

Gauggel, S., & Hoop, M. (2004). Goal-setting as a motivational technique for neurorehabilitation. In W. M. Cox & E. Klinger (Eds.), *Handbook of motivational counseling: Concepts, approaches, and assessment* (pp. 440–455). New Jersey, NJ: John Wiley & Sons Inc..

Glodich, A. M., & Allen, J. G. (1998). Adolescents exposed to violence and abuse: A review of the group therapy literature. *Journal of Child and Adolescent Group Therapy*, *8*(3), 135–154. https://doi.org/10.1023/A:1022988218910.

Hallgren, K. A., Dembe, A., Pace, B. T., Imel, Z. E., Lee, C. M., & Atkins, D. C. (2018). Variability in motivational interviewing adherence across sessions, providers, sites, and research contexts. *Journal of Substance Abuse Treatment*, *84*, 30–41. https://doi.org/10.1016/j.jsat.2017.10.011.

Ingersoll, K. S., Wanger, C. C., & Gharib, S. (2002). *Motivational groups for community substance abuse programs* (2nd ed.). Mid-Atlantic ATTC.

Lincourt, P., Kuettel, T. J., & Bombardier, C. H. (2002). Motivational interviewing in a group setting with mandated clients: A pilot study. *Addictive Behaviors*, *27*(3), 381–391. https://doi.org/10.1016/S0306-4603(01)00179-4.

Magill, M., Apodaca, T. R., B^3, B., Gaume, J., Hoadley, A., Gordon, R. E. F., et al. (2018). A meta-analysis of motivational interviewing process: Technical, relational, and conditional process models of change. *Journal of Consulting and Clinical Psychology*, *86*(2), 140–157. https://doi.org/10.1037/ccp0000250.

Magill, M., Apodaca, T. R., Barnett, P. B., & Monti, P. M. (2010). The route to change: Within-session predictors of change plan completion in a motivational interview. *Journal of Substance Abuse Treatment*, *38*(3), 299–305. https://doi.org/10.1016/j.jsat.2009.12.001.

Martino, S., & Moyers, T. B. (2008). Motivational interviewing with dually diagnosed patients. In H. Arkowitz, H. A. Westra, W. R. Miller, & S. Rollnick (Eds.), *Motivational interviewing in the treatment of psychological problems* (pp. 277–303). New York, NY: The Guilford Press.

Miller, W. R. (1995). *Motivational enhancement therapy with drug abuse* Retrieved fromhttp://www.motivationalinterview.org/clinical/METDrugAbuse.PDF.

Miller, W. R. (2008). SAMSHA center for substance abuse treatment, enhancing motivation for change in substance abuse treatment. In *Treatment Improvement Protocol (TIP) Series 35*.

Miller, W. R., & Rollnick, S. (2002). *Motivational interviewing: Preparing people for change* (2nd ed.). New York, NY: The Gilford Press.

Miller, W. R., & Wilbourne, P. L. (2002). Mesa Grande: A methodological analysis of clinical trials of treatments for alcohol use disorders. *Addiction*, *97*(3), 265–277. https://doi.org/10.1046/j.1360-0443.2002.00019.x.

Morgenstern, J., Kuerbis, A., Houser, J., Levak, S., Amrhein, P., Shao, S., et al. (2017). Dismantling motivational interviewing: Effects on initiation of behavior change among problem drinkers seeking treatment. *Psychology of Addictive Behaviors*, *31*(7), 751–762. https://doi.org/10.1037/adb0000317.

Morgenstern, J., & Longabaugh, R. (2000). Cognitive-behavioral treatment for alcohol dependence: A review of evidence for its hypothesized mechanisms of action. *Addiction*, *95*(10), 1475–1490.

Muuss, R. (1996). *Theories of adolescence* (6th ed.). New York: The McGraw-Hill.

National Institute on Alcohol Abuse and Alcoholism (NIAAA). (1999). *Alcohol alert*. 43.

National Survey of Substance Abuse Treatment Services, The N-SSATS Report. (2009). *Data on substance abuse treatment facilities [data file]* Retrieved fromhttp://www.atforum.com/addiction-resources/documents/nssats2k9web.pdf.

National Survey of Substance Abuse Treatment Services, The N-SSATS Report. (2010). *Clinical or therapeutic approaches used by substance abuse treatment facilities [data file]* Retrieved fromhttp://oas.samhsa.gov/2k10/238/238ClinicalAp2k10.htm.

Noonan, W. C., & Moyers, T. B. (1997). Motivational interviewing: A review. *Journal of Substance Misuse*, *2*, 8–16.

O'Brien, W. B., & Perfas, F. B. (2005). The therapeutic community. In J. H. Lowinson, P. Ruiz, R. B. Millman, & J. C. Langrod (Eds.), *Substance abuse: A comprehensive textbook.* (4th ed.)(pp. 609–615). Philadelphia, PA: Lippincott Williams & Wilkins.

Pita, D. D. (1992). *Addictions counseling*. New York, NY: Crossroad.

Prochaska, J. O., & DiClemente, C. C. (1982). Transtheortical therapy: Toward a more integrative model of change. *Psychotherapy: Theory, Research, and Practice*, *19*, 161–173.

Prochaska, J. O., & DiClemente, C. (1984). *The trans-theoretical approach: Crossing traditional boundaries of therapy*. Homewood, Illinois, IL: Down Jones-Irwin.

Prochaska, J., & DiClemente, C. (1992). The transtheoretical approach. In Norcross and Goldfried (Eds.), *Handbook of psychotherapy integration* (pp. 300–334). New York, NY: Basic Books.

Prochaska, J., DiClemente, C., & Norcross, J. (1998). Stages of change: Prescriptive guidelines for behavioral medicine and psychotherapy. In Koocher, Norcross, Hill (Eds.), *Psychologists desk reference* (pp. 230–236). New York, NY: Oxford University Press.

Prochaska, J., Norcross, J., & DiClemente, C. (2006). *Positive workplace alliance: Changing for good* Retrieved fromhttp://www.positiveworkplace.com/files/Abstract%20Change%20Prochaska.pdf.

Project MATCH Research Group. (1997). Matching alcoholism treatments to client heterogeneity; project MATCH post-treatment outcomes. *Journal of Studies on Alcohol*, *58*, 7–29.

Rani, S., Maheshwari, S. K., & Arora, K. (2014). Motivational enhancement therapy for substance abusers: A quasi experimental study. *Delhi Psychiatry Journal*, *17*(2), 349–356.

Rasekh, B., Mon Saw, Y., Azimi, S., Kariya, T., Yamamoto, E., & Hamajima, N. (2018). Associations of treatment completion against drug addiction with motivational interviewing and related factors in Afghanistan. *Nagoya Journal of Medical Science*, *80*(3), 329–340. https://doi.org/10.18999/nagjms.80.3.329.

Rawson, R. A., & McCann, M. J. (2011). *The matrix model of intensive outpatient treatment: A guideline developed for the behavioral health recovery management project* Retrieved fromhttp://www.bhrm.org/guidelines/The_Matrix_Model_Of_Intensive_Outpatient_Treatment.pdf.

Resnicow, K., Baskin, M. L., Rahotep, S. S., Periasamy, S., & Rollnick, S. (2004). Motivational interviewing in health promotion and behavioral medicine. In W. M. Cox & E. Klinger (Eds.), *Handbook of motivational counseling: Concepts, approaches, and assessment* (pp. 457–476). New Jersey, NJ: John Wiley & Sons Inc..

Resnicow, K., Dilorio, C., Soet, J. E., Borrelli, B., Hecht, J., & Ernst, D. (2002). Motivational interviewing in health promotion: It sounds like something is hanging. *Health Psychology*, *21*(5), 444–451.

Riper, H., Andersson, G., Hunter, S. B., de Wit, J., Berking, M., & Cuijpers, P. (2014). Treatment of comorbid alcohol use disorders and depression with cognitive-behavioural therapy and motivational interviewing: A meta-analysis. *Addiction*, *109*(3), 394–406. https://doi.org/10.1111/add.12441.

Rotgers, F., & Nguyen, T. A. (2006). Substance abuse. In P. J. Bieling, R. E. McCabe, M. M. Antony, & M. M (Eds.), *Cognitive-behavioral therapy in groups* (pp. 298–323). New York, NY: The Guilford Press.

Rounsaville, B. J., Carroll, K. M., & Back, S. (2005). Individual psychotherapy. In J. H. Lowinson, P. Ruiz, R. B. Millman, & J. C. Langrod (Eds.), *Substance abuse: A comprehensive textbook.* (4th ed.)(pp. 654–671). Philadelphia, PA: Lippincott Williams & Wilkins.

Royce, J. E., & Scratchley, D. (1996). *Alcoholism and other drug problems*. New York, NY: The Free Press.

Satre, D. D., Delucchi, K., Lichtmacher, J., Sterling, S. A., & Weisner, C. (2013). Motivational interviewing to reduce hazardous drinking and drug use among depression patients. *Journal of Substance Abuse Treatment*, *44*(3), 323–329. https://doi.org/10.1016/j.jsat.2012.08.008.

Saunders, B., Wilkinson, C., & Phillips, M. (1995). The impact of a brief motivational intervention with opiate users attending a methadone programme. *Addiction*, *90*(3), 415–424.

Schroer, B. M., Fuhrmann, A., & Jong-Meyer, R. D. (2004). Systematic motivational counseling in groups: Clarifying motivational structure during psychotherapy. In W. M. Cox & E. Klinger (Eds.), *Handbook of motivational counseling: Concepts, approaches, and assessment* (pp. 239–258). New Jersey, NJ: John Wiley & Sons Inc..

Sharma, M., & Branscum, P. (2010). Is alcoholics anonymous effective? *Journal of Alcohol and Drug Education*, *54*(3), 3–6.

Siu, A. M. H., Ko, F. S. L., & Mak, S. K. (2018). Outcome evaluation of a short-term hospitalization and community support program for people who abuse ketamine. *Frontiers in Psychiatry*, *17*(9), 313. https://doi.org/10.3389/fpsyt.2018.00313.

Spirito, A., Hernandez, L., Cancilliere, M. K., Graves, H. R., Rodriguez, A. M., Operario, D., et al. (2017). Parent and adolescent motivational enhancement intervention for substance-using, truant adolescents: A pilot randomized trial. *Journal of Clinical Child & Adolescent Psychology*, *47*(Suppl. 1), S467–S479.

Stephen, R. S., Roffman, R. A., & Curtrain, L. (2000). Comparison of extended versus brief treatments for marijuana use. *Journal of Consulting and Clinical Psychology*, *68*(5), 898–908. https://doi.org/10.1037/0022-006X.68.5.898.

Stewart, J. L., Christner, R. W., & Freeman, A. (2007). An introduction to cognitive-behavior group therapy with youth. In R. W. Christner, J. L. Stewart, & A. Freeman (Eds.), *Handbook of cognitive-behavior group therapy with children and adolescents: Specific settings and presenting problems* (pp. 3–21). New York, NY: Routledge.

Stotts, A. L., DiClemente, C. C., & Dolan-Mullen, P. (2001). One-to-one: A motivational intervention for resistant pregnant smokers. *Addictive Behaviors*, *27*(2), 275–292. https://doi.org/10.1016/S0306-4603(01)00174-5.

Tillitski, C. J. (1990). A meta-analysis of estimated effect sizes for group versus individual versus control treatments. *International Journal of Group Psychotherapy*, *40*(2), 215–224.

Tomlin, K. M., & Richardson, H. (2004). *Motivational Interviewing and stages of change: Integrated best practices for substance abuse professionals*. Center City, Minnesota, MN: Hazeldon.

Treasure, J. L., Katzaman, M., Schmidt, U., Troop, N., Todd, G., & de Silva, P. (1999). Engagement and outcome in the treatment of bulimia nervosa: First phase of sequential design comparing motivation enhancement therapy and cognitive behavioral therapy. *Behavior Research and Therapy*, *37*(5), 405–418.

Treasure, J., & Schmidt, U. B. (2008). Motivational interviewing in the management of eating disorders. In H. Arkowitz, H. A. Westra, W. R. Miller, & S. Rollnick (Eds.), *Motivational interviewing in the treatment of psychological problems* (pp. 194–224). New York, NY: The Guilford Press.

Velasquez, M. M., Maurer, G. G., Crouch, C., & DiClemente, C. C. (2001). *Group treatment for substance abuse: A stage-of- change therapy manual*. New York: The Gilford Press.

Wagner, C. C., & Sanchez, F. P. (2002). The role of values in motivational interviewing. In W. R. Miller & S. Rollnick (Eds.), *Motivational interviewing: Preparing people for change*. (2nd ed.)(pp. 284–298). New York, NY: Guilford Press.

Walters, S. T., Ogle, R., & Martin, J. E. (2002). Perils and possibilities of group –based motivational interviewing. In M. R. Miller & S. Rollinck (Eds.), *Motivational interviewing* (pp. 377–390). New York, NY: The Gilford Press.

Washton, A. M. (2005). Group therapy with outpatients. In J. H. Lowinson, P. Ruiz, R. B. Millman, & J. C. Langrod (Eds.), *Substance abuse: A comprehensive textbook*. (4th ed.)(pp. 671–679). Philadelphia, PA: Lippincott Williams & Wilkins.

Webb, C., Scudder, M., Kaminer, Y., & Kadden, R. (2003). The motivational enhancement therapy and cognitive behavioral therapy supplement: 7 sessions of cognitive behavioral therapy for adolescent cannabis users. In Vol. 2. *Cannabis Youth Treatment (CYT) series*. DHHS Pub. No. (SMA) 03-3818.

Westra, H. A., & Dozois, J. A. (2008). Integrating motivational interviewing into the treatment of anxiety. In H. Arkowitz, H. A. Westra, W. R. Miller, & S. Rollnick (Eds.), *Motivational interviewing in the treatment of psychological problems* (pp. 26–56). New York, NY: The Guilford Press.

Witkiewitz, K., Hartzler, B., & Donovan, D. (2010). Matching motivation enhancement treatment to client motivation: Re-examining the project MATCH motivation matching hypothesis. *Addiction*, *105*(8), 1403–1413. https://doi.org/10.1111/j.1360-0443.2010.02954.x.

Zuckoff, A., Swartz, H. A., & Grote, N. K. (2008). Motivational interviewing as a prelude to psychotherapy of depression. In H. Arkowitz, H. A. Westra, W. R. Miller, & S. Rollnick (Eds.), *Motivational interviewing in the treatment of psychological problems* (pp. 109–144). New York, NY: The Guilford Press.

Index

Note: Page numbers followed by *f* indicate figures and *t* indicate tables.

B

C

D

E

F

N

O

P

9780128169797